Student Problem Set with Solutions

Po-Chin Lin

Fifth Edition

BASIC ELECTRIC CIRCUIT ANALYSIS

DAVID E. JOHNSON
Professor Emeritus, Louisiana State University
Science and Mathematics Division
Birmingham-Southern College

JOHN L. HILBURN
Microcomputer Systems Inc.

JOHNNY R. JOHNSON
Professor Emeritus, Louisiana State University
Department of Mathematics
University of North Alabama

PETER D. SCOTT
Associate Professor
Department of Electrical and Computer Engineering
Department of Biophysical Sciences, School of Medicine
State University of New York at Buffalo

Prentice Hall, Englewood Cliffs, New Jersey 07632

Acquisition Editor: ***Alan Apt***
Project Manager: ***Joanne E. Jimenez***
Supplements Editor: ***Alice Dworkin***
Cover Design: ***Amy Rosen***
Manufacturing Coordinator: ***Alan Fischer***

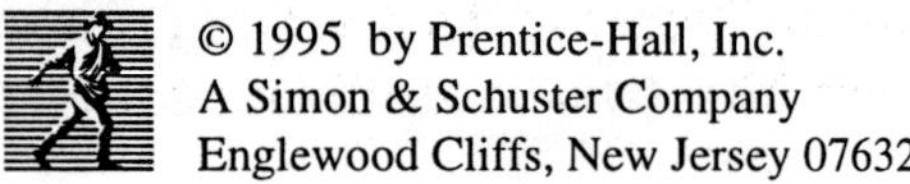

A Simon & Schuster Company
Englewood Cliffs, New Jersey 07632

Printed in the United States of America

10 9 8 7 6 5 4 3 2 1

ISBN 0-13-320250-X

PRENTICE-HALL INTERNATIONAL (UK) LIMITED, LONDON
PRENTICE-HALL OF AUSTRALIA PTY. LIMITED, SYDNEY
PRENTICE-HALL CANADA INC. TORONTO
PRENTICE-HALL HISPANOAMERICANA, S.A., MEXICO
PRENTICE-HALL OF INDIA PRIVATE LIMITED, NEW DELHI
PRENTICE-HALL OF JAPAN, INC., TOKYO
SIMON & SCHUSTER ASIA PTE. LTD., SINGAPORE
EDITORA PRENTICE-HALL DO BRASIL, LTDA., RIO DE JANEIRO

Chapter 1
Introduction

1.1 Definitions and Units

1.1 The time to read/write data from/to a modern computer's semiconductor memory is in the range of 10 nano seconds. The same operations performed on a magnetic disk require time in the range of milli seconds. What is the speed ratio of the two storage devices.

1.2 A charge of 1 coulomb is moved between two points of space in 10s. If the potential difference existing between the two points is said to be 5 V, what is the average power required in order to perform the task described above.

1.3 If 7 J of work are required to move $\frac{1}{3}$C of negative charge from point a to point b , find v_{ab} and v_{ba}.

1.2 Charge and current

1.4 Let $f(t)$ in the graph shown be the current $i(t)$ in milliamperes entering an element terminal as a function of time. Find (a) the charge that enters the terminal between 0 and 10s and (b) the rate at which the charge is entering at $t = 2$s and $t = 4$s.

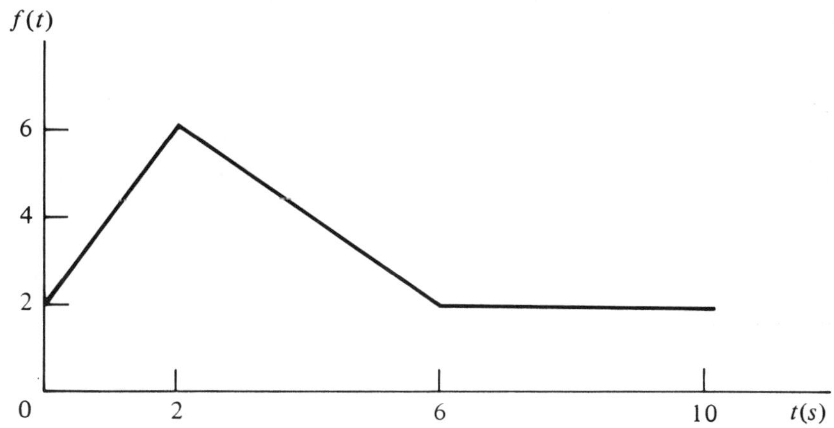

Problem 1.4

1.5 An ac current given by $i = te^{-t}$ flows into a terminal. Find the total charge entering the terminal between $t = 0$ and $t = 10$s.

1.3 Voltage, energy and power

1.6 The voltage across an element is 6V and the charge entering the positive terminal is as shown. Find the total charge and total energy delivered to the element between 1 and 10ms.

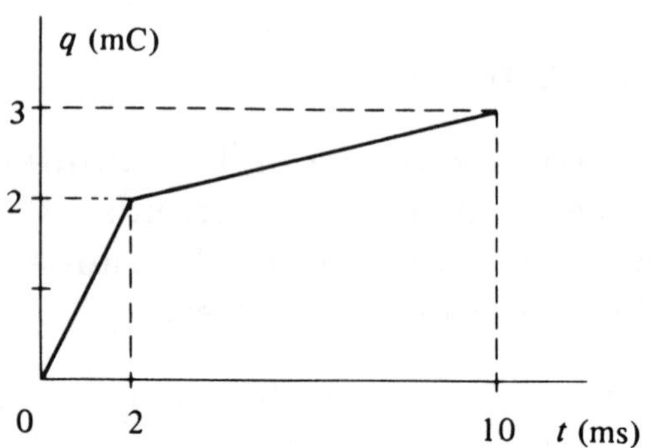

Problem 1.6

1.7 If the graph of Prob.1.6 is the current iA versus ts, and $\nu = 10\frac{di}{dt}$V, find the instant power at 1s and 4s.

1.8 A current given by $i = t+1$A flow into a two-terminal element with potential difference of 5 Volts between its two terminals as shown. Does the element absorb energy or deliver energy ? How much energy does it absorb or deliver between time t=0 and t=5s ?

1.9 A current of 2A flows through a conducting wire of 1 millimeter in diameter. What is the charge density coulomb/cm^{2}sec seen at a given cross section of the wire. How many electrons pass through the given cross section per cm^2 per second ?

1.10 The power delivered to an element is $P = 4te^{-t}$W and the charge entering the positive terminal of the element is $q = \frac{1}{2}t^2$C. Find (a) the voltage across the element and (b) the energy delivered to the element between 0 and 100ms.

1.11 An element has $v(t) = e^{-t}\cos 3t$V for $t \geq$0s. If we wish this element to supply net energy of $\frac{1}{2}t^2$J for $t \geq$0, what must $i(t)$ be ?

1.12 Let the current entering the positive terminal of an element be i=0 for $t <$0, and $i = 6\sin(2t)$A for $t >$0. (a) If the voltage is $\nu = 4\frac{di}{dt}$V, show that the energy delivered to the element is nonnegative for all time.(The element is passive.) (b) Repeat part (a) if $\nu = 2\int_0^t i dt$V.

1.13 If a current i entering the positive terminal of an element with $v(t) = t - 1$V and $i(t) = t^2 + 2t + 1$A, find the power dissipated by the element and the total energy supplied to the element between times t=0 and t=4.

1.14 Let the current entering the positive terminal of an element be $i = 0$ for $t < 0$, and $i = 3\cos(4t)$A for $t > 0$. If the voltage is $\nu = 2\frac{di}{dt}$V, find the energy delivered to the element for all time.

1.15 If the current entering the positive terminal of an element is $i = \frac{1}{6}t - 20$mA and the voltage is $\nu = 12$V, find the energy delivered to the element in 5 min.

1.16 The charge entering the positive terminal of an element is $q = -2e^{-4t}$C. Find the power delivered to the element as a function of time and the energy delivered to the element between 0 and 1s if (a) $\nu = 6i$, (b) $\nu = 3\frac{di}{dt}$, (c) $\nu = 2\int_0^t i dt - 4$.

1.17 The charge stored in a fully charged car battery depends on temperature. If $q_0(T) = 10^{-3}T^2 + \frac{3}{4}T + 20$ where $q_0(T)$ is the stored charge in kilocoulombs with temperature T in Farenheit, how many times can we attemp to start the car in $100°$F environment ? $32°$F environment ? Assume each attemp lasts 10s and draws 50A current from the battery.

1.4 Passive and active elements

1.18 The element shown satisfies $v = |t| i$. Is this element active or passive ? Assume passive sign convention.

1.19 The power absorbed by a circuit element is shown. At what time is the net energy a maximum ? What is the maximum value of net energy ? Is this element passive or active ? Assume $p(t)$=0 for $t < 0$.

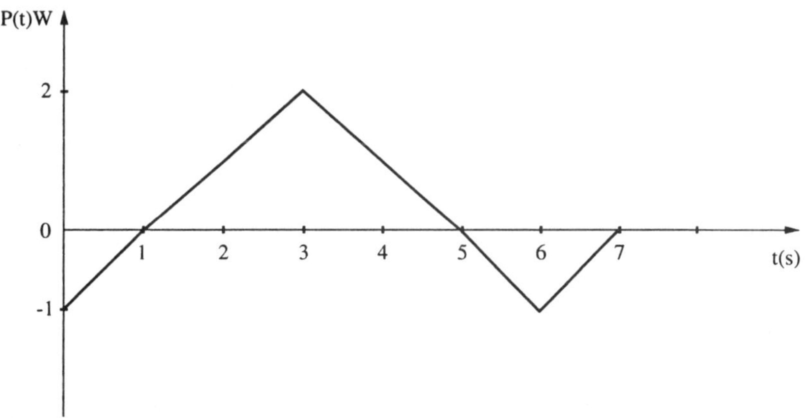

Problem 1.19

1.20 A current i, i=0 for t <0, entering the positive terminal of an element with a constant potential of 5V between its two terminals is shown in the graph. What type of device this element is ? Passive ? or Active ?

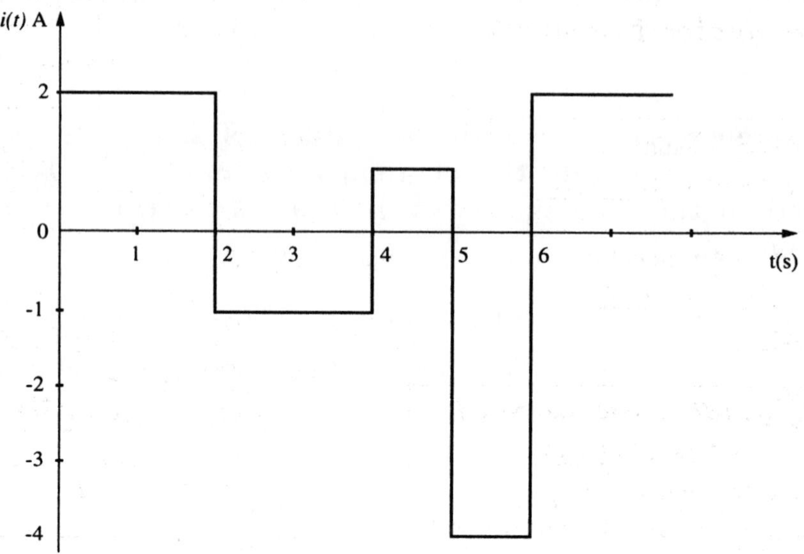

Problem 1.20

1.1 10 nano seconds $= 10\cdot 10^{-9}$ sec.
1 milli second $= 10^{-3}$ sec
$10\cdot 10^{-9}/10^{-3} = 10^{5}$ times
semiconductor memory is about 10^5 times faster than magnetic disks.

1.2 $1\,C \times 5\,V = 5$ Joules
5 Joules / 10 sec = $\underline{0.5 \text{ watts}}$

1.3 By passive sign convention
$V_{ab} = 7J/-\frac{1}{3}C = \underline{-21\ V}$
$V_{ba} = \underline{21\ V}$

1.4 (a) $q_t = \int_0^{10} i(t)\,dt$ = area under $i(t)$
$= 2\times 10 + \frac{1}{2}(6)(4) = \underline{32\,mC}$
(b) rate $= i(t)$, from the graph
$i(2) = \underline{6mA}$; $i(4) = \frac{1}{2}(2+6) = \underline{4\,mA}$

1.5 $q = \int_0^{10} i\,dt = \int_0^{10} te^{-t}\,dt$
$= \int_0^{10} e^{-t}dt - te^{-t}\Big|_0^{10} = \underline{1-11e^{-10}C}$

1.6 From the graph
$q_T = q(10ms) - q(1\text{-}ms) = 3-1 = \underline{2\ mC}$
$$i = \frac{dq}{dt} = \begin{cases} \frac{2-0}{2-0} = 1A & 0 \le t \le 2ms \\ \frac{3-2}{10-2} = \frac{1}{8}A & 2ms \le t \le 10ms \end{cases}$$
$W_T = \int_{10^{-3}}^{10(10^{-3})} v i\,dt$
$= \int_{10^{-3}}^{2(10^{-3})} (6)(1)dt + \int_{2(10^{-3})}^{10(10^{-3})} (6)(\frac{1}{8})\,dt$
$= (6+6)10^{-3} = \underline{12\ mJ}$

1.7 $$v = \begin{cases} 10\,V & 0 \le t \le 2s \\ 10/8\,V & 2s \le t \le 10s \end{cases}$$
$P(1) = v(1)\,i(1) = (10)(1) = \underline{10W}$
$P(4) = (\frac{10}{8})(2+\frac{2}{8}) = \underline{2.813\ W}$

1.8 The element absorbs energy.
$E = vC = 5\cdot\int_0^5 (t+1)dt$
$= 5\cdot[(\frac{1}{2}t^2+t)|_0^5] = 5(\frac{1}{2}\cdot 25+5) = \underline{87.5J}$

1.9 2A = 2 coulomb/second
$1\ mm = 10^{-1}$ cm
area of cross section $= \pi r^2 = \pi(\frac{10^{-1}cm}{2})^2$
charge density $= \dfrac{2 \text{ coulomb/second}}{\pi(10^{-2}/4)\ cm^2}$
$= \underline{254.64 \dfrac{\text{Coulombs}}{cm^2\cdot sec}}$

$\dfrac{254.64}{1.60\times 10^{-19}} = \underline{1.59\times 10^{21} \dfrac{\text{electrons}}{cm^2\cdot sec}}$

1.10 (a) $q = \frac{1}{2}t^2$
$i = dq/dt = t$
$v = P/i = 4te^{-t}/t = \underline{4e^{-t}V}$
(b) $E = \int_0^{0.1} 4te^{-t}dt$
$= 4[\int_0^{0.1} e^{-t}dt - te^{-t}|_0^{0.1}]$
$= \underline{4-4.4e^{-0.1}\ J}$

1.11 $E = \int_0^t p(t)dt$
$\frac{dE}{dt} = p(t) = t = vi$
$i = t/v = t/e^{-t}\cos 3t = \underline{te^{t}/\cos 3t\ \ A}$

1.12 (a) $v = 4\frac{di}{dt} = 4(12\cos 2t)$
$W = \int_{-\infty}^{t} v i\,dt = \int_0^t (48\cos 2t)(6\sin 2t)\,dt$
$= \underline{72\sin^2 2t \ge 0}$

(b) $v = 2\int_0^t i\,dt = 2\int_0^t 6\sin 2t$
$= -6\cos 2t + 6V$
$W = \int_0^t (-6\cos 2t+6)(6\sin 2t)dt$
$= -9\sin^2 2t - 18\cos 2t + 18$
$= 9(\cos^2 2t - 1 - 2\cos 2t + 2)$
$= 9(\cos^2 2t - 2\cos 2t + 1)$
$= \underline{9(\cos 2t - 1)^2 \ge 0}$

1.13 $P = vi = (t-1)(t^2+2t+1) = \underline{t^3+t^2-t-1\ W}$
$E = \int_0^4 p\,dt = \int_0^4 t^3+t^2-t-1\,dt$
$= \frac{1}{4}t^4+\frac{1}{3}t^3-\frac{1}{2}t^2-t\,|_0^4 = \underline{73\frac{1}{3}\ J}$

1.14 $v = 2(3)(4)(-\sin 4t)\ V$

$W = \int_0^t (-24\sin 4t)(3\cos 4t)\,dt$

$= -\frac{72}{8}\sin^2 4t = \underline{-9\sin^2 4t\ J}$

1.15 $W = \int_0^{5(60)} v i\,dt = \int_0^{300} (12)(\frac{t}{6} - 20\text{ mA})\,dt$

$= (300)^2 - (240)(300)$

$= \underline{18\ J}$

1.16 $i = \frac{dq}{dt} = 8e^{-4t}\ A$

(a) $P = vi = 6i^2 = 6(8e^{-4t})^2 = \underline{384e^{-8t}\ W}$

$W = \int_0^1 P\,dt = \int_0^1 384e^{-8t}\,dt = \underline{48(1-e^{-8})\ J}$

(b) $v = 3\frac{di}{dt} = 3(-32e^{-4t}) = \underline{-96e^{-4t}\ V}$

$P = (-96e^{-4t})(8e^{-4t}) = \underline{-768e^{-8t}\ W}$

$W = \int_0^1 -768e^{-8t}\,dt = \underline{96(e^{-8}-1)\ J}$

(c) $v = 2\int_0^t 8e^{-4t}\,dt - 4 = \underline{-4e^{-4t}\ V}$

$P = (-4e^{-4t})(8e^{-4t}) = \underline{-32e^{-8t}\ W}$

$W = \int_0^1 -32e^{-8t}\,dt = \underline{4(e^{-8}-1)\ J}$

1.17 At each attemp

$q = 50 \cdot 10 = 500\ C$

At 100°F

$q_o(T) = 10^{-3}(100)^2 + (\frac{3}{4})100 + 20 = 105\times10^3\ C$

$\frac{105\times10^3}{500} = \underline{210\text{ attemps}}$

At 32°F

$q_o(t) = 10^{-3}(32)^2 + \frac{3}{4}(32) + 20 = 45.0\times10^3\ C$

$\frac{45.0\times10^3}{500} = \underline{90\text{ attemps}}$

1.18 $P = vi = (|t| \cdot i)\,i = |t|\,i^2 \geq 0$

Hence $E(t) = \int_{-\infty}^{t} P\,dt = \int_{-\infty}^{t} |t|\,i^2\,dt > 0$

This is a <u>passive element</u>

1.19 At t=5 the net energy is a maximum

The maximum value of net energy is

$\frac{1}{2}\times2\times4 - \frac{1}{2}\times1\times1 = \underline{\frac{7}{2}\ J}$

This is an active element because at $0<t<1$ the net energy is <u>negative.</u>

1.20 Total energy delivered to the element at time $t=6$ is

$E = \int_0^6 v i\,dt = 5\int_0^6 i\,dt = -5\ J$

Hence this element is <u>active</u>

Chapter 2
Resistive Circuits

2.1 Kirchhoff's Laws

2.1 Find currents i_1 and i_2.

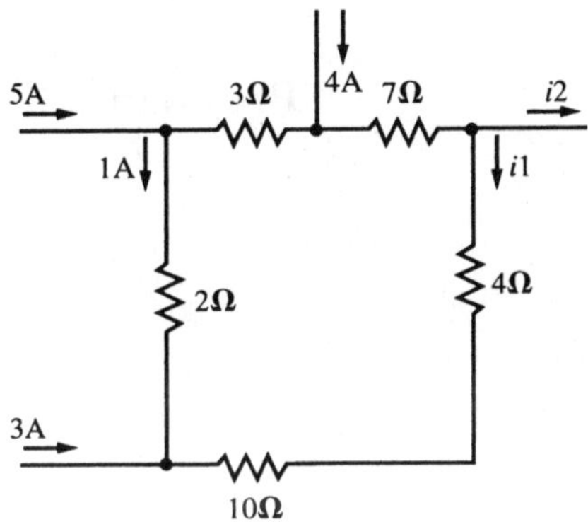

Problem 2.1

2.2 Write KVL around each loop of the network shown below and find v_1, v_2

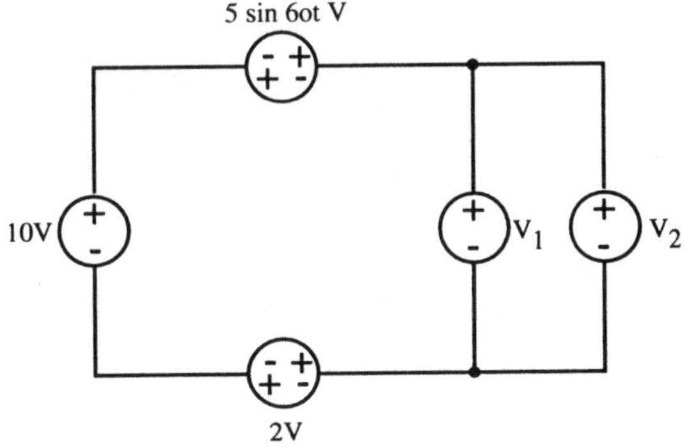

Problem 2.2

2.3 Find ν.

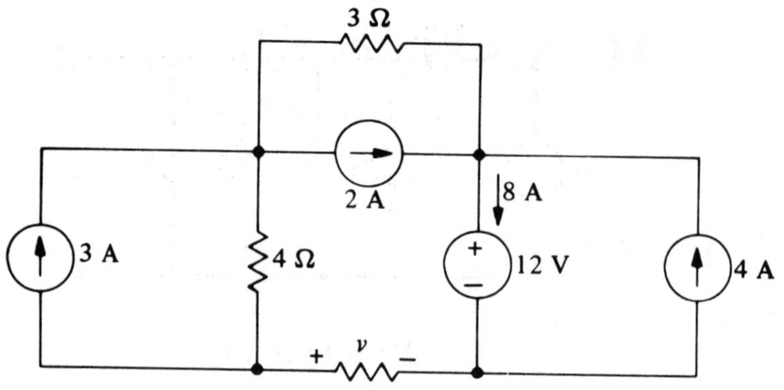

Problem 2.3

2.2 Ohm's Law

2.4 Find i and ν_{ab}.

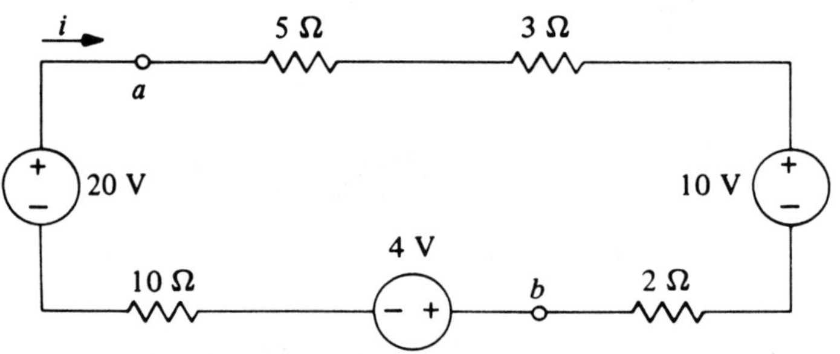

Problem 2.4

2.5 If $\nu_x = 10$V, find R_y, ν_y, and ν_z.

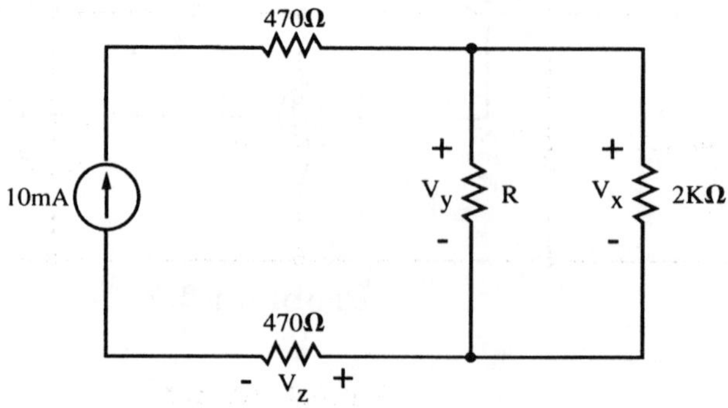

Problem 2.5

2.6 Find the values of R_1, R_2, R_3, R_4, and v_0.

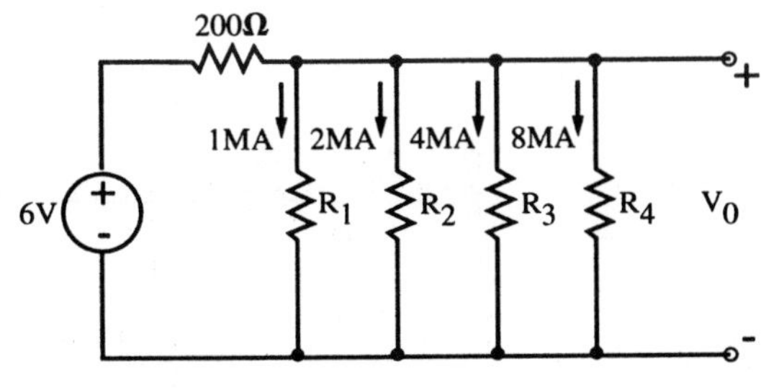

Problem 2.6

2.7 Find i_1, $i2$, and v_{ba}.

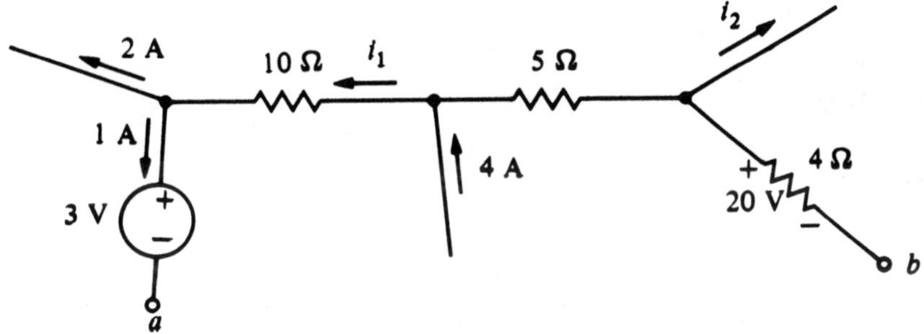

Problem 2.7

2.3 Parallel resistance and current division

2.8 Find i and the power deliver to the 3Ω resister.

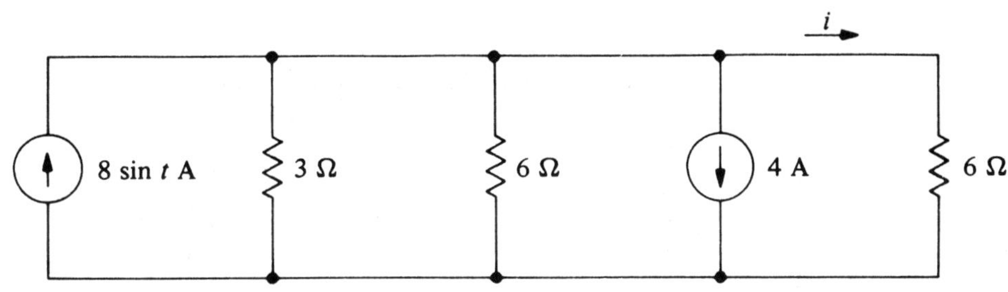

Problem 2.8

2.9 Show that ν_0 is the weighted sum of voltages ν_1, ν_2, ... ν_N with weights $\frac{G_1}{G_t}$, $\frac{G_2}{G_t}$, ... $\frac{G_i}{G_t}$, ... $\frac{G_N}{G_t}$ repeatively,where $G_i = \frac{1}{R_i}$, $G_t = \sum_{i=1}^{N} G_i$ for the circuit shown below.

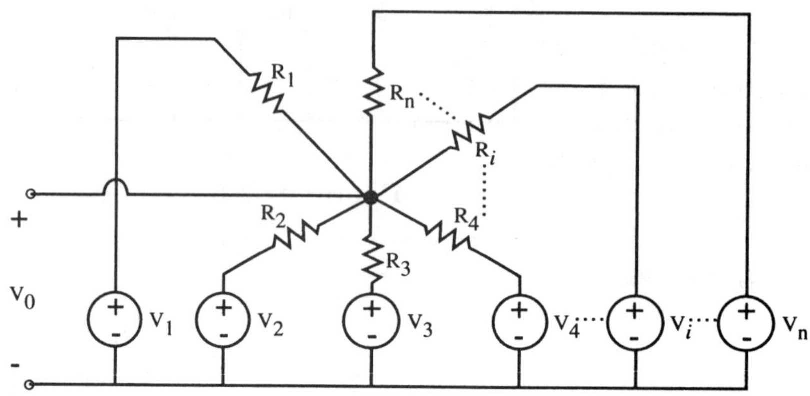

Problem 2.9

2.4 Series resistance and voltage division

2.10 A voltage divider is to be connstructed with a 60V source and a number of 10 KΩ resistors. Find the minimum number of resistors required if the output voltage is (a) 40V and (b) 30V.

2.11 The regular axial lead carbon composition resistors have maximim power dissipation rated as $\frac{1}{4}$W, $\frac{1}{2}$W, 1W, and 2W. (a) A 200Ω carbon composition resistor is connected to a source of 12 V. Find the required power rating of the resistor. (b) If a 100Ω resistor of the same type is inserted in series in the circuit, find the power ratings of the two resistors.

2.12 (a) If all resistors in a voltage divider are scaled by the same factor, how does this affect the voltages across them? current through the power dissipated? (b) If the conductances of all resistors in a current divider are scaled by the same factor α, how does this affect the current through the voltages across them? power?

2.13 Design a voltage divider which delivers 5V, 3.3V both with a common negative terminal. The voltage divider is to be powered by a 9V battery. Use 10KΩ resistor only in the divider circuit. Your design shall have a total power consumption less than 10mW when there is no load attached to it. The voltage outputs have to be accurate within ± 10% of the specified values. Draw the circuit diagram.

2.5 Analysis examples

2.14 (a) Find the equivalent resistance looking in the terminals a-b if terminals c-d are open, and if terminals c-d are shorted together. (b) Find the equivalent resistances looking into terminals c-d if terminals a-b are open and if a-b are shorted together.

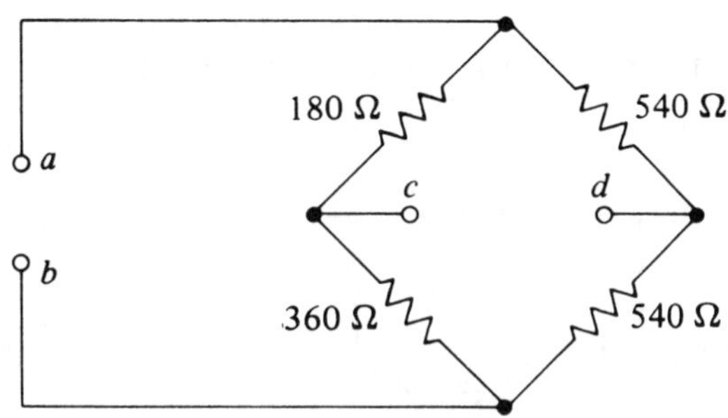

Problem 2.14

2.15 (a) Design a voltage source of 3V output using a 5V voltage source and 1KΩ resistors. Use as few resistors as possible. (b) Design another one with a 3.1V output. Determine the error in your design.

2.16 A current divider consists of 10 resistors in parallel. Nine of them have equal resistance of 30KΩ and the tenth is a 10KΩ resistor. Find the equivalent resistance of this current divider, and if the total current entering the divider is 20mA, find the current in the tenth resistor.

2.17 Find R_{eq} and the resulting values of i and ν if a 50V battery is connected to the open terminals.

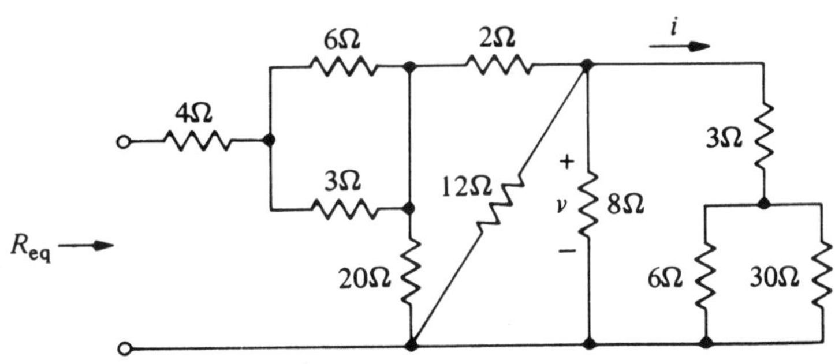

Problem 2.17

2.18 A ladder resistive network is shown as below. Find R_{eq}.

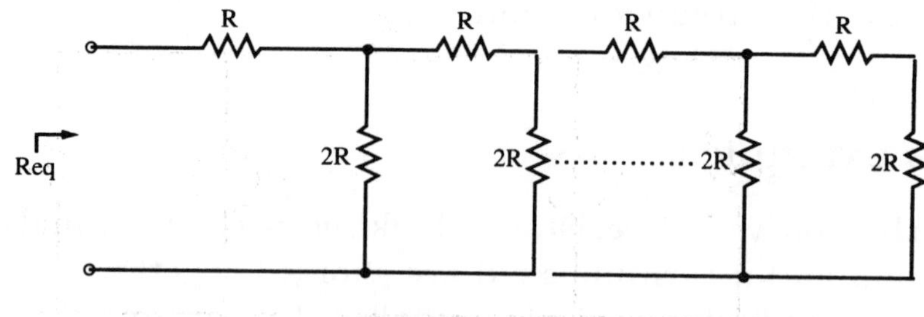

Problem 2.18

2.19 A resistive network with 2 single-pole, double-throw toggle switches (s1, s2) is constructed as shown below. Find the output voltage V_0 for each of the four combinations of switches s1 and s2.

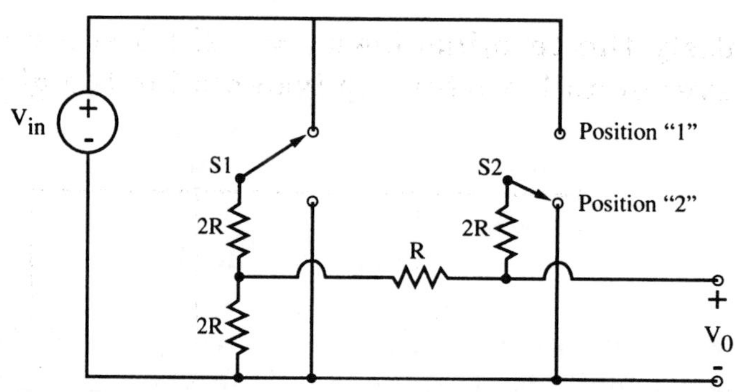

Problem 2.19

2.6 Thevenin-Norton equivalent

2.20 Suppose the terminal laws for subcircuits N_1 and N_2 are $v = -3i + 4$ and $v = -5i + 8$ respectively. Find the Thevenin equivalent if N_1 and N_2 are placed in series. Assume passive sign convention.

2.21 Use equivalent circuits to find the voltage v_1 and current i.

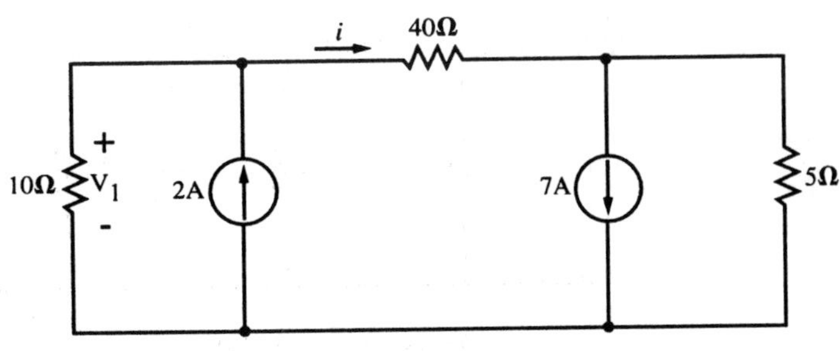

Problem 2.21

2.22 Suppose the terminal laws for subcircuits N_1 and N_2 are $v = -\frac{1}{2}i + 4$ and $v = -\frac{1}{9}i + 8$ respectively. Find the Thevenin equivalent if N_1 and N_2 are placed in parallel.

2.23 N_1 and N_2 satisfy the terminal laws $v = -2i + 5$ and $v = -3i + 7$ respectively. Find the Thevenin and Norton equivalents for the given circuit.

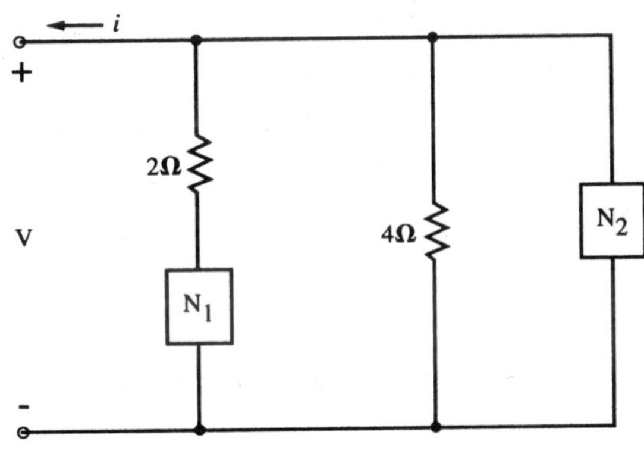

Problem 2.23

2.24 Find the Thevenin equivalent of the given circuit.

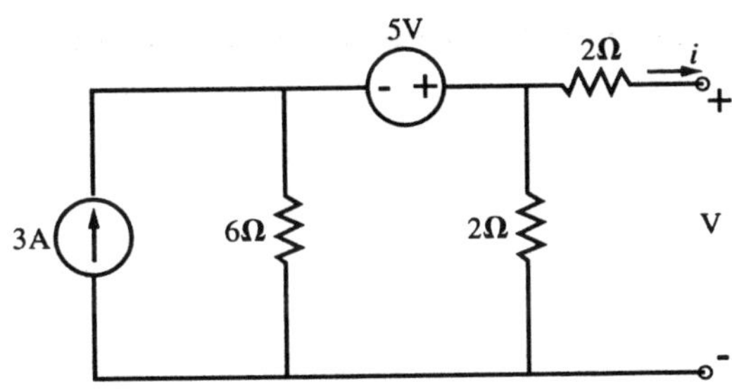

Problem 2.24

2.25 Find a value of R such that $\nu = 4V$.

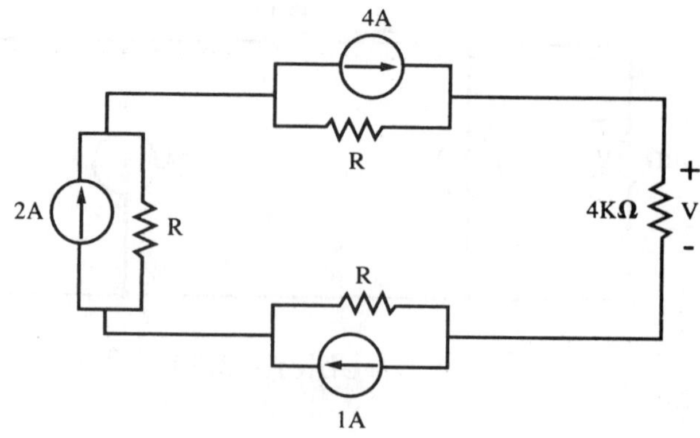

Problem 2.25

2.26 Find ν by using Thevenin equivalent seen by the 4Ω resistor.

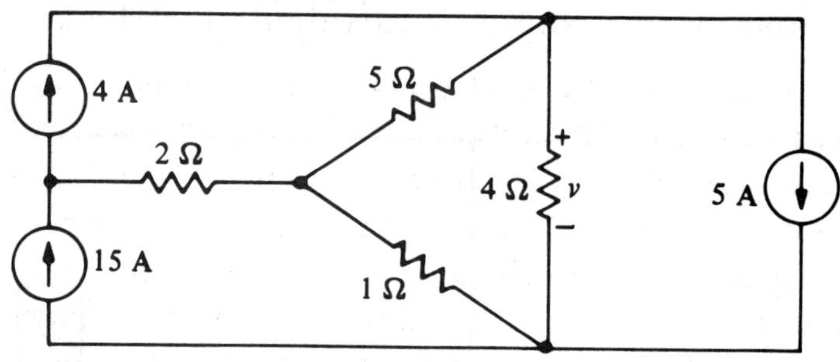

Problem 2.26

2.27 Find the Thevenin and Norton equivalents.

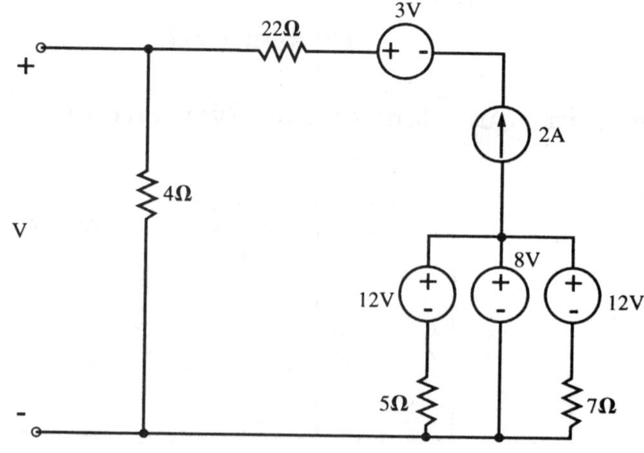

Problem 2.27

2.28 Find i and v_{ab}.

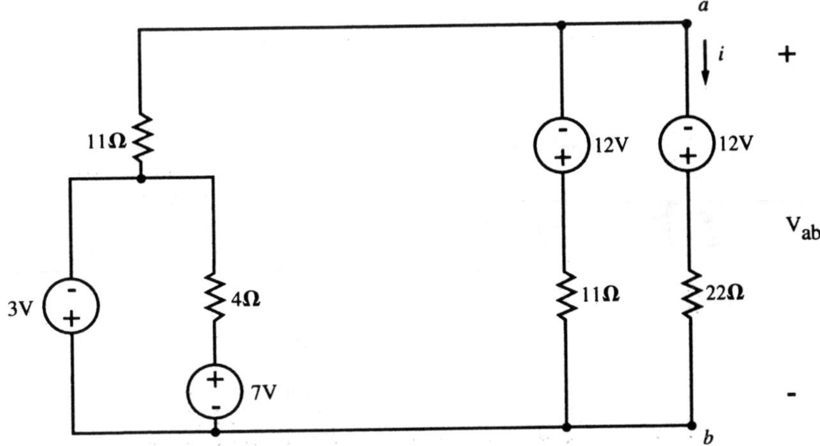

Problem 2.28

2.29 Ohm's law states that the voltage across a resistor is directly proportional to the current flowing through the resistor. Other devices may possess such linear voltage-current characteristic when they operate over a certain range of voltage and current. (The resistor-like operating range is usually called the Ohmic region of such devices). The voltage-current characteritic of a silicon junction diode is shown. Find the Thevenin equivalent circuit of a junction diode within its ohmic region indicated.

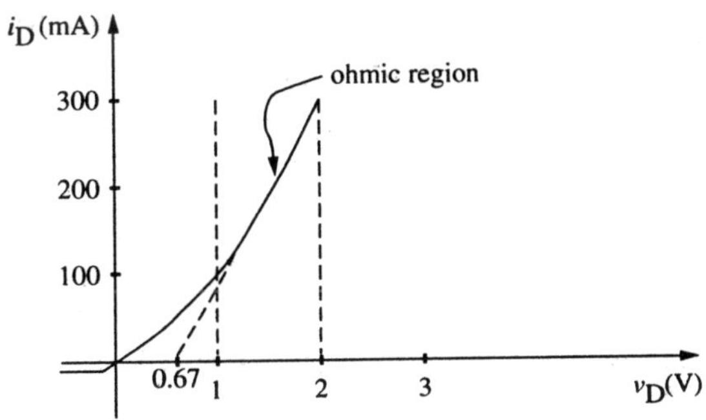

Problem 2.29

2.30 **Find the resistance range of carbon composition resistor with color bands of (a) brown, black, red, silver, (b) red, violet, yellow, silver and (c) blue, gray, silver, gold.**

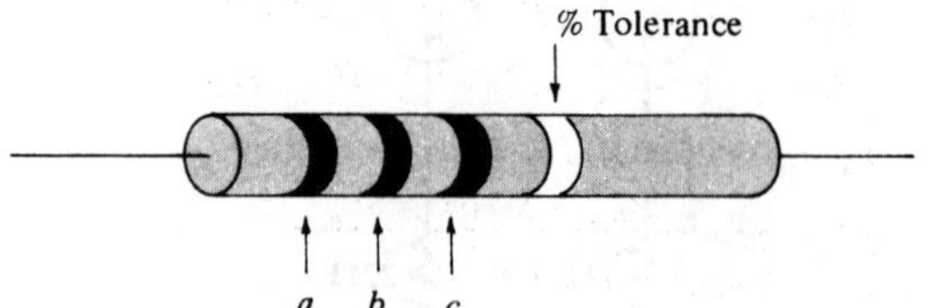

Color Code for Carbon Resistors

Bands a, b, and c			
Color	Value	Color	Value
Silver*	−2	Yellow	4
Gold*	−1	Green	5
Black	0	Blue	6
Brown	1	Violet	7
Red	2	Gray	8
Orange	3	White	9
	% Tolerance Band		
Gold	±5%		
Silver	±10%		

*These colors apply to band c only.

Problem 2.30

2.1 $i_1+1+3=0$, $i_1=-4A$

$i_0' = i_0+4 = (5-1)+4 = 8A$

$i_0' = i_1+i_2$, $i_2 = 8-(-4) = 12A$

2.2 By KVL around left loop

$10+5\sin 60t - V_1 - 2 = 0$, $V_1 = 8+5\sin 60t\ V$

By KVL around right loop

$V_1 = V_2 = 8+5\sin 60t\ V$

By KVL around outter loop

$10+5\sin 60t - V_2 - 2 = 0$

2.3 By KCL; $i_{3\Omega} = 8A-2A-4A = 2A$ to the right

$i_{4\Omega} = 2A-3A+2A = 1A$ up

By KVL; $v = (4)(1)+(3)(2)+12 = 22V$

2.4 By KVL

$(5+3+2+10)i+10V+4V-20V=0$

$i = \frac{6V}{20\Omega} = 0.3A$

$V_{ab} = (5+3+2)i+10V = 13V$

2.5 By Ohm's law; $V_Z = 10\cdot 10^{-3}\cdot 470 = 4.7V$

$V_X = I_X\cdot 2\cdot 10^3 = 10V$, $I_X = \frac{10}{2K} = 5mA$

By KCL, $I_Y = 10 - I_X = 5mA$

By KVL, $V_Y = V_X = 10V$

$R_Y = \frac{V_Y}{I_Y} = \frac{10}{5\cdot 10^{-3}} = 2K\Omega$

2.6 $v_o = 6-200(1+2+4+8)\cdot 10^{-3} = 6-200\cdot 15\cdot 10^{-3} = 3V$

$R_1 = \frac{3}{1\times 10^{-3}} = 3K\Omega$

$R_2 = \frac{3}{2\times 10^{-3}} = 1.5K\Omega$

$R_3 = \frac{3}{4\times 10^{-3}} = 0.75K\Omega$

$R_4 = \frac{3}{8\times 10^{-3}} = 0.375K\Omega$

2.7 By KCL

$i_1 = 2+1 = 3A$

$i_3 = 4-i_1 = 1A$

$i_4 = \frac{20V}{4\Omega} = 5A$

2A, 1A, i_1, i_3, i_2, 10Ω, 5Ω, 4A, i_4, 20V, a, $-V_{ba}+$, b

2.7 Cont.

$i_2 = i_3 - i_4 = 1-5 = -4A$

By KVL, $V_{ba} - 3 - 10i_1 + 5i_3 + 20 = 0$

$\therefore V_{ba} = 3+10(3)-5(1)-20 = 8V$

2.8 Equivalent current source has current $8\sin t - 4A$, up

By current division,

$i = \frac{(1/6)(8\sin t - 4)}{(1/3)+(1/6)+(1/6)}$

$= 2\sin t - 1\ A$

$i_{3\Omega} = \frac{1/3\,(8\sin t-4)}{1/3+1/6+1/6} = 4\sin t - 2$

$P_{3\Omega} = (i_{3\Omega})^2(3) = 12(2\sin t-1)^2 W$

2.9 By KCL, $\sum_{i=1}^{N} i_i = \sum_{i=1}^{N} \frac{(V_i - V_o)}{R_i} = 0$

$\sum_{i=1}^{N} \frac{V_i}{R_i} = V_o \sum_{i=1}^{N} \frac{1}{R_i}$,

$\sum_{i=1}^{N} G_i V_i = V_o \sum_{i=1}^{N} G_i = V_o G_t$

where $G_t = \sum_{i=1}^{N} G_i$

$V_o = \frac{\sum_{i=1}^{N} G_i V_i}{G_t} = \sum_{i=1}^{N} \left(\frac{G_i}{G_t}\right) V_i$

2.10 R_1 and R_2 are the voltage divider resistances with output taken across R_2, by voltage division

$\frac{v_{out}}{60} = \frac{R_2}{R_1+R_2}$

(a) $v_{out} = 40V$, $\therefore 40(R_1+R_2) = 60R_2$ or $4R_1 = 2R_2$. Thus $R_2 = 2R_1$ and three $10\,k\Omega$ resistors are required using two in series for R_2.

(b) $V_{out} = 30V$, $\therefore R_1+R_2 = 2R_2$ or $R_1 = R_2$. Two $10k\Omega$ resistors are required.

2.11 (a) $P = vi = \frac{v^2}{R} = \frac{144}{200}$

$= 0.72\,W$

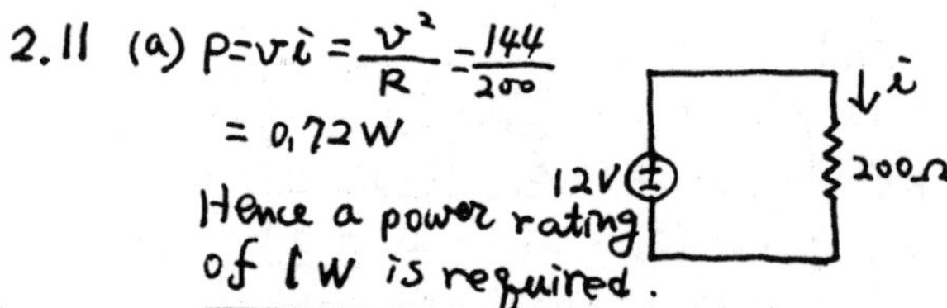

Hence a power rating of 1 W is required.

(b) $i = \frac{12}{400+100} = 0.04A$

$P_{R_1} = V_{R_1} i_{R_1} = i_{R_1}^2 \cdot R_1$

$= (40\times10^{-3})^2 \times 200$

$= 320\times10^{-3}\,W$

Hence a power rating of ½ W is required for resistor R_1.

$P_{R_2} = V_{R_2} i_{R_2} = i_{R_2}^2 \cdot R_2 = (40\cdot10^{-3})^2 \cdot 100$

$= 160\cdot10^{-3}\,W$

Hence the required power rating of resistor R_2 is ¼ W.

2.12 (a) Let $R_1, R_2, \cdots R_i$ be the resistances before scaling and $R_1', R_2', \cdots R_i'$ be the resistances after scaling. $R_i' = \alpha R_i$.

$V = \left(\sum_{i=1}^{N} R_i\right) I = \left(\sum_{i=1}^{N} \alpha R_i\right)\left(\frac{1}{\alpha} I\right)$

$= \left(\sum_{i=1}^{N} R_i'\right)(I')$

Hence $I' = \frac{1}{\alpha} I$,

$v_i' = R_i' I_i' = (\alpha R_i)\left(\frac{1}{\alpha} I_i\right) = R_i I_i = V_i$

$P_i' = I_i'^2 R_i' = \left(\frac{1}{\alpha} I_i\right)^2 (\alpha R_i) = \frac{1}{\alpha} I_i^2 R_i = \frac{1}{\alpha} P_i$

where I', V_i' and P_i' are the current, voltage and power dissipated by resistor R_i after scaling.

(b) Similarly, $I = \left(\sum_{i=1}^{N} G_i\right) V$

$= \left(\sum_{i=1}^{N} \alpha G_i\right)\left(\frac{1}{\alpha} V\right) = \left(\sum_{i=1}^{N} G_i'\right) V'$

Hence
$$\begin{cases} V_i' = \frac{1}{\alpha} V_i \\ I_i' = G_i' V_i' = (\alpha G_i)\left(\frac{1}{\alpha} V_i\right) = G_i V_i = I_i \\ P_i' = I_i'^2 R_i' = I_i^2 \left(\frac{1}{\alpha} R_i\right) = \frac{1}{\alpha} I_i^2 R_i = \frac{1}{\alpha} P_i \end{cases}$$

2.13

9V, R_1, R_2, R_3; + 5V −; + 3.3V

2.13 Cont.

$$\begin{cases} \frac{R_2+R_3}{R_1+R_2+R_3}(9) = 5 \\ \frac{R_3}{R_1+R_2+R_3}(9) = 3.3, \end{cases} \Rightarrow 9R_2 = 1.7(R_1+R_2+R_3)$$

or $0.233(R_1+R_3) = R_2 \cdots (1)$

The other constraint

$P = \frac{V^2}{R} = \frac{81}{R_1+R_2+R_3} \leq 0.01 \cdots (2)$

From ① and ②, choose $R_1 = R_3 = 10k\Omega$

$R_2 = 0.233(20k) = 4.66k\Omega$

Let $R_2 = 10k\Omega \,//\, 10k\Omega = 5k\Omega \cong 4.66k\Omega$

Error in this design;

$5 - v_{o1} = 5 - \frac{10+5}{10+10+5}(9) = -0.4V$

$3.3 - v_{o2} = 3.3 - \frac{10}{25}(9) = -0.3V$

Both are within the 10% specified value.

2.14 (a) c-d open:

$R_{ab} = \frac{(180+360)(540+540)}{180+360+540+540} = 360\Omega$

c-d shorted:

$R_{ab} = \frac{(180)(540)}{180+540} + \frac{(360)(540)}{360+540} = 351\Omega$

(b) a-b open

$R_{cd} = \frac{(360+540)(180+540)}{360+540+180+540} = 400\Omega$

a-b shorted:

$R_{cd} = \frac{(180)(360)}{360+180} + \frac{540}{2} = 390\Omega$

2.15 (a) $v_o = \frac{R_2}{R_1+R_2}(5) = 3$, $2R_2 = 3R_1$

choose $R_1 = 1k\Omega$, $R_2 = 1.5k\Omega$

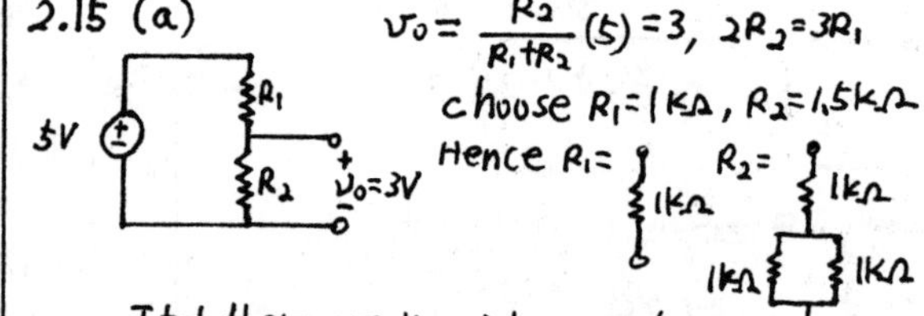

Total there are 4 resistor used.

(b) For arbitrary v_o, use bridge topology.

i →, i_1, 5V, $R_1 = 2k\Omega$, $R_3 = 2k\Omega$, $R_2 = 3k\Omega$, $R_4 = (3+\Delta X)k\Omega$, $v_o = 3.1V$

properly choose $R_1 = 2k\Omega$, $R_2 = 3k\Omega$, $R_3 = 2k\Omega$ and $R_4 = (3+\Delta X)k\Omega$ shuch that the bridge is only slightly unbanlanced.

Resistance seen by 5V voltage source is R_{eq},

$R_{eq} = 5K // (5+\Delta X) k\Omega = \frac{5(5+\Delta X)}{10+\Delta X} k\Omega$

$i = \frac{5}{R_{eq}} = \frac{5}{\frac{5(5+\Delta X)}{10+\Delta X}} = \frac{10+\Delta X}{5+\Delta X}$ mA

$i_1 = \frac{5}{10+\Delta X} \times \frac{10+\Delta X}{5+\Delta X} = \frac{5}{5+\Delta X}$ mA

$v_o = 3,1 = (3+\Delta X)\left(\frac{5}{5+\Delta X}\right)$, $15+5\Delta X = 15.5+3,1\Delta X$

$\Delta X = \frac{0,5}{1,9}$. So one can choose $\Delta X = \frac{1}{3}$

Use this design $R_1 = 2k\Omega$ = 1kΩ, 1kΩ (in series), $R_2 = 3k\Omega$ = 1kΩ, 1kΩ, 1kΩ (in series)

$R_3 = 2k\Omega$ = 1kΩ, 1kΩ (in series)

and $R_4 = 3\frac{1}{3}k\Omega$ = (1kΩ // 1kΩ // 1kΩ) in series with 1kΩ, 1kΩ, 1kΩ

Total of 13 resistors are used.

Error in this design:

$v_o = (3+\frac{1}{3})\frac{5}{5+\frac{1}{3}} = \frac{\frac{10}{3}\times 5}{\frac{16}{3}} = 3,125 V$

error $= (3,125 - 3,1)/3,1 \cong \underline{0,8\,\%}$

2.16 R_1 = equivalent resistance of 9 equal R's $= 30/9 = 10/3$ kΩ

R_p = divider resistance $= \frac{(10/3)(10)}{10/3+10} = \underline{2,5k\Omega}$

By current division,

$i_{10} = \frac{10/3}{10/3+10}(20mA) = \underline{5mA}$, current in the tenth resistor.

2.17 $6\Omega // 30\Omega = 5\Omega$, $12\Omega // 8\Omega = 4.8\Omega$

$6\Omega // 3\Omega = 2\Omega$. Equivalent circuit shown:

4Ω 2Ω 2Ω 3Ω i 5Ω
$R_{eq} \rightarrow$ 20Ω v_1 4.8Ω v

6Ω 2Ω
$R_{eq} \rightarrow$ 20Ω v_1 v 3Ω

$\frac{(3+5)4.8}{4.8+3+5} = 3\Omega$

$R_{eq} = 6 + \frac{20(2+3)}{20+2+3} = \underline{10\Omega}$

6Ω, 50V, v_1, 4Ω

$v_1 = \frac{4}{6+10}(50) = 20V$

$v = \frac{3}{2+3}(v_1) = \underline{12V}$, $i = \frac{12V}{(3+5)\Omega} = \underline{1,5A}$

2.18

R 0 R 1 R R; 2R 2R 2R ··· 2R 2R; $R_{00'}$, $R_{11'}$; 0', 1'

Let $R_{11'} = X$, $R_{00'} = (R+X) // 2R = \frac{2R(X+R)}{2R+X+R}$

Since the ladder has infinite elements

$R_{00'} = R_{11'}$, or $2RX + 2R^2 = X^2 + 3RX$

$X^2 + RX - 2R^2 = 0$, $X = R$ or $-2R$

∵ R has to be positive, $X = R$

$R_{eq} = R + R_{00'} = R + R = \underline{2R\ \Omega}$

2.19 (a) Both S1 and S2 are at position "0"

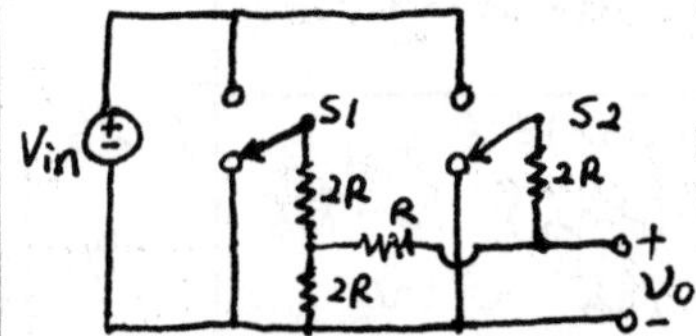

Vin is in an open loop and there is no current in the network, $\underline{V_o = 0}$

(b) Both S1 and S2 are at position "1"

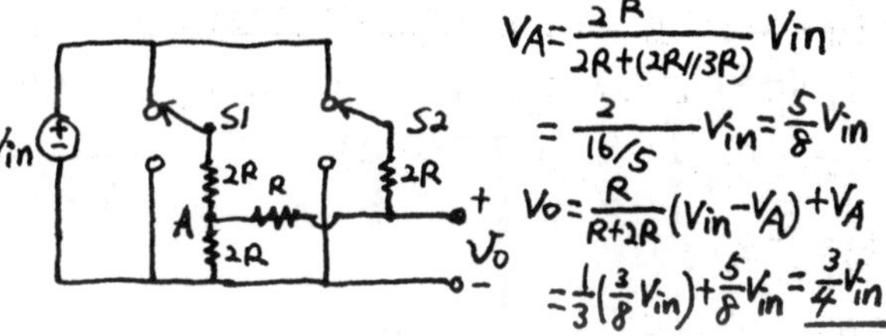

$V_A = \frac{2R}{2R+(2R//3R)}V_{in}$

$= \frac{2}{16/5}V_{in} = \frac{5}{8}V_{in}$

$V_o = \frac{R}{R+2R}(V_{in} - V_A) + V_A$

$= \frac{1}{3}(\frac{3}{8}V_{in}) + \frac{5}{8}V_{in} = \frac{3}{4}V_{in}$

(c) S1 is at position "0" and S2 at "1".

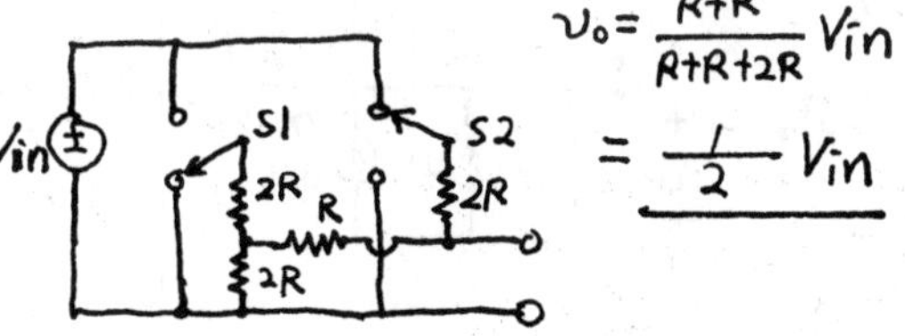

$v_o = \frac{R+R}{R+R+2R}V_{in}$

$= \underline{\frac{1}{2}V_{in}}$

(d) S1 is at position "1" and S2 at "0"

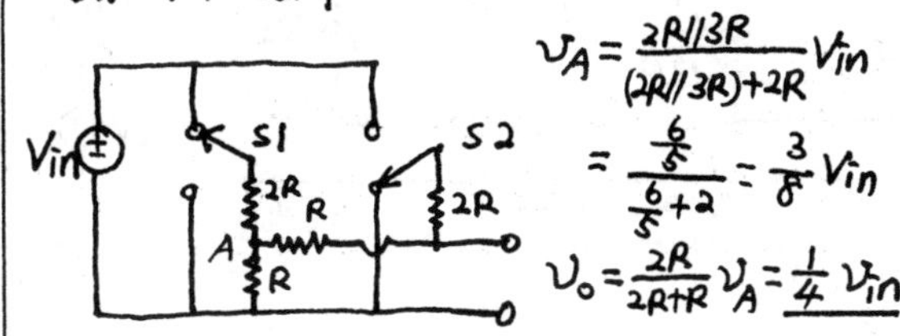

$v_A = \frac{2R//3R}{(2R//3R)+2R}V_{in}$

$= \frac{\frac{6}{5}}{\frac{6}{5}+2} = \frac{3}{8}V_{in}$

$V_o = \frac{2R}{2R+R}v_A = \frac{1}{4}v_{in}$

2.20 $v_1=-3i_1+4,\ v_2=-5i_2+8$

v_1, v_2 in series, the terminal law of the circuit

$v=v_1+v_2=-3i_1-5i_2+12,\ i_1=i_2=i$

$v=-8i+12$

The Thevenin equivalent:

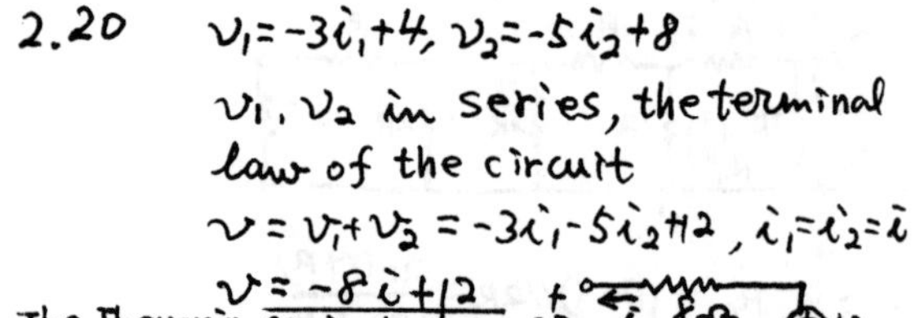

2.21 By Thevenin-Norton equivalence the network is:

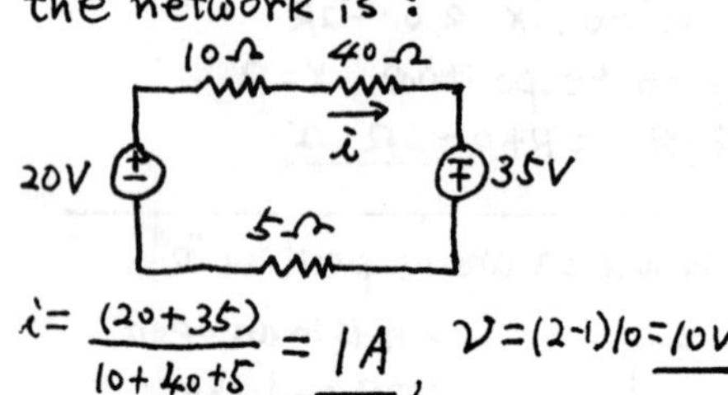

$i=\frac{(20+35)}{10+40+5}=1A$, $v=(2-1)10=10V$

2.22 $v_1=-\frac{1}{2}i_1+4,\ i_1=2(4-v_1)=8-2v_1$

$v_2=-\frac{1}{9}i_2+8,\ i_2=9(8-v_2)=72-9v_2$

$i=i_1+i_2=80-2v_1-9v_2,\ v_1=v_2$

$i=80-11v$

The Norton equivalent: (80 A source in parallel with 11)

2.23 $v_1=-2i_1+5$

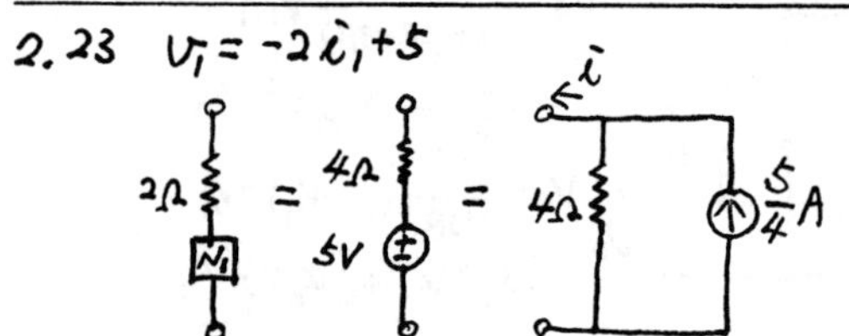

$v_2=-3i_2+7$

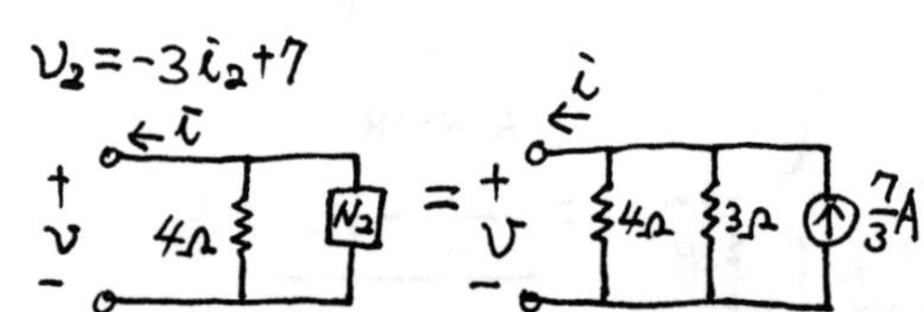

Hence the Norton

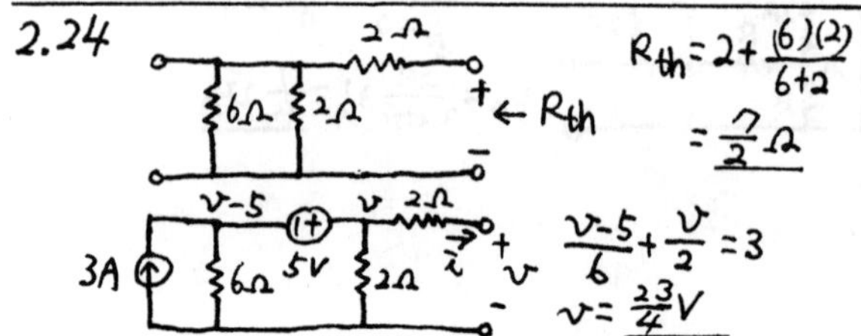

2.24

$R_{th}=2+\frac{(6)(2)}{6+2}=\frac{7}{2}\Omega$

$\frac{v-5}{6}+\frac{v}{2}=3$

$v=\frac{23}{4}V$

2.24 Cont.

Thevenin equivalent circuit

($\frac{23}{4}V$ source in series with $\frac{7}{2}\Omega$)

2.25 By the Thevenin equivalent Circuit

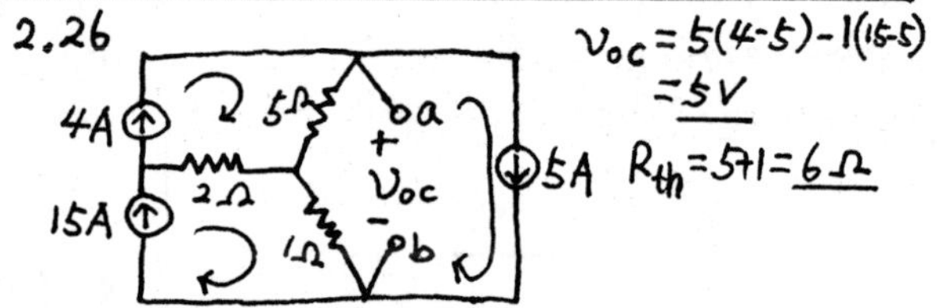

$(4R+2R+R)$

$i=\frac{4}{4\times10^3}=1mA$

$v=4V,\ \frac{7R}{3R+4\times10^3}=10^{-3},$

$7R=3\times10^{-3}R+4$

By approximation, $7R=4,\ R=\frac{4}{7}\Omega$

2.26

$v_{oc}=5(4-5)-1(15-5)=-5V$

$R_{th}=5+1=6\Omega$

From the Thevenin equivalent circuit, voltage division gives $v=\frac{4}{4+6}(5)=2V$

2.27 The 2A current source will force i_{NT} to be 2A. Hence the equivalent circuits are:

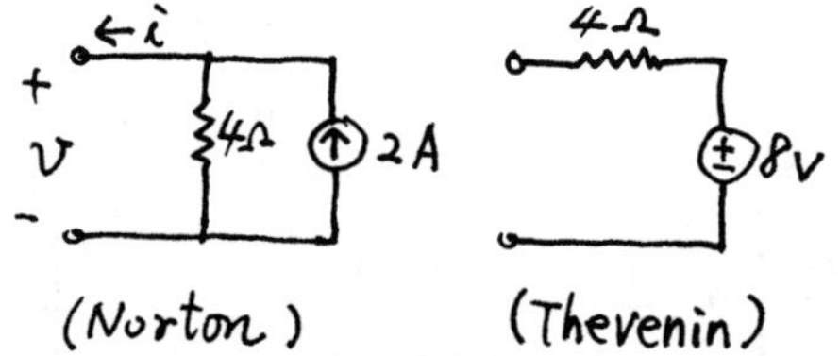

(Norton) (Thevenin)

2.28

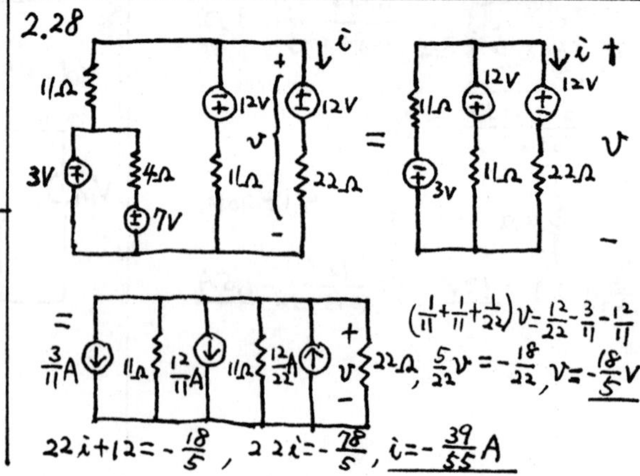

$\left(\frac{1}{11}+\frac{1}{11}+\frac{1}{22}\right)v=\frac{12}{22}-\frac{3}{11}-\frac{12}{11}$

$\frac{5}{22}v=-\frac{18}{22},\ v=-\frac{18}{5}V$

$22i+12=-\frac{18}{5},\ 22i=-\frac{78}{5},\ i=-\frac{39}{55}A$

2.29 Within the indicated region, a diode has a resistance of $\frac{(2-1)}{(0.3-0.1)} = 5\Omega$

$\because$ the straight line intercept the x axis at $v \simeq 0.67V$, $i = 0.2V + 0.67$.

The Thevenin equivalent circuit of a junction diode is

0.67V $R_{th} = 5\Omega$ v_{th} + -

2.30 (a) $R = (1 \times 10 + 0) \times 10^2 \pm 10\% = 1000 \pm 100$

resistance range = 900–1100 Ω

(b) $R = (2 \times 10 + 7) \times 10^4 \pm 10\% = 270k\Omega \pm 27k\Omega$

resistance range = 243–297 $k\Omega$

(c) $R = (6 \times 10 + 8) \times 10^{-2} \pm 5\% = 680 \pm 34\ m\Omega$

resistance range = 646–714 $m\Omega$

Chapter 3
Dependent Sources and Op Amps

3.1 Circuits with dependent sources

3.1 Find the voltage v across the 1Ω resistor.

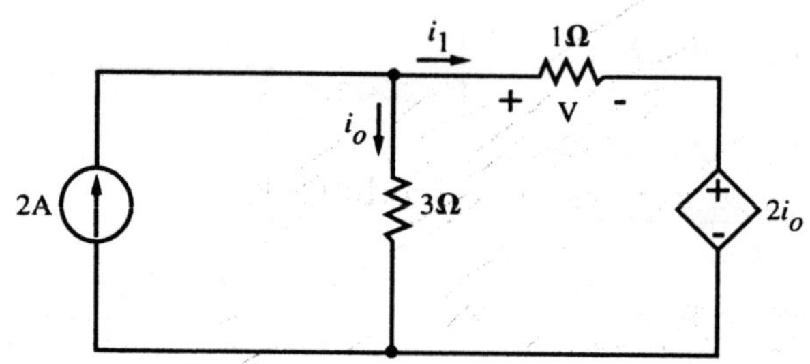

Problem 3.1

3.2 Find the power delivered to the 6Ω resistor.

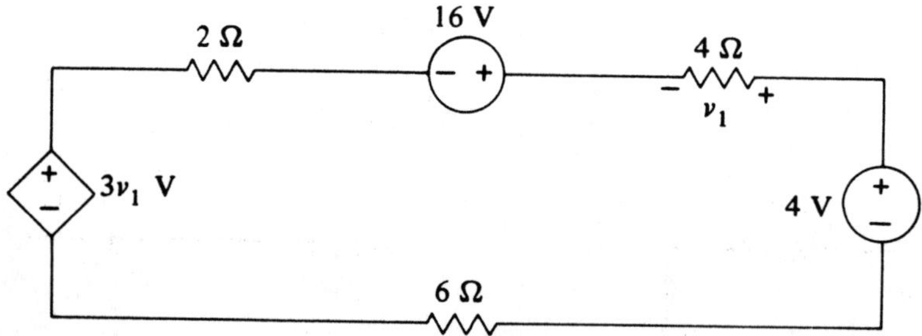

Problem 3.2

3.3 Find v.

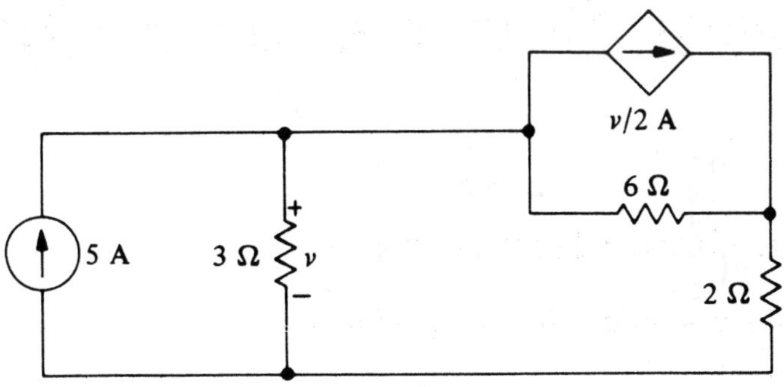

Problem 3.3

3.4 Find the voltage v_{CE} across terminals C-E.

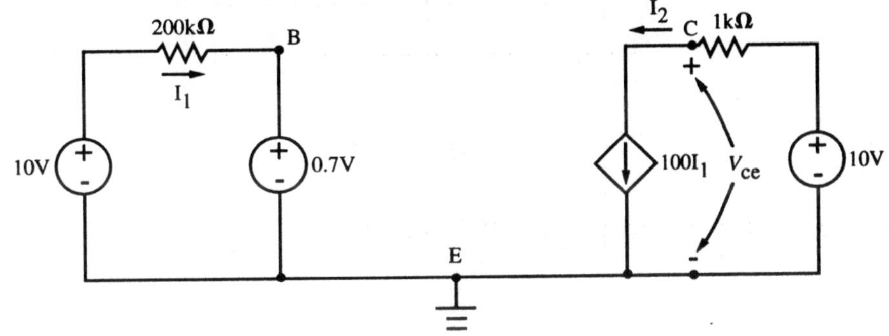

Problem 3.4

3.5 Find v.

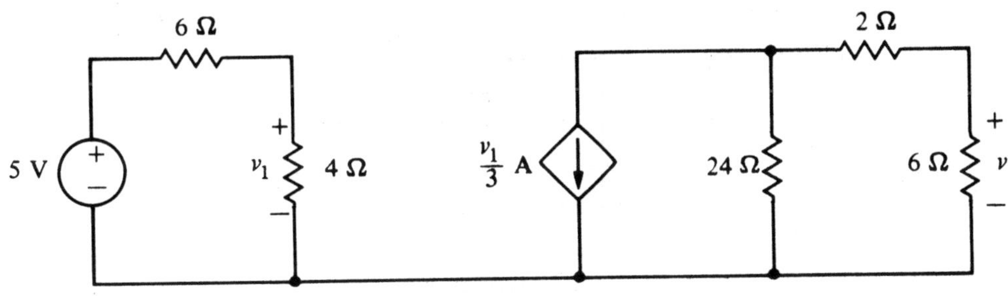

Problem 3.5

3.6 In Prob.3.5 if the voltage v equals to 4V, what is the transconductance of the dependent current source?

3.7 **Find v.**

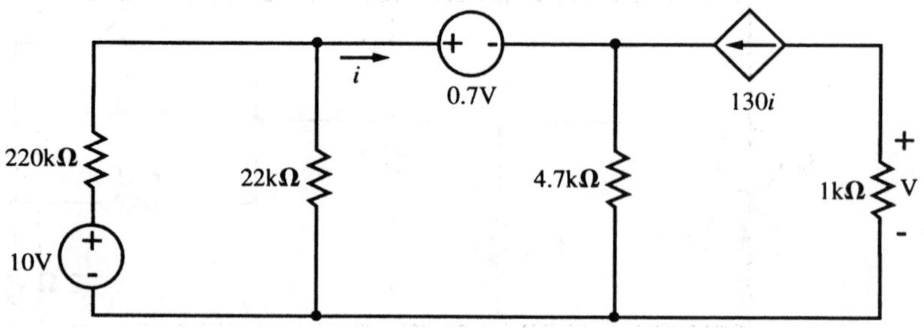

Problem 3.7

3.8 **Find the Thevenin equivalent circuit.**

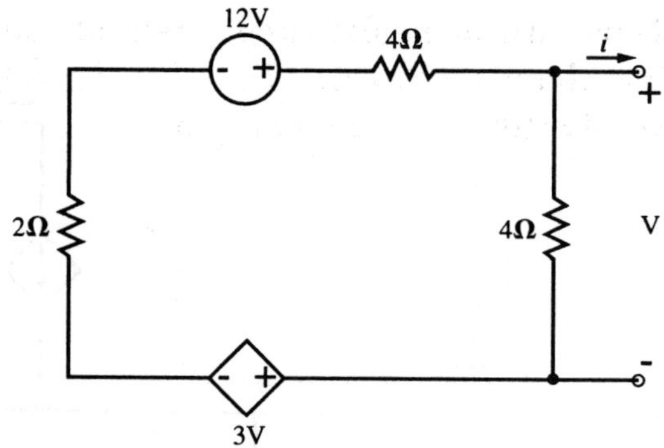

Problem 3.8

3.9 **Find the Thevnin and Norton equivalent circuits.**

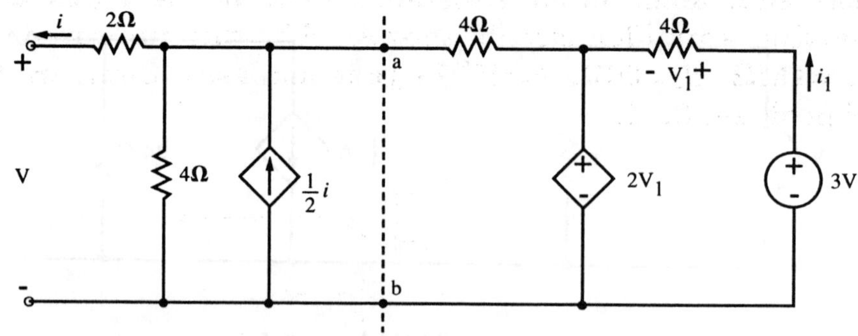

Problem 3.9

3.10 Let $R_1 = R_2 = \frac{1}{10}\Omega$, $i_g = 3\text{A}$. Find the resistance seen by the source.

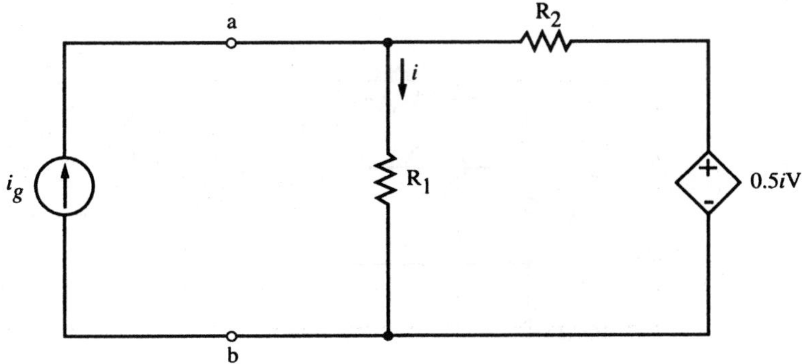

Problem 3.10

3.2 Operational amplifiers

3.11 Find the equivalent input resistance seen at the input terminals of the inverting amplifier circuit shown. Use the improved op amp model with $R_i = 1\text{M}\Omega$, $R_o = 30\Omega$, $A = 10^5$ in your analysis.

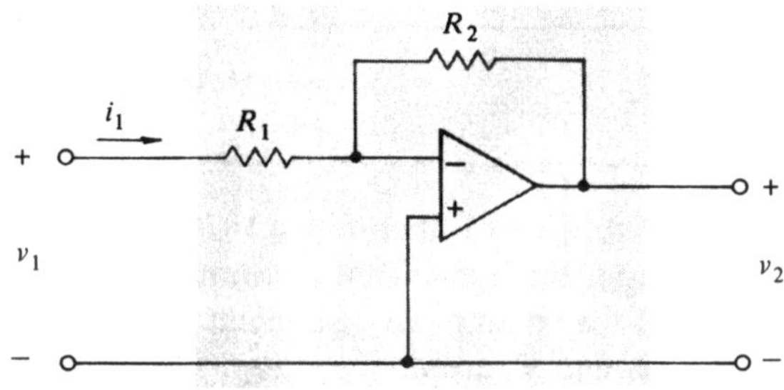

Problem 3.11

3.12 Find the equivalent input resistance seen at the input terminals of the noninverting amplifier circuit shown. Use the improved op amp model with $R_i = 1\text{M}\Omega$, $R_o = 30\Omega$, $A = 10^5$ in your analysis. Compare the result with that of problem 3.11.

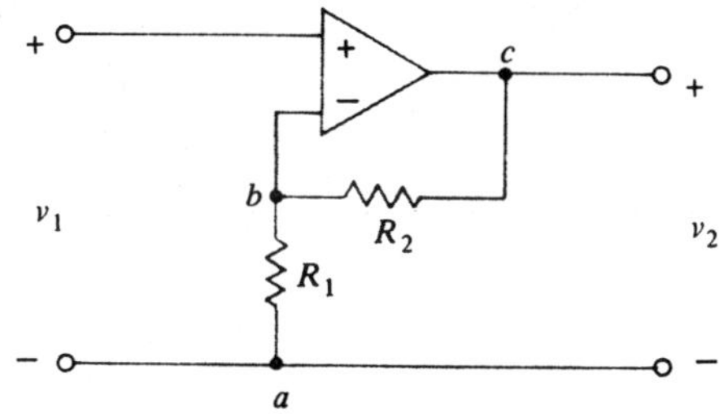

Problem 3.12

3.13 **Find k in the voltage transfer function $v_O = kv_i$.**

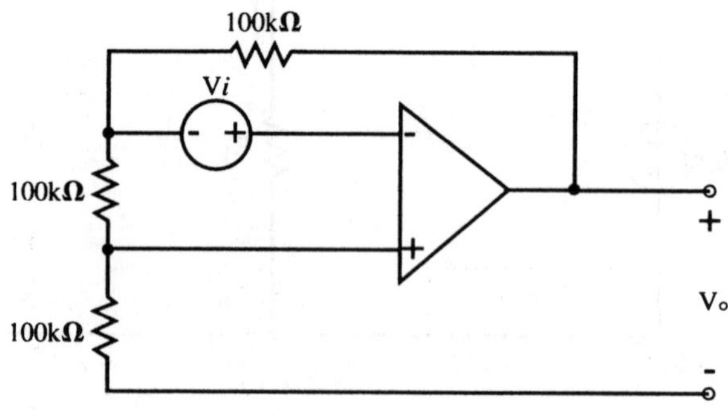

Problem 3.13

3.14 **Show that $v_O = -i_{in}R$ for the given circuit. This simple amplifier circuit can be used as a current-to-voltage converter. By Ohm's law, a single resistor also behaves as a current-to-voltage converter. What is the advantage of such amplifier-based current-to-voltage converter over a resistive one?**

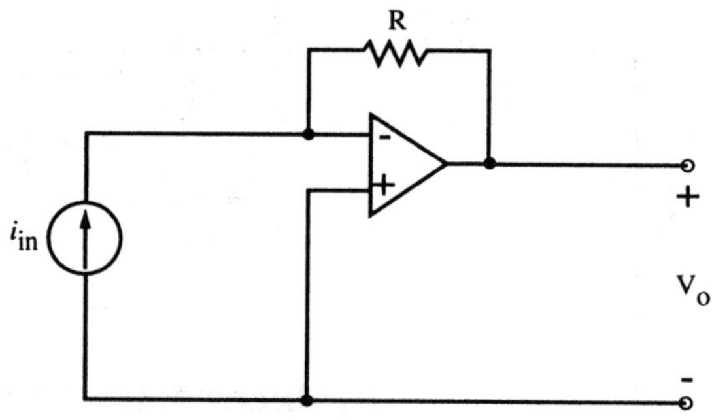

Problem 3.14

3.15 **Find v.**

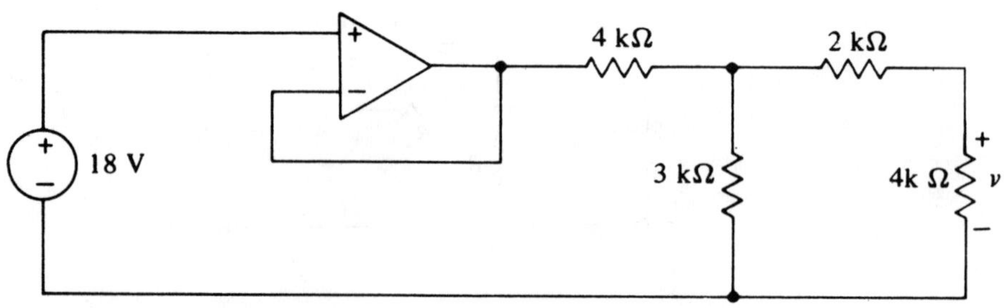

Problem 3.15

3.16 Find v_1 in (a) and v_2 in (b).

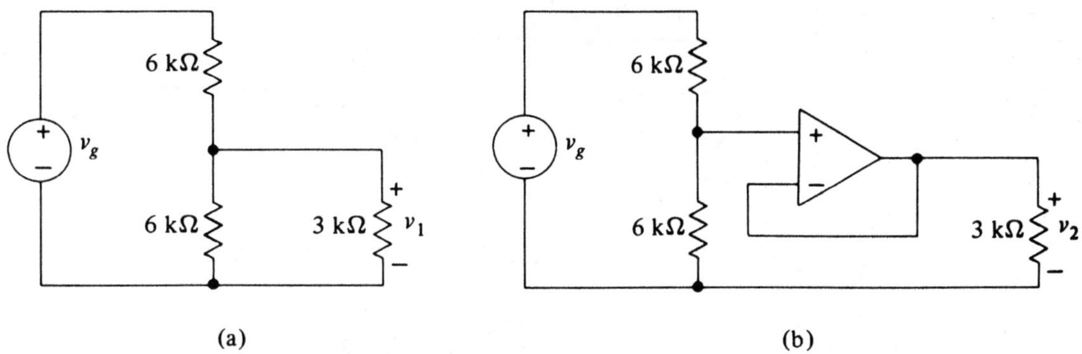

Problem 3.16

3.17 Compute the power delivered to the 60KΩ resistor with and without the indicated voltage follower (a and b shorted together).

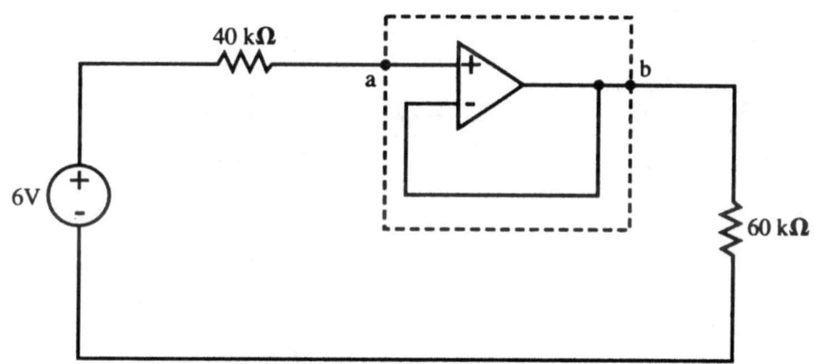

Problem 3.17

3.18 (a) Using an inverting amplifier, design a circuit with $v_O = -8v_i$. (b) Design another one with $v_O = -\frac{1}{8}v_i$.

3.19 If we wish to deliver $\frac{1}{4}$mW of power to the 25KΩ load resistor. What is the value of R_F?

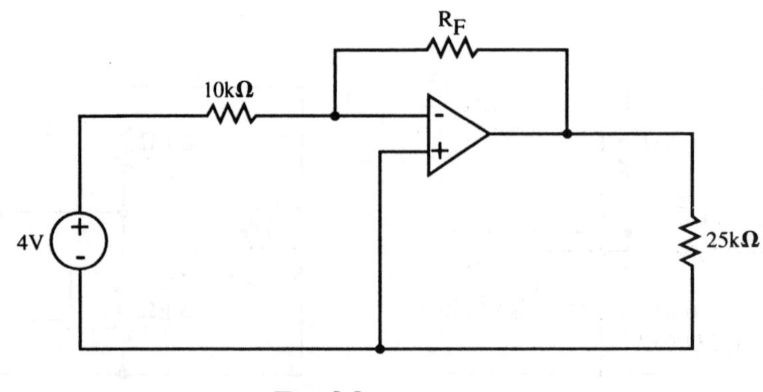

Problem 3.19

3.20 Find v_O.

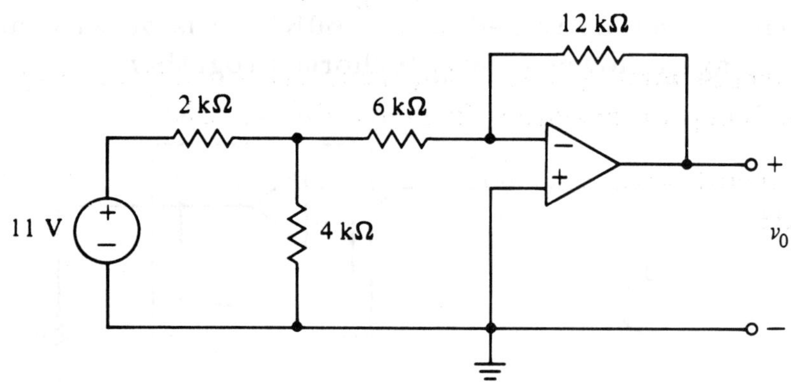

Problem 3.20

3.21 Connect a resistance of R=5KΩ across terminals a-b and find the resulting current i_{ab} if v_1=1V, v_2=3V, R_0=4KΩ, R_1=1KΩ, and R_2=2KΩ.

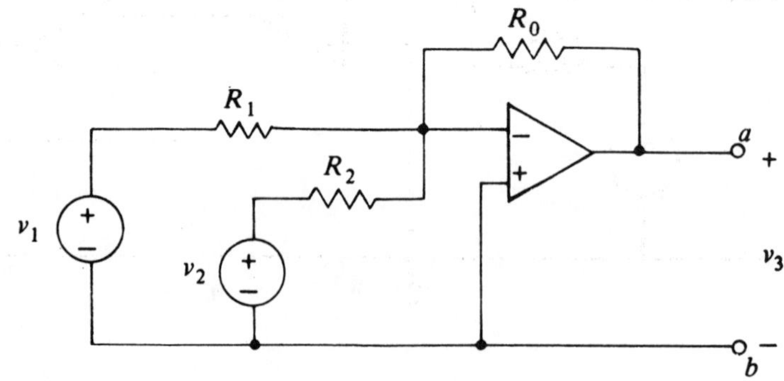

Problem 3.21

3.22 **Design an amplifier circuit with $v_O = 125v_i$ by cascading three identical amplifier stages. Keep all resistors in range 5KΩ-500KΩ.**

3.23 **Find i.**

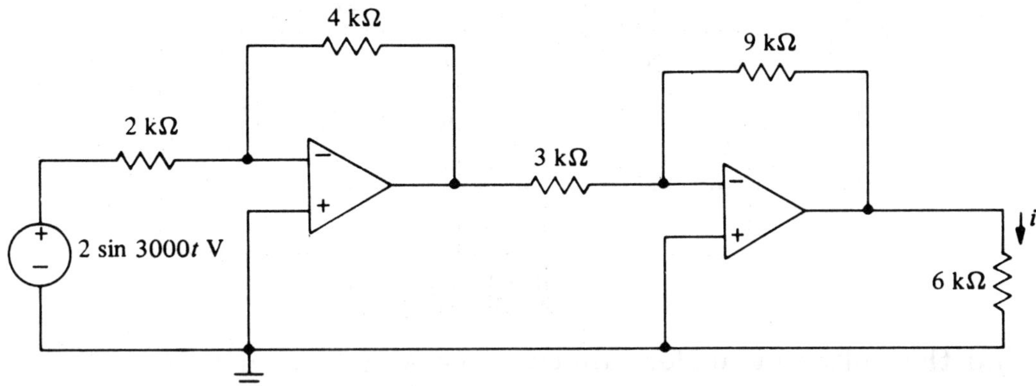

Problem 3.23

3.24 **Design a circuit with $v_O = 7v_1 + 3v_2$ using cascaded inverting amplifier building blocks. Keep all resistors in range 5KΩ-500KΩ.**

3.25 **Design a circuit with $v_O = -4v_1 - 5v_2 + 2v_3 + v_4$. Keep all resistors in range 5KΩ-500KΩ.**

3.26 **Find v.**

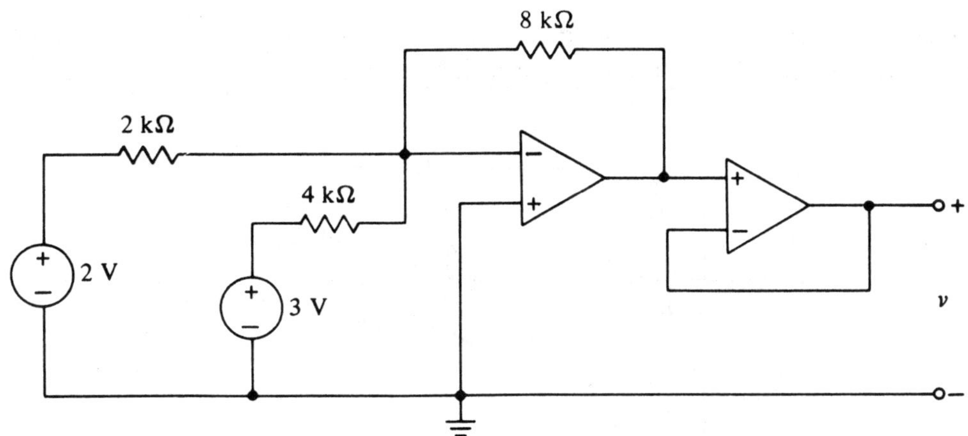

Problem 3.26

3.27 **Find** v.

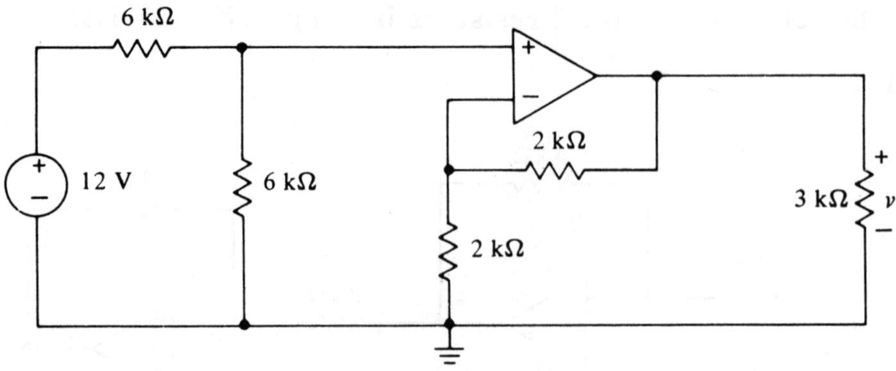

Problem 3.27

3.28 **Find the voltage transfer function** $v_O = k_1v_1 + k_2v_2 + k_3v_3$.

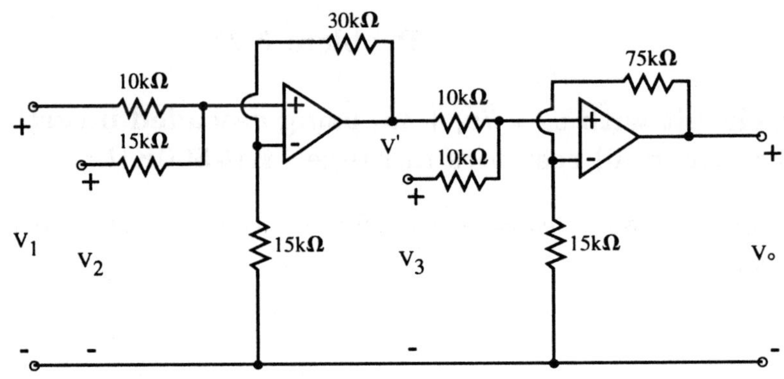

Problem 3.28

3.29 **Find the voltage transfer function** $v_3 = k_1v_1 + k_2v_2$ **when** R=**47KΩ**

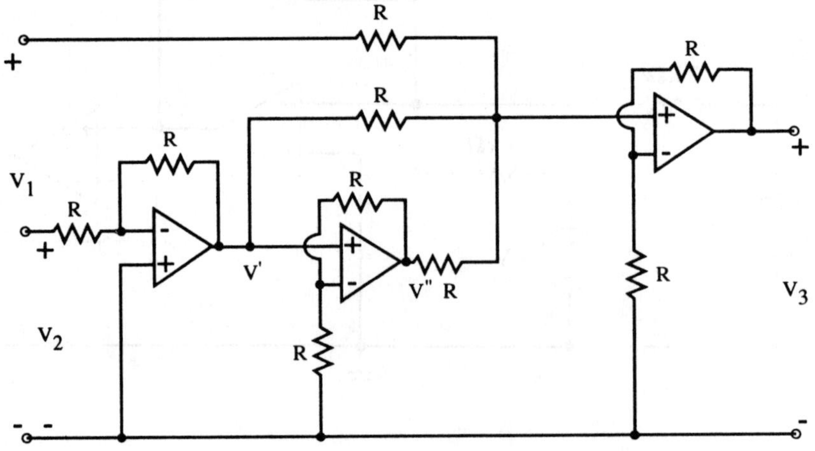

Problem 3.29

3.30 Show that $v_O = \frac{R_2}{R_1}(v_1 - v_2)$ for the circuit shown with $R_1 = R_3$, $R_2 = R_4$.

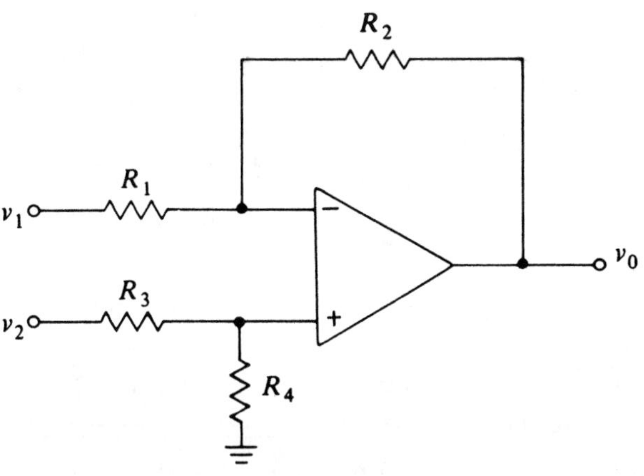

Problem 3.30

3.31 Show that $v_O = (1 + \frac{2R_2}{R_1})\frac{R_4}{R_3}(v_2 - v_1)$.

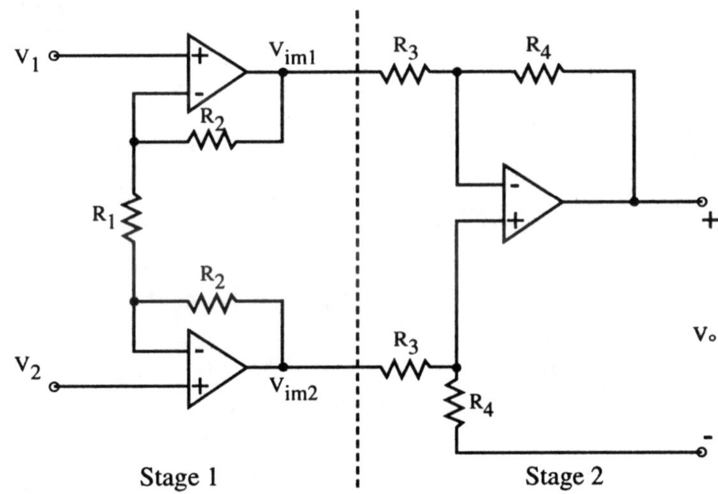

Problem 3.31

3.1 By KVL, $3i_o = v + 2i_o$, $v = i_o$

By KCL, $2 = i_o + i_1 = i_o + v = i_o + i_o = 2i_o$

$i_o = 1A$, $v_o = \underline{1V}$

3.2 i leaves the + terminal of the 16-V source: $i = -\frac{v_1}{4}$,

By KVL, $16 + v_1 - 4 + 6\frac{v_1}{4} + 3v_1 + 2\frac{v_1}{4} = 0$

$v_1 = \frac{12}{-3/2 - 1/2 - 4} = -2V$, $i = -\frac{2}{4} = -\frac{1}{2}$

$P_{6\Omega} = 6i^2 = \underline{1.5\ W}$

3.3 Current downward in 3-Ω resistor is $\frac{v}{3}$. By KCL the current down in 2-Ω resistor is $5 - \frac{v}{3}$. Current to the right in 6Ω resistor is, by KCL, $5 - \frac{v}{3} - \frac{v}{2} = 5 - \frac{5v}{6}$. By KVL around the loop containing only resistors.

$v - 6(5 - 5v/6) - 2(5 - v/3) = 0$

$\therefore v = \underline{6V}$

3.4 $I_1 = \frac{10 - 0.7}{220\times10^3} = 44\times10^{-6}A$

$I_2 = 100 I_1 = 44\times10^{-3}A$

$V_{CE} = 10 - 10^3 \times 44\times10^{-3} = 10 - 4.4 = \underline{5.6V}$

3.5 By voltage division,

$v_1 = \frac{4}{6+4}(5) = 2V$. Let i be the current entering + terminal of v.

Then by current division

$i = \frac{24}{24+6+2}\left(-\frac{v_1}{3}\right) = -\frac{1}{2}A$

$\therefore v = 6\left(-\frac{1}{2}\right) = \underline{-3V}$

3.6 By voltage division, $v_1 = \frac{4}{10}(5) = 2V$

$g v_1 \times \left(\frac{24}{24+8}\right)(6) = 4$, $2g\left(\frac{3}{4}\right)(6) = 4$, $g = \frac{4}{9}$

3.7 $220K // 22K = 20K\Omega$

$\frac{22}{220+20}(5) = 0.45V$

By the Thevenin equivalent, the circuit is simplified as:

20 KΩ, 0.91 V, i, 0.7V, i_2, $130i$, 4.7KΩ, 1KΩ, + v −

3.7 Cont.

By KCL around the left loop,

$0.91 = 2\times10^3 i + 0.7 + i(130+1)\times4.7\times10^3$

$i = 0.0157\times10^{-3} = 1.57\ uA$

$v = 130 i \times 10^3 = \underline{0.204\ mV}$

3.8 Let the loop current i_ℓ, $i_\ell = \frac{v}{4}$

KVL: $v + 3v + \frac{6}{4}v = 12$, $v = 2V$

$R_{th} = (4+2)//4 = \frac{12}{5}\Omega$

The Thevenin equivalent circuit is

3.9 By KVL, $2v_1 + v_1 = 3V$, $v_1 = 1V$

By Thevenin-Norton equivalence, the circuit given can be simplified as:

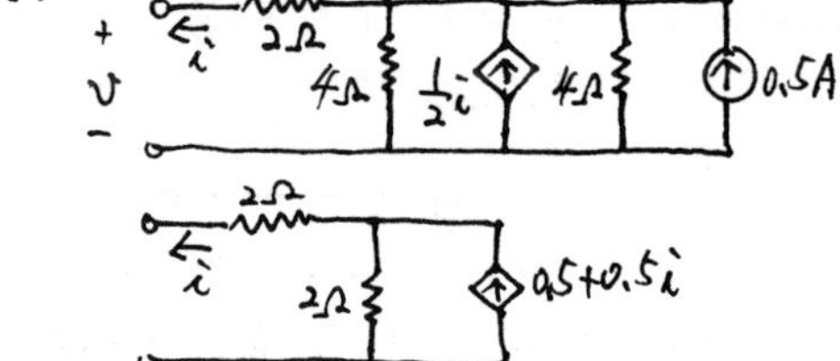

= 2Ω, i, 2Ω, $0.5 + 0.5i$

$i_n = \frac{1}{2}(0.5 + 0.5i) = 0.25 + 0.25i$, $R_n = 4\Omega$

The Norton and Thevenin equivalents are:

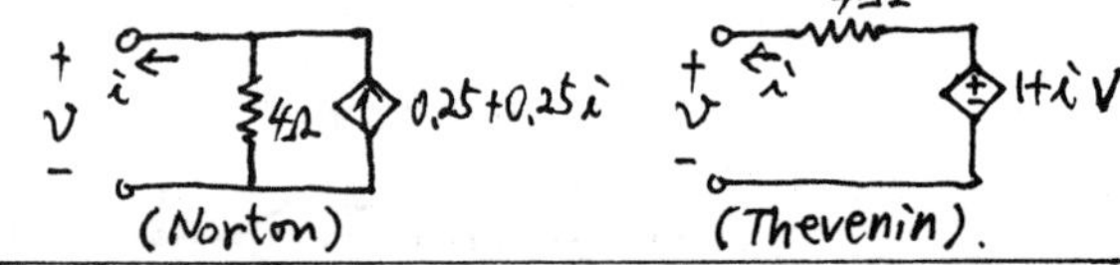

(Norton) (Thevenin).

3.10 $i = \frac{(\frac{1}{10})3}{\frac{1}{10} + \frac{1}{10} - 0.5} = -1A$, $v_{a-b} = (-1)\left(\frac{1}{10}\right) = -\frac{1}{10}V$

$R_{a-b} = -\frac{1}{10}/3 = -\frac{1}{30}\Omega$

3.11 Use the improved Op-amp model

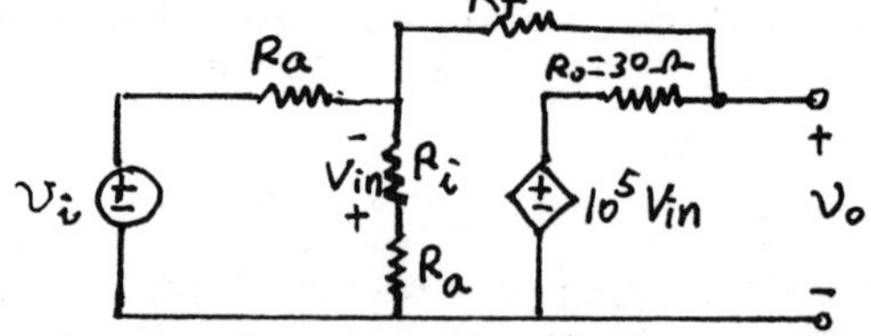

By the Thevenin equivalent

R_{eq}, R_a, R_f, $R_i = 1M\Omega$, R_a', $R_o = 30$

$R_{eq} = R_a + (1M + R_a')//(R_f + 30)$

assume R_a, R_a' and R_f are in the range of 1k ~ 100k

$\underline{R_{eq} \cong R_a + R_f + 30}$

3.12

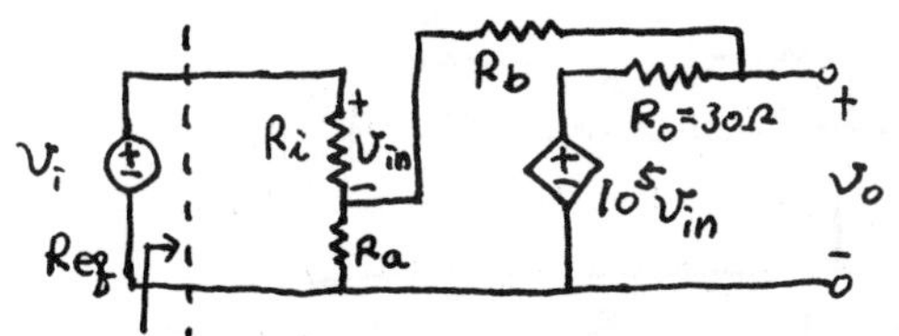

By the Thevenin equivalent circuit.

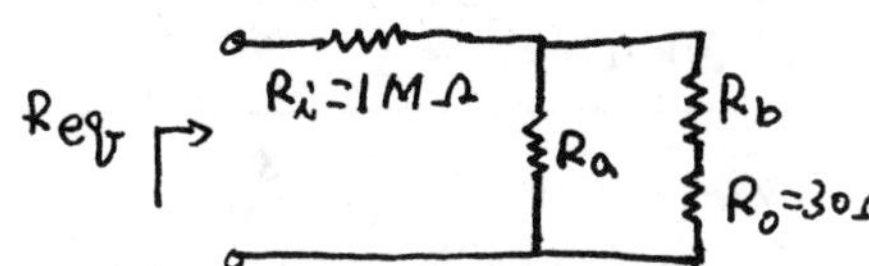

$R_{eq} = R_i + R_a // (R_b + 30)$. Assume R_a, R_b are in the range of $1K-100k\Omega$.

$R_{eq} \cong R_i$ which is in the range of $\underline{1M\Omega}$

3.13

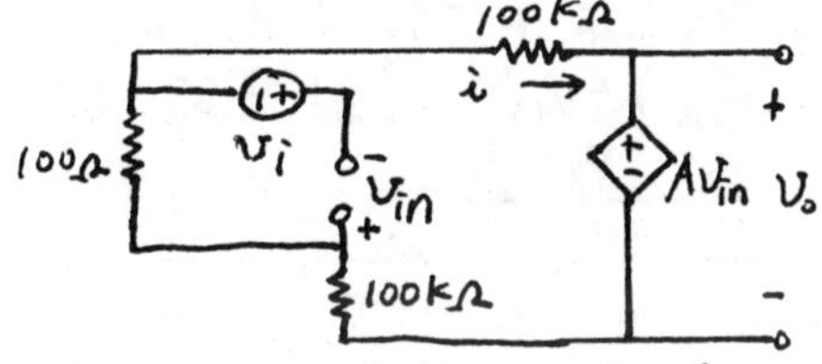

Apply KVL around the right loop

$v_i + v_{in} + 100 \cdot 10^3 i + v_o + 100 \cdot 10^3 i = 0$ and

$v_{in} = \frac{v_o}{A}$, $\quad v_i + \frac{v_o}{A} + 200 \cdot 10^3 i + v_o = 0$ —(1)

Apply KCL around the left loop

$100 i = v_i + v_{in} = v_i + \frac{v_o}{A}$ —(2)

Substitute (2) into (1)

$v_i + \frac{v_o}{A} + 2000 v_i + 2000 \frac{v_o}{A} + v_o = 0$

$2001 v_i + (\frac{1}{A} + \frac{2000}{A} + 1) v_o = 0$

As A becomes arbitrarily large,

$\frac{1}{A} + \frac{2000}{A} + 1 \cong 1$, $2001 v_i + v_o = 0$, $k = \frac{v_o}{v_i} = \underline{-2001}$

3.14

R

i_{in} $\quad v_{in}$ $\quad Av_{in}$ $\quad v_o$

By KVL, $v_{in} + i_{in} R + A v_{in} = 0$, and $v_o = A v_{in}$

$(\frac{1+A}{A}) v_o = -i_{in} R$

As A is arbitrarily large, $\underline{v_o = -i_{in} R}$

Consider a single resistor current-to-voltage converter. When a load resistance R_L is connected, the effective resistance seen by the input current source is

$R // R_L = \frac{RR_L}{R+R_L}$. Hence $\underline{v_o = -i_{in} \frac{RR_L}{R+R_L}}$

3.14 Cont.

Compare $v_o = -i_{in} \frac{RR_L}{R+R_L}$ with the result using an inverting amplifier. There is a voltage $-i_{in}(R - \frac{RR_L}{R+R_L}) = -i_{in}(\frac{R^2}{R+R_L})$

This voltage drop is significant when R_L is not large enough relative to R

3.15 Output of voltage follower is 18V

By voltage division

$$v_{3k\Omega} = \frac{(3)(2+4)/(3+2+4)}{4+2}(18) = 6V$$

$$v = \frac{4}{2+4} v_{3k\Omega} = \underline{4V}$$

3.16 (a) By voltage division

$$v_1 = \frac{(\frac{6 \cdot 3}{6+3})(v_g)}{\frac{6 \cdot 3}{6+3} + 6} = \underline{v_g/4}$$

(b) The current into the noninverting terminal is zero. Therefore the 6-kΩ resistors carry the same current and constitute a voltage divider. Input voltage for voltage follower is $v_g/2$, $\therefore \underline{v_2 = v_g/2}$

3.17 Without the voltage follower

$$i = \frac{6}{40k + 100k} = 60 \times 10^{-6} A$$

$$P = i^2 R = (60 \times 10^{-6})^2 (60 \times 10^3) = \underline{0.216\ mW}$$

With the voltage follower

$$i = \frac{6}{60 \cdot 10^3} = 100 \times 10^{-6} A$$

$$P = i^2 R = (100 \times 10^{-6})^2 \times (60 \times 10^3) = \underline{0.6\ mW}$$

3.18 (a)

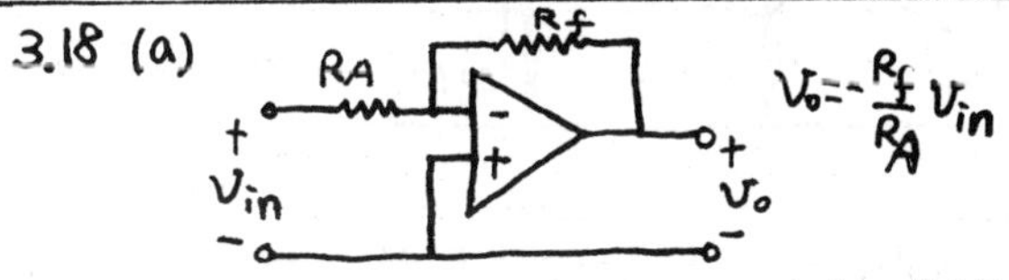

Choose $R_f = 80k\Omega$, $R_A = 10k\Omega$, $\underline{v_o = -8 v_{in}}$

(b) Choose $R_f = 10k\Omega$, $R_A = 80k\Omega$

$\underline{v_o = -\frac{1}{8} v_{in}}$

3.19 $p = vi = \frac{v^2}{R} = \frac{v^2}{25\times10^3} = \frac{1}{4\times10^3}$

$v^2 = \frac{25\times10^3}{4\times10^3} = \frac{25}{4}$

∵ This is an inverting amplifier

$v = -\frac{5}{2}, \quad v = -\frac{R_F}{10\times10^3}(4) = -\frac{5}{2}$

$R_F = \frac{10^4\times5}{8} = \underline{6.25k\Omega}$

3.20 Since the op amp input voltage is zero, the 6-kΩ and 4 kΩ resistors have same voltages (v_1). By voltage division

$v_1 = \frac{6(4)/(6+4)}{2+(6)(4)/(6+4)}(11) = 6V$

Since the op amp input current are zero, the 6-kΩ and 12-kΩ have the same current (i_1) to the right.

$i_1 = \frac{6V}{6k\Omega} = 1mA$

KVL around the loop containing v_o

$v_o + 12(1) + 0 = 0$ or $v_o = \underline{12V}$

3.21 $v_3 = -R_0\left(\frac{v_1}{R_1} + \frac{v_2}{R_2}\right) = -4\left(\frac{1}{1} + \frac{3}{2}\right) = -10V$

$i_{ab} = \frac{v_3}{R} = -\frac{10V}{5k\Omega} = \underline{-2\ mA}$

3.22 One can design a three stage amplifier with voltage gain of 5 for each stage. A non-inverting amplifier can be used as basic building block.

$v_o = \left(1 + \frac{R_f}{R_a}\right)v_i, \quad 1 + \frac{R_f}{R_a} = 5, \quad R_f = 4R_a$

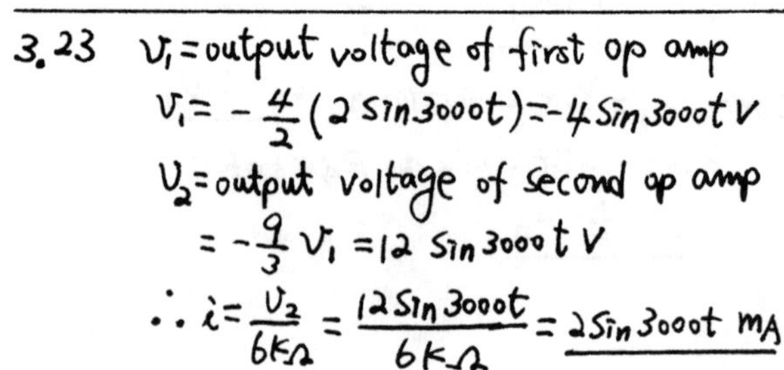

3.23 v_1 = output voltage of first op amp

$v_1 = -\frac{4}{2}(2\sin 3000t) = -4\sin 3000t\ V$

v_2 = output voltage of second op amp

$= -\frac{9}{3}v_1 = 12\sin 3000t\ V$

$\therefore i = \frac{v_2}{6k\Omega} = \frac{12\sin 3000t}{6k\Omega} = \underline{2\sin 3000t\ mA}$

3.24

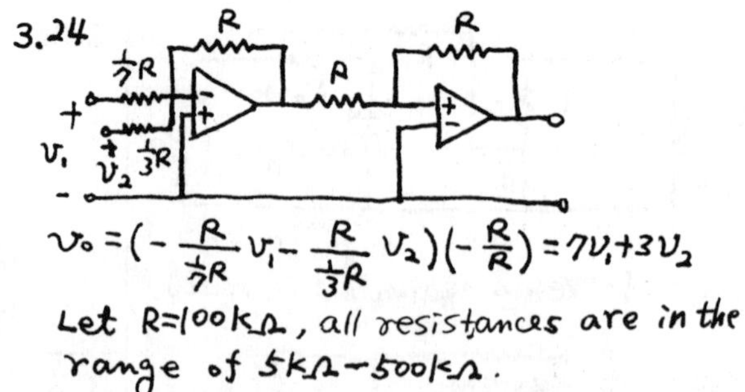

$v_o = \left(-\frac{R}{\frac{1}{7}R}v_1 - \frac{R}{\frac{1}{3}R}v_2\right)\left(-\frac{R}{R}\right) = 7v_1 + 3v_2$

Let R=100kΩ, all resistances are in the range of 5kΩ–500kΩ.

3.25

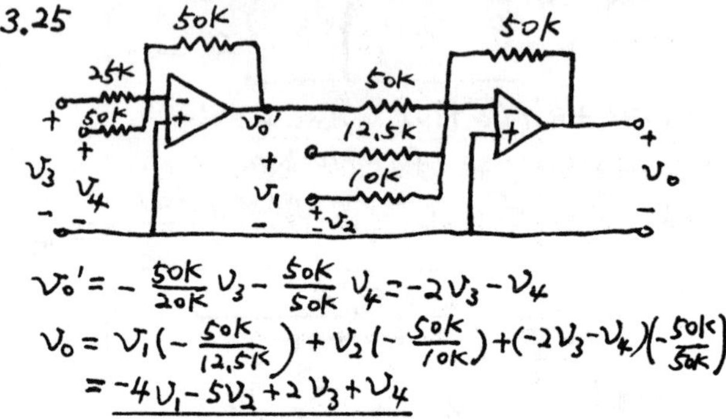

$v_o' = -\frac{50k}{20k}v_3 - \frac{50k}{50k}v_4 = -2v_3 - v_4$

$v_o = v_1\left(-\frac{50k}{12.5k}\right) + v_2\left(-\frac{50k}{10k}\right) + (-2v_3 - v_4)\left(-\frac{50k}{50k}\right)$

$= \underline{-4v_1 - 5v_2 + 2v_3 + v_4}$

3.26 v = output of the first op amp

$= -8\left(\frac{2}{2} + \frac{3}{4}\right) = \underline{-14V}$

3.27 Current in op amp terminals is zero. By voltage division, the VCVS input voltage is

$v_1 = \frac{6}{6+6}(12) = 6V$

Therefore $v = \left(1 + \frac{2}{2}\right)v_1 = \underline{12V}$

3.28 $v' = \left(\frac{10k//15k}{10k}v_1 + \frac{10k//15k}{15k}v_2\right)\left(1 + \frac{30k}{15k}\right)$

$= (6v_1 + 4v_2)(3) = 18v_1 + 12v_2$

$v_o = \left(\frac{10k//10k}{10k}v' + \frac{10k//10k}{10k}v_3\right)\left(1 + \frac{75k}{15k}\right)$

$= \left(\frac{1}{2}v' + \frac{1}{2}v_3\right)(6)$

$= 3v' + 3v_3 = \underline{54v_1 + 36v_2 + 3v_3}$

3.29 $v' = -\frac{R}{R}v_2 = -v_2$

$v'' = \left(1 + \frac{R}{R}\right)v' = -2v_2$

$v_3 = \left(1 + \frac{R}{R}\right)v_1 + v'\left(1 + \frac{R}{R}\right) + v''\left(1 + \frac{R}{R}\right)$

$= 2v_1 + (-v_2)(2) + (-2)(v_2)(2)$

$= \underline{2v_1 - 6v_2}$

3.30 By superposition principle

$$v_o = v_{o1}\Big|_{v_2 \text{ is killed}} + v_{o2}\Big|_{v_1 \text{ is killed}}$$

$$v_{o1} = \left(1+\frac{R_2}{R_1}\right)\left(\frac{R_2}{R_1+R_2}v_1\right) = \frac{R_1+R_2}{R_1}\frac{R}{R_1+R_2}v_1$$

$$= \frac{R_2}{R_1}v_1$$

$$v_{o2} = -\frac{R_2}{R_1}v_2$$

Hence $v_o = v_{o1}+v_{o2} = \frac{R_2}{R_1}v_1 - \frac{R_2}{R_1}v_2$

$$= \left(\frac{R_2}{R_1}\right)(v_1 - v_2)$$

3.31 Separate the circuit into two stages and analyze them separately.

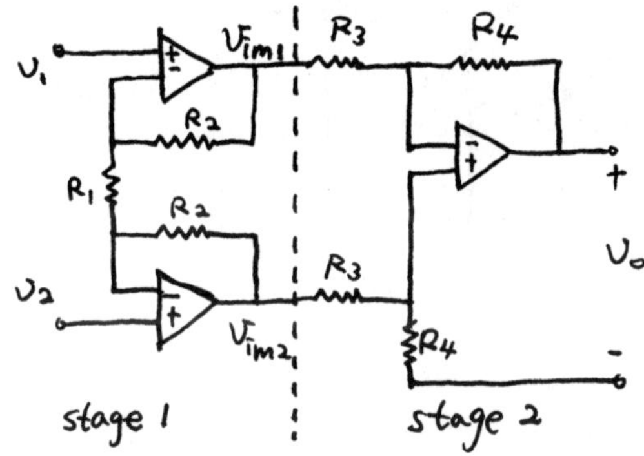

For stage 2, using the result of prob(3.30)

$$v_o = \frac{R_4}{R_3}(v_{im2} - v_{im1})$$

For stage 1, write its op amp model as

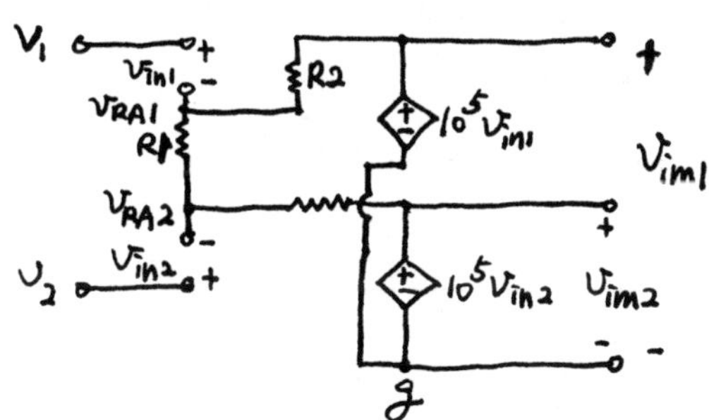

All voltages are with respect to the common negative terminal g except v_{in1} and v_{in2}.

3.31 Cont.

By voltage division

$$\begin{cases} v_{RA1} = \dfrac{R_1+R_2}{R_2+R_1+R_2}(v_{im1}-v_{im2}) \\ v_1 = v_{in1} + v_{RA1} \\ v_{in1} = v_{im1}/A \end{cases}$$

$$\Rightarrow v_1 = \frac{v_{im1}}{A} + \frac{R_1+R_2}{R_1+2R_2}(v_{im1}-v_{im2}) \quad \text{—}\langle 1\rangle$$

and

$$\begin{cases} v_{RA2} = \dfrac{R_2}{R_2+R_1+R_2}(v_{im1}-v_{im2}) \\ v_2 = v_{in2} + v_{RA2} \\ v_{in2} = v_{im2}/A \end{cases}$$

$$\Rightarrow v_2 = \frac{v_{im2}}{A} + \frac{R_2}{R_1+2R_2}(v_{im1}-v_{im2}) \quad \text{—}\langle 2\rangle$$

$\because$ A is arbitrarily large, $\frac{v_{im1}}{A} = \frac{v_{im2}}{A} \cong 0$

From ⟨1⟩ and ⟨2⟩

$$v_1 - v_2 = \frac{R_1}{R_1+2R_1}(v_{im1} - v_{im2})$$

$$v_{im2} - v_{im1} = \frac{R_1+2R_2}{R_1}(v_2 - v_1)$$

Use the result of stage 2, $v_o = \frac{R_4}{R_3}(v_{im2} - v_{im1})$

$$v_o = \left(\frac{R_4}{R_3}\right)\left(\frac{R_1+2R_2}{R_1}\right)(v_2-v_1) = \left(1+\frac{2R_2}{R_1}\right)\left(\frac{R_4}{R_3}\right)(v_2-v_1)$$

Chapter 4
Analysis methods

4.1 Linearity and Proportionality

4.1 Express v as a linear combination of the three source functions v_{g1}, v_{g2} and i_{g1}.

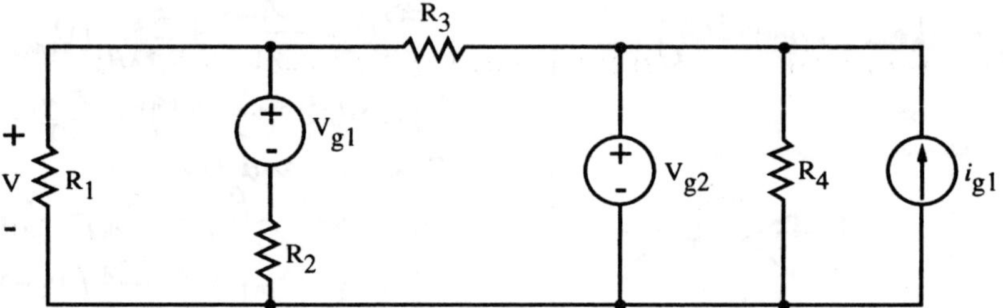

Problem 4.1

4.2 Use proportionality to find i_1 and i_2.

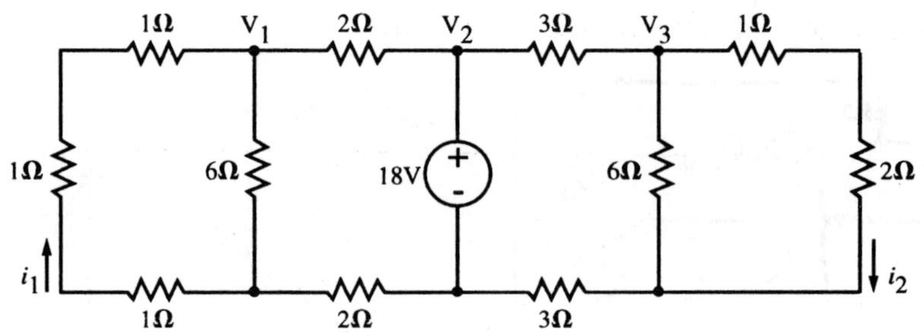

Problem 4.2

4.3 Use proportionality to find v.

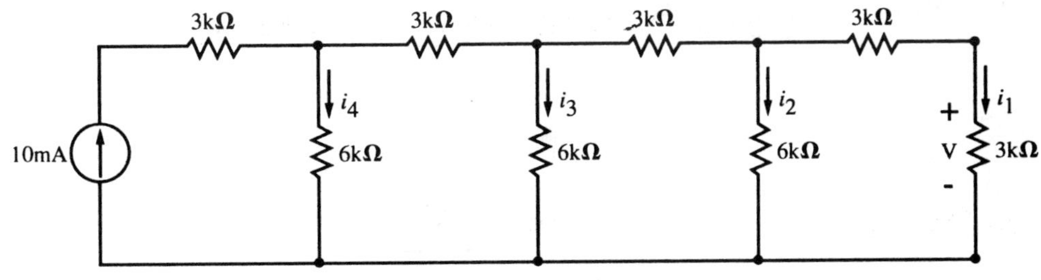

Problem 4.3

4.2 Superposition

4.4 Find I by superposition. Check using Thevenin-Norton transformations.

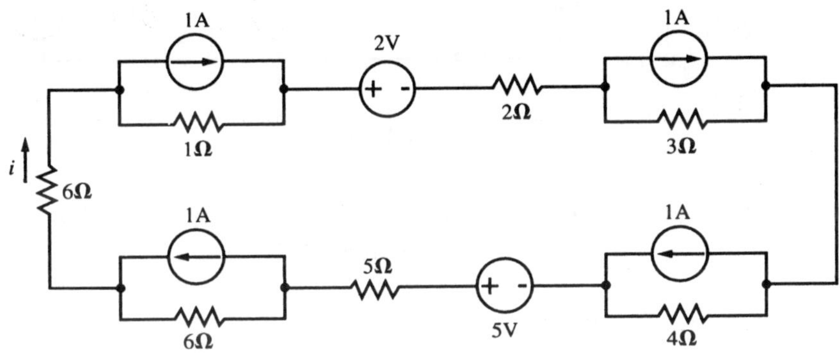

Problem 4.4

4.5 Find the current i of the circuit shown.

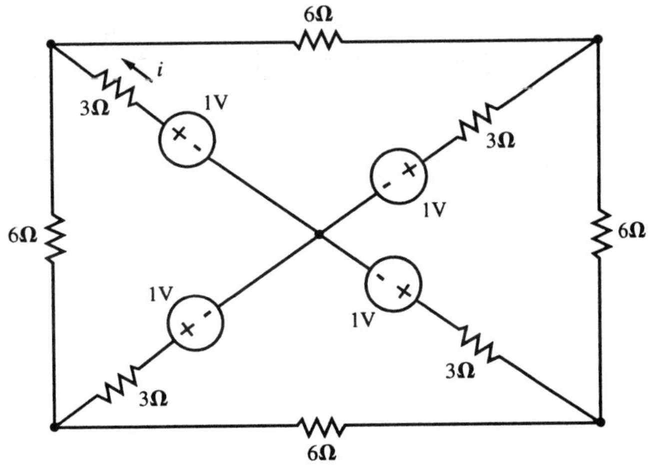

Problem 4.5

4.3 Nodal Analysis

4.6 Using nodal analysis, find i.

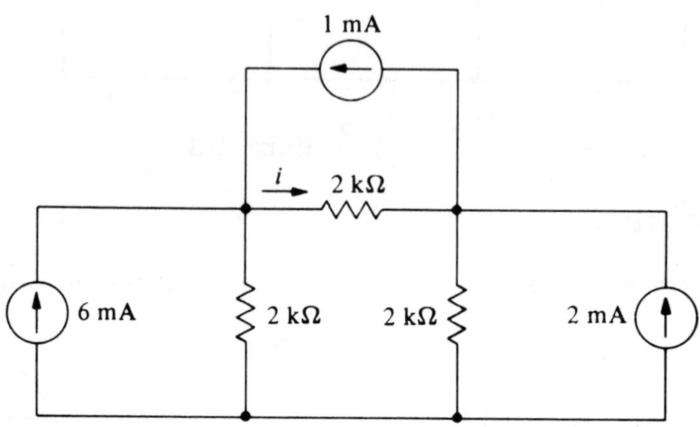

Problem 4.6

4.7 Using nodal analysis, find v_1 and v_2, if R_1=2Ω, R_2=1Ω, R_3=4Ω, i_{g1}=14A, and i_{g2}=7A.

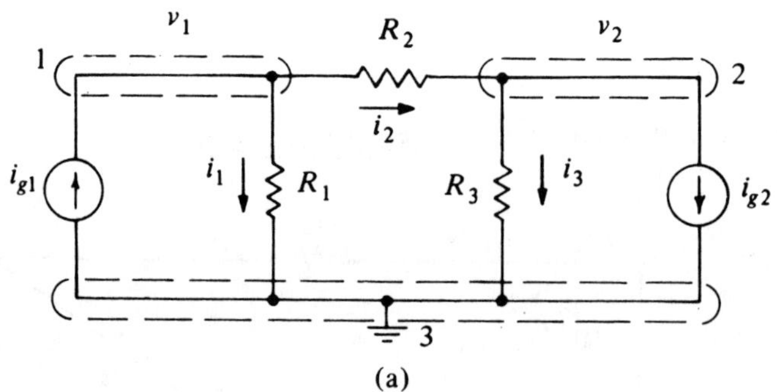

(a)

Problem 4.7

4.8 Using nodal analysis, find v_1, v_2 and v_3.

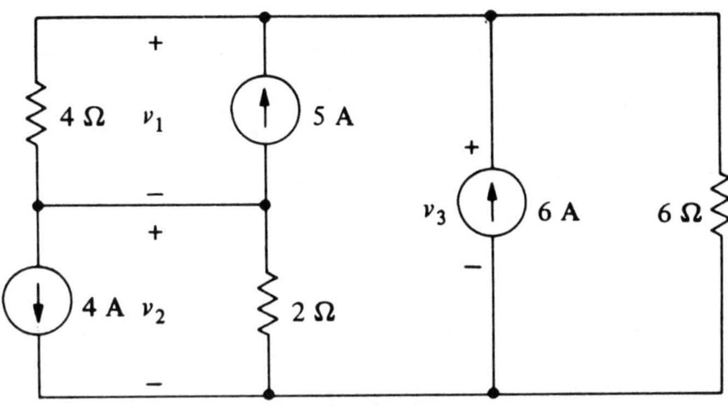

Problem 4.8

4.9 Using nodal analysis, find i_1.

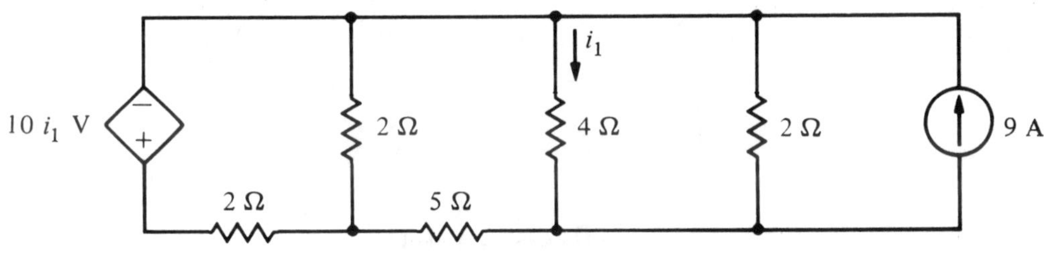

Problem 4.9

4.10 Find v and i using nodal analysis.

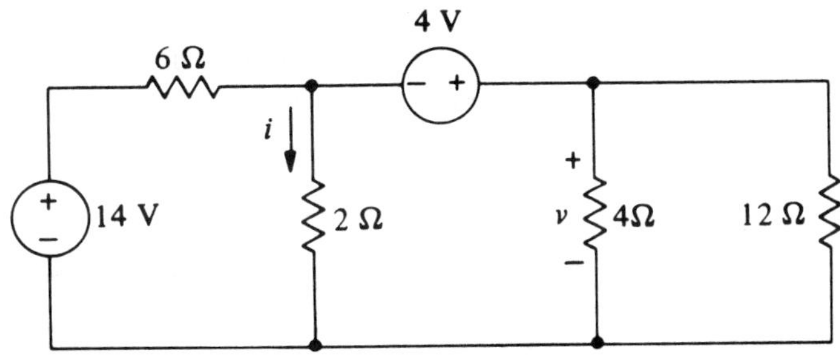

Problem 4.10

4.11 **Using nodal analysis, find v.**

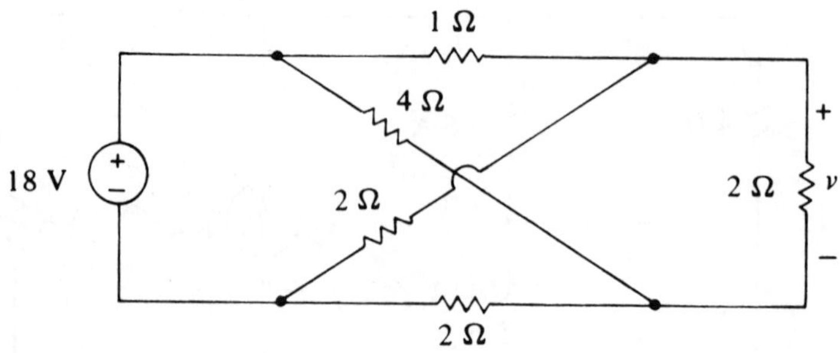

Problem 4.11

4.12 **Using nodal analysis, find i_1.**

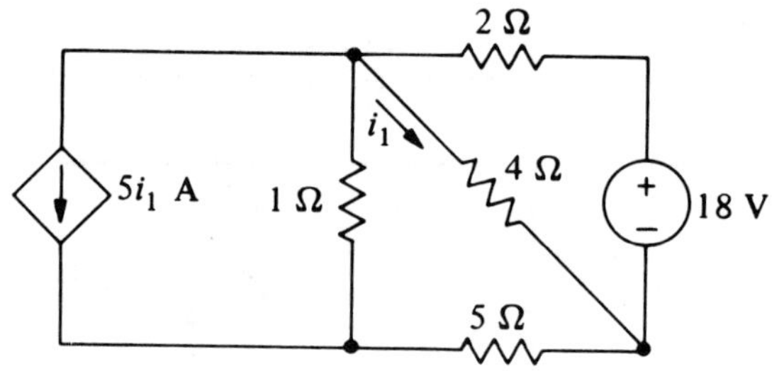

Problem 4.12

4.13 **Using nodal analysis, find v_2 if G_1=2s, G_2=0.5s, G_3=0.25s, G_4=1s, G_5=1s, β= 5, and v_g=4V.**

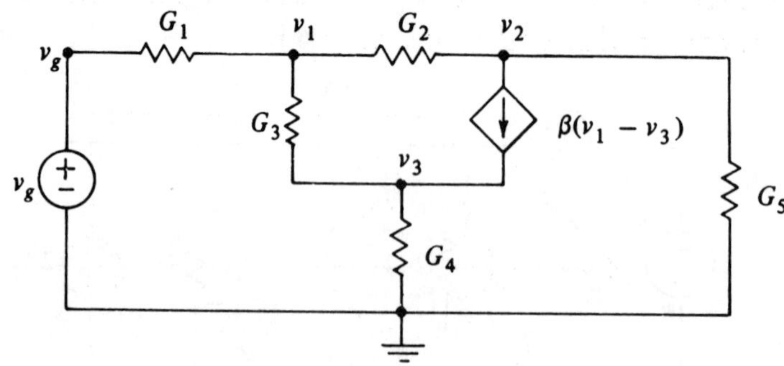

Problem 4.13

4.14 Find the equivalent resistance using nodal equations.

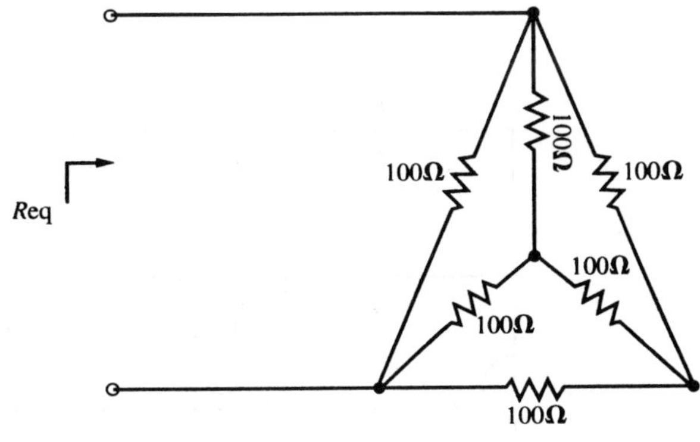

Problem 4.14

4.5 Mesh analysis

4.15 Find the equivalent resistance using mesh equations.

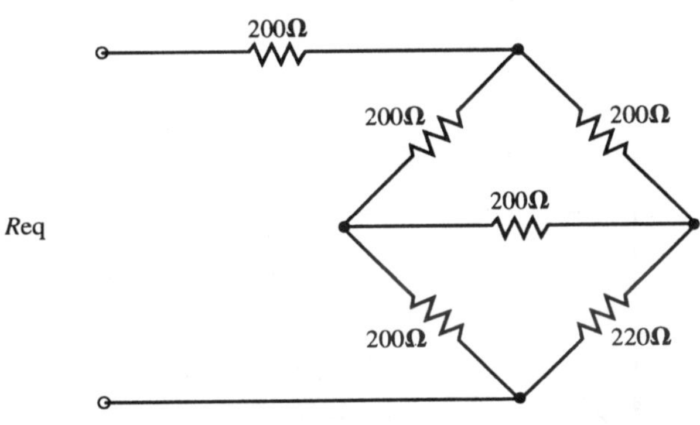

Problem 4.15

4.16 (a)Find v_g using superposition and mesh equations in terms of v, i, R and ΔR. Assume $\Delta R << R$ and $iR = v$V. (b)Compare the resulted v_g when the bridge is perfectly balanced (i.e.,ΔR=0).

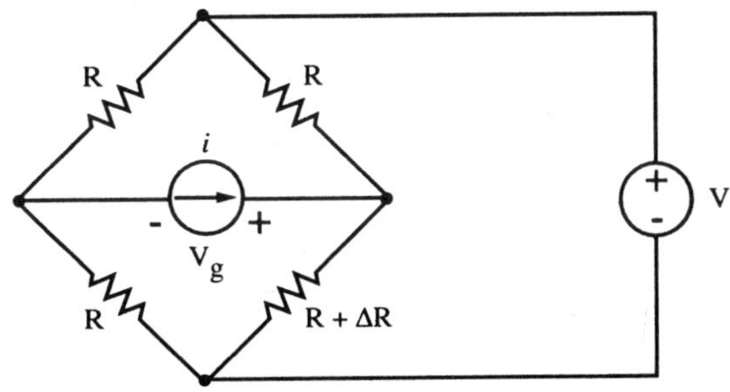

Problem 4.16

4.17 Using superposition and mesh equations to find v.

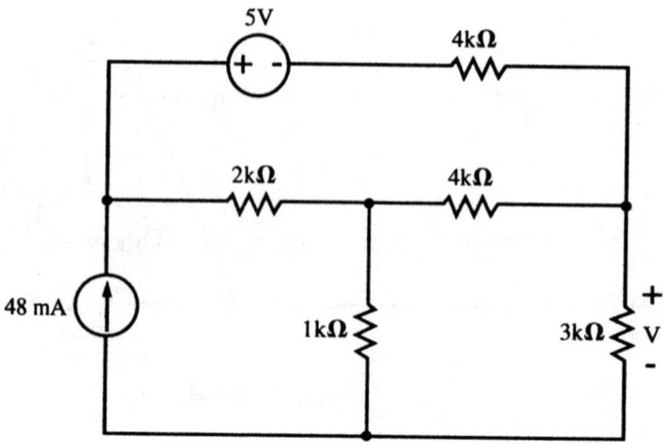

Problem 4.17

4.18 Find v_1 using mesh analysis.

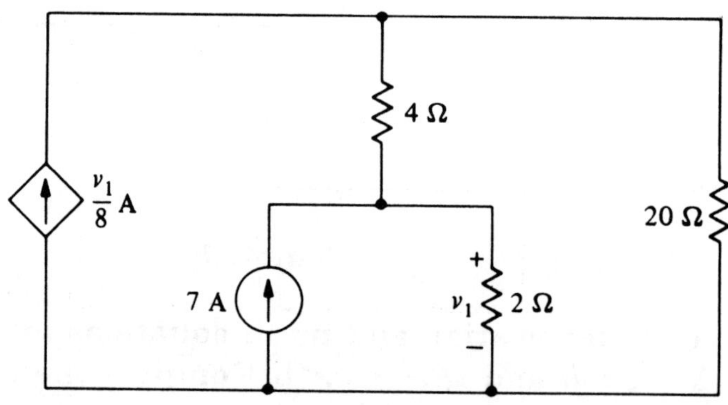

Problem 4.18

4.19 Find the power delivered to the 3KΩ resistor using mesh analysis.

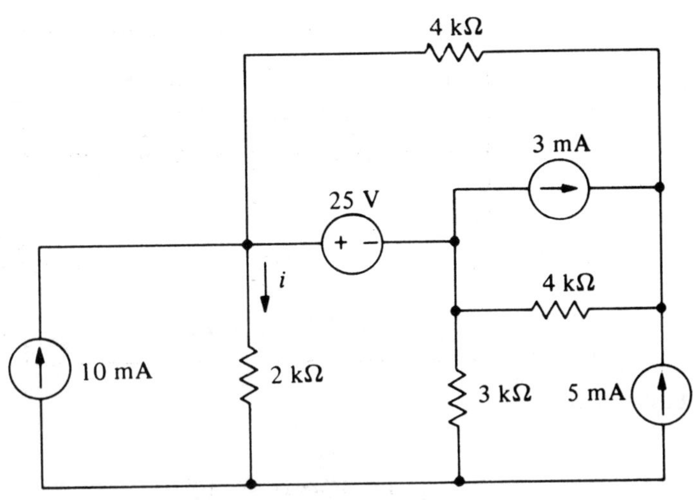

Problem 4.19

4.20 If i_{g1}=2A,i_{g2}=8A, v_{g3}=24V, R_1=1Ω, R_2=3Ω, R_3=4Ω, find the power delivered to R_3.

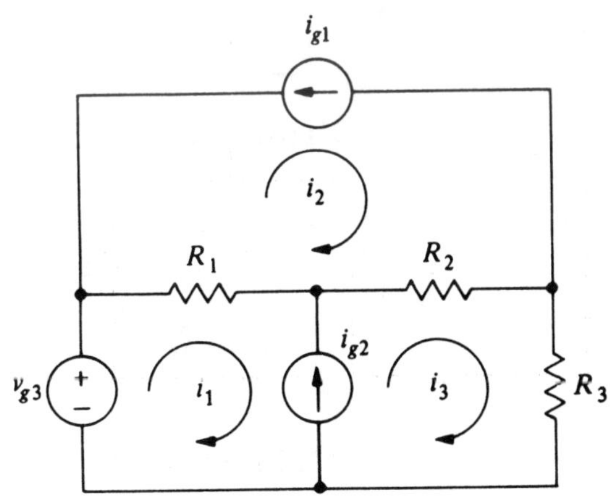

Problem 4.20

4.7 The virtual short principle for Op-amps.

4.21 Find V_{out}.

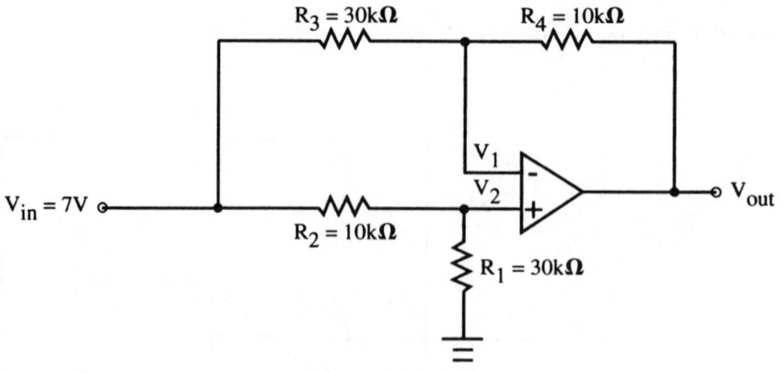

Problem 4.21

4.22 Find ν_{out} in terms of ν_{in}.

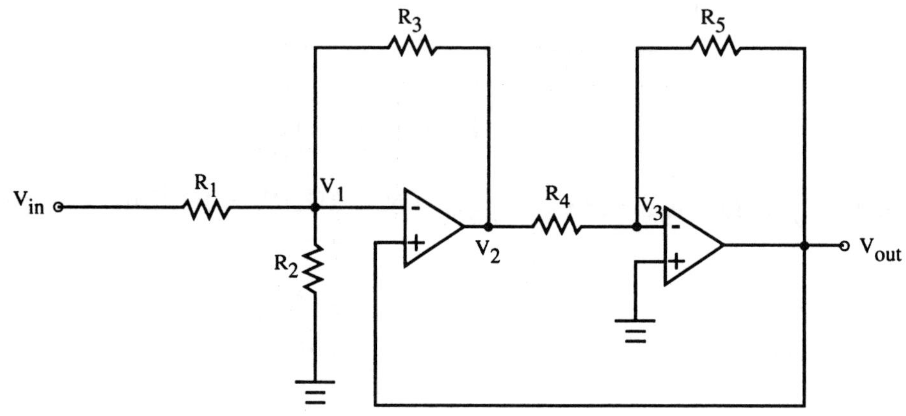

Problem 4.22

4.23 Find the current in the 5KΩ resistor if ν_g=5V.

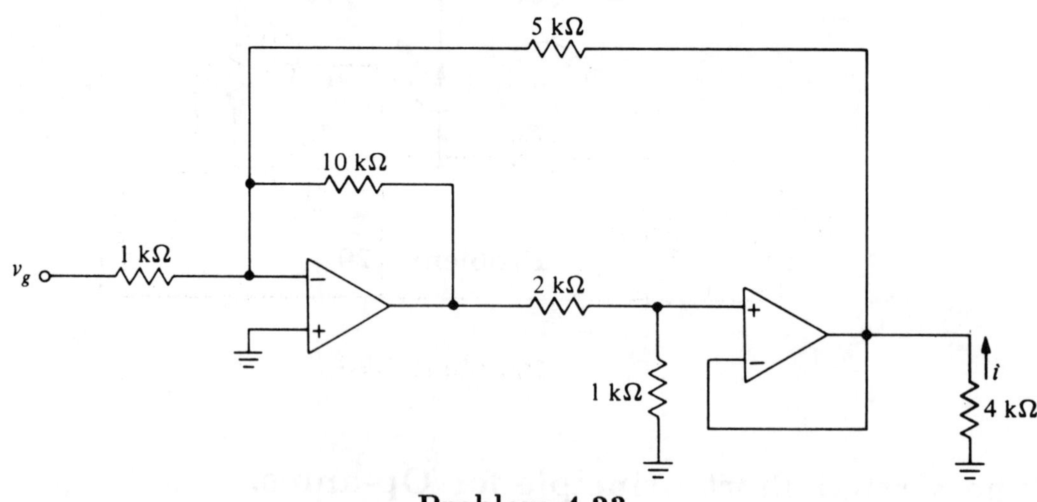

Problem 4.23

4.24 (a) Give the conditions that (1)superposition principle (2)KCL (3)KVL hold. (b) An Op-amp is modeled as a three-terminal device as shown. Does it violate KCL ? Give your explanation.

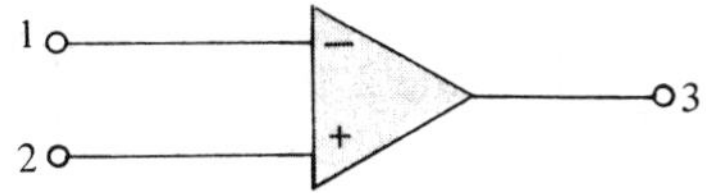

Problem 4.24

4.25 (a) Give the operating condition of the circuit shown that superposition principle holds . (b) Assume superposition principle holds, use it and virtual short principle to find v_O in terms of v_1, and v_2.

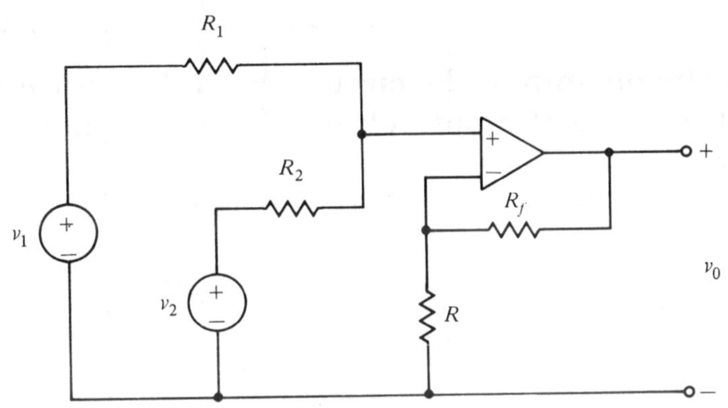

Problem 4.25

4.26 Find current gain $\frac{I_2}{I_1}$ by Nodal analysis.

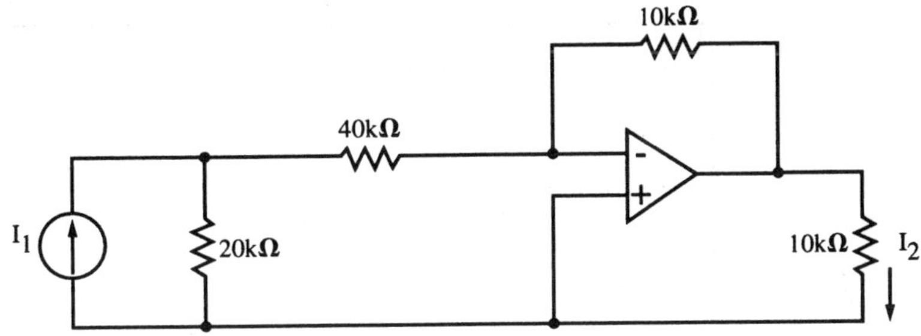

Problem 4.26

4.27 Find voltage v_2 in terms of v_1.

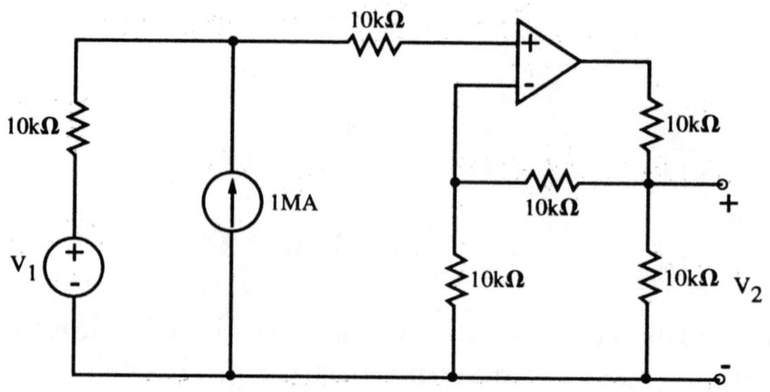

Problem 4.27

4.28 Assume that the op amp in the circuit shown has output saturation voltage ±10V. Find the range of input voltage v_{in} to guarantee the linear operation of the circuit.

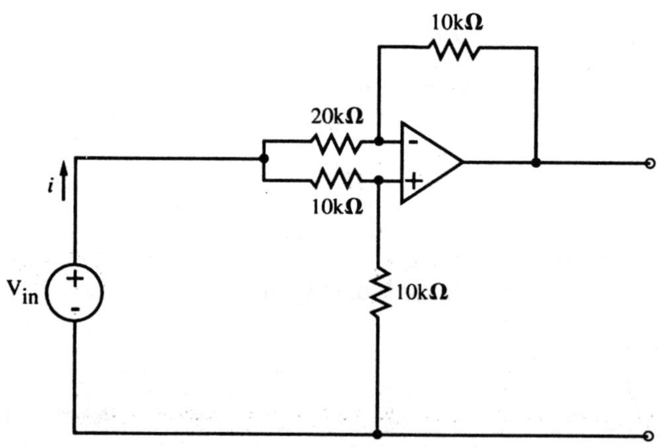

Problem 4.28

4.29 Find the voltage transfer ratio $\frac{v_{out}}{v_{in}}$.

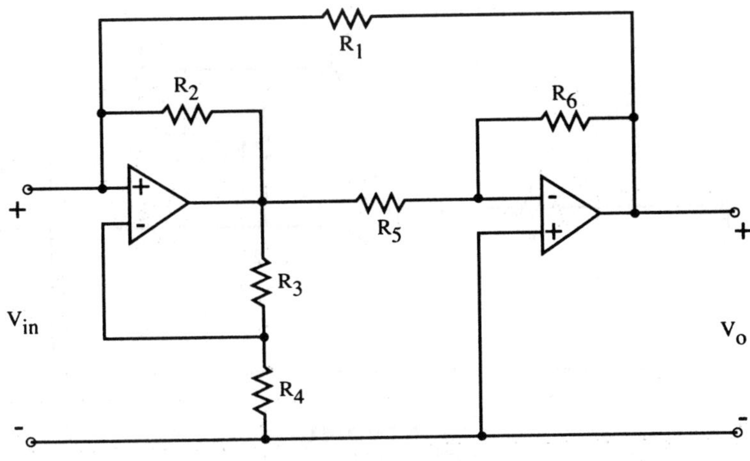

Problem 4.29

4.30 For a two-port network shown, find the transresistance of the two-port network.

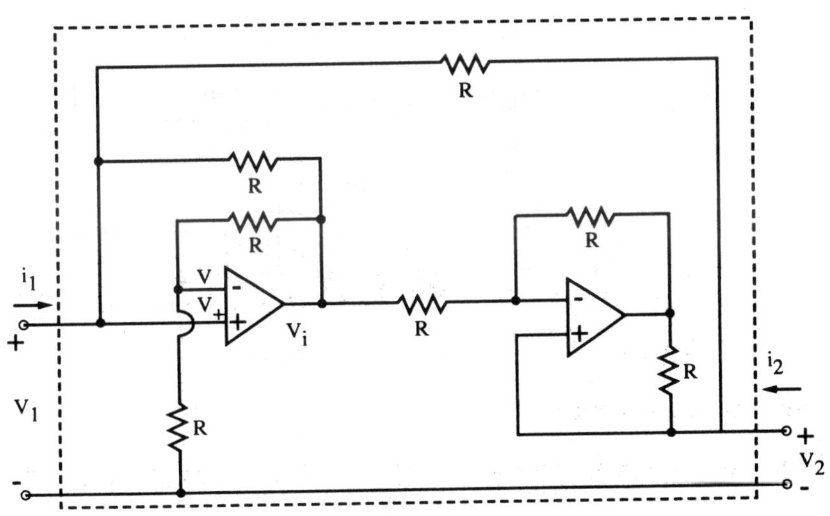

Problem 4.30

4.31 Use the virtual short and open principles to find the voltage transfer ratio $\frac{v_{out}}{v_{in}}$.

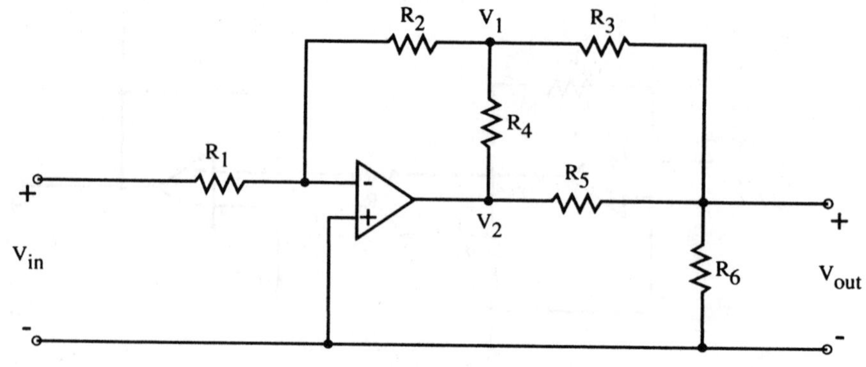

Problem 4.31

4.32 Find v_2 for G_1=0.25s, G_2=0.5s, G_3=1s, G_4=2s, G_5=0.5s and $v_1 = 8\cos(1000t)$V by (a) appling nodal analysis at points a and b . (b) appling nodal analysis at points a and c. Which answer is the correct one?

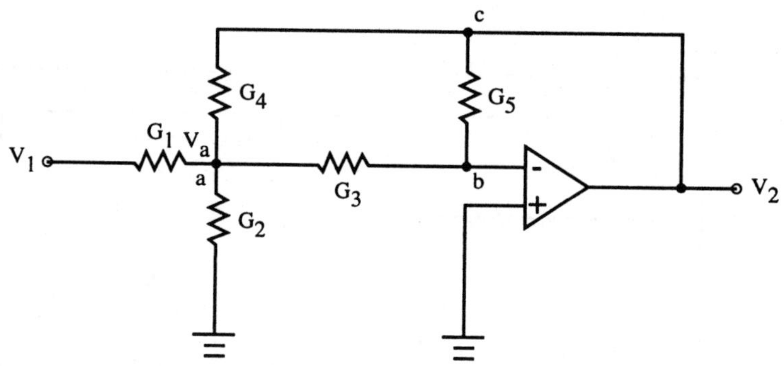

Problem 4.32

4.8 Computer-aided circuit analysis using SPICE

4.33 Find the equivalent resistance of Prob.4.15 using SPICE.

4.34 Use SPICE to solve Prob.4.13.

4.35 Use SPICE to solve Prob.4.23.

4.36 Solve Prob.4.30 using SPICE with R=20KΩ, $v_1 = 5u(t)$V.

4.1 The Norton equivalent circuit of the problem is:

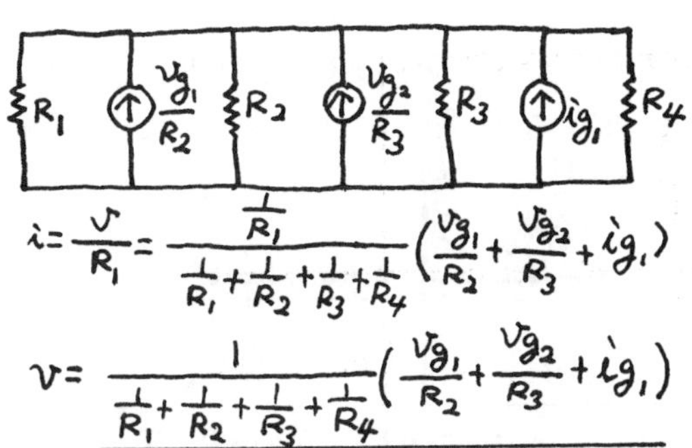

$$i = \frac{v}{R_1} = \frac{\frac{1}{R_1}}{\frac{1}{R_1}+\frac{1}{R_2}+\frac{1}{R_3}+\frac{1}{R_4}}\left(\frac{v_{g1}}{R_2}+\frac{v_{g2}}{R_3}+i_{g1}\right)$$

$$v = \frac{1}{\frac{1}{R_1}+\frac{1}{R_2}+\frac{1}{R_3}+\frac{1}{R_4}}\left(\frac{v_{g1}}{R_2}+\frac{v_{g2}}{R_3}+i_{g1}\right)$$

4.2 Let $i_1 = 1A$, $v_1 = (1+1+1)\cdot 1 = 3V$

By voltage division, the supposed value of the voltage source has to be $-\frac{2+2+(3//6)}{(3//6)}v_1 = -3v_1 = -9V$

Hence the scaling factor is $\frac{18}{-9} = -2$

$i_1 = -2 \times 1 = \underline{-2A}$

Similarly, let $i_2 = 1A$, $v_3 = (1+2)\cdot 1 = 3V$.

By voltage division, the supposed value of voltage source has to be

$\frac{3+3+(3//6)}{(3//6)}v_3 = \frac{8}{2}\times 3 = 12\ V,$

The scaling factor is $\frac{18}{12} = \frac{3}{2}$

$i_2 = \frac{3}{2}\times 1 = \underline{1.5A}$

4.3 Assume $i_1 = 1\ mA$, $i_1 = i_2 = \frac{1}{2}i_3 = \frac{1}{4}i_4$ then $i_4 = 4\ mA$.

Given the current source of 10 mA, $2i_4 = 10\ mA$ or $i_4 = 5\ mA$

The scaling factor for i_1 is

$\frac{5}{4} = 1.25$, $v = 1.25 \times 1 \times 3 = \underline{3.75\ V}$

4.4 (a) By superposition

$i_1 = 1 \times \frac{1}{1+(2+3+4+5+6)} = \frac{1}{27}$

$i_2 = \frac{2}{1+2+3+4+5+6} = \frac{2}{27}$, $i_3 = \frac{3}{27}$

$i_4 = \frac{4}{27}$, $i_5 = \frac{5}{27}$, $i_6 = \frac{6}{27}$

$I = i_1 + i_2 + i_3 + i_4 + i_5 + i_6 = \underline{\frac{7}{9}A}$

4.4 Cont.

By Thevenin–Norton transformation

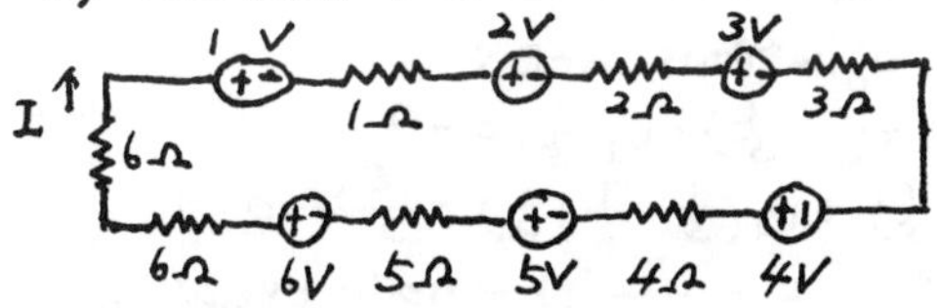

$$I = \frac{1+2+3+4+5+6}{1+2+3+4+5+6+6} = \frac{7}{9}A$$

4.5 By superposition, $i = i_1 + i_2 + i_3 + i_4$.

$\because$ Symmetrical, $i_1 = i_2 = i_3 = i_4 = \frac{1}{4}i$

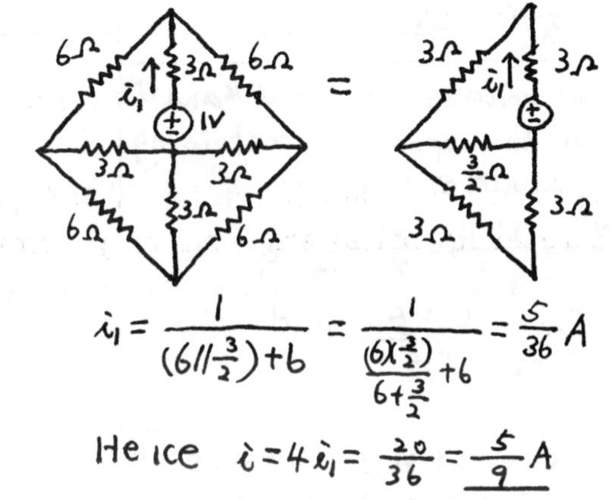

$$i_1 = \frac{1}{(6//\frac{3}{2})+6} = \frac{1}{\frac{(6\times\frac{3}{2})}{6+\frac{3}{2}}+6} = \frac{5}{36}A$$

Hence $i = 4i_1 = \frac{20}{36} = \underline{\frac{5}{9}A}$

4.6 Let the node voltages be v_1 and v_2 at the terminals of the 2kΩ resistor. The node equations are:

$(\frac{1}{2}+\frac{1}{2})v_1 - \frac{1}{2}v_2 = 6+1$ or $2v_1 - v_2 = 14$

$-\frac{1}{2}v_1 + (\frac{1}{2}+\frac{1}{2})v_2 = 2-1$ or $-v_1 + 2v_2 = 2$

Adding, $3v_2 = 18$ and $v_2 = 6V$,

$v_1 = \frac{14+v_2}{2} = 10V$, $i = \frac{v_1 - v_2}{2k} = \underline{2\ mA}$

4.7 $(\frac{1}{2}+1)v_1 - v_2 = 14$ or $3v_1 - 2v_2 = 28$

$-v_1 + (1+\frac{1}{4})v_2 = -7$ or $-4v_1 + 5v_2 = -28$

$v_1 = \begin{vmatrix} 28 & -2 \\ -28 & 5 \end{vmatrix} \Big/ \begin{vmatrix} 3 & -2 \\ -4 & 5 \end{vmatrix} = \frac{140-56}{15-8} = \underline{12V}$

$v_2 = \begin{vmatrix} 3 & 28 \\ -4 & -28 \end{vmatrix} \Big/ 7 = \frac{-84+112}{7} = \underline{4V}$

4.8 $(\frac{1}{4}+\frac{1}{6})v_3 - \frac{1}{4}v_2 = 6+5$ or $-3v_2 + 5v_3 = 132$

$(\frac{1}{2}+\frac{1}{4})v_2 - \frac{1}{4}v_3 = -5-4$ or $3v_2 - v_3 = -36$

adding, $4v_3 = 96$ and $v_3 = \underline{24V}$,

$v_2 = \frac{-36+v_3}{3} = \underline{-4V}$

$v_1 = v_3 - v_2 = \underline{28V}$

4.9 With v_1 and v_2 as node voltages, at terminals of 5Ω resistor. The node equations are:

$$v_1\left(\frac{1}{4}+\frac{1}{2}+\frac{1}{5}\right)-v_2\left(\frac{1}{5}\right)=-9$$

$$v_1\left(-\frac{1}{5}\right)+v_2\left(\frac{1}{5}+\frac{1}{2}+\frac{1}{2}\right)-\frac{10i_1}{2}=0$$

Since $i_1=-\frac{v_1}{4}$, the node equations become:

$19v_1-4v_2=-180$, $21v_1+24v_2=0$

Adding six times the first equation

$135v_1=-1080$, $v_1=-8V$

$i_1=-\frac{(-8V)}{4}=\underline{2A}$

4.10 The node voltages are 14, $v-4$ and v. KCL for the supernode containing the $4V$ source yields

$$\frac{v-4-14}{6}+\frac{v-4}{2}+\frac{v}{4}+\frac{v}{12}=0 \quad \text{or}$$

$12v=60$, $\therefore v=\underline{5V}$ and

$i=(v-4)/2=\underline{0.5A}$

4.11 With v_1 and v_2 as node voltages at terminals of 2Ω resistor. The node equations are

$$\begin{cases} v_1\left(\frac{1}{2}+\frac{1}{2}+1\right)-\frac{v_2}{2}=\frac{18}{1} \\ -\frac{v_1}{2}+v_2\left(\frac{1}{4}+\frac{1}{2}+\frac{1}{2}\right)=\frac{18}{4} \end{cases} \text{or} \begin{cases} 4v_1-v_2=36 \\ -2v_1+5v_2=18 \end{cases}$$

Adding twice the second equation

$9v_2=72$ or $v_2=8V$, $v_1=\frac{36+v_2}{4}=11V$

$v=v_1-v_2=\underline{3V}$

4.12 With v_1 and v_2 as node voltages at terminals of 1Ω resistor and $i_1=v_1/4$. The node equations are

$v_1\left(1+\frac{1}{4}+\frac{1}{2}+\frac{5}{4}\right)-v_2=\frac{18}{2}$ or $12v_1-4v_2=36$

$-v_1\left(1+\frac{5}{4}\right)+v_2\left(\frac{1}{5}+1\right)=0$ or $-45v_1+24v_2=0$

Adding 6 times the first equation

$27v_1=216$ or $v_1=8V$ $\therefore i_1=8/4=\underline{2A}$

4.13 The node equations are

$v_1(2+0.5+0.25)-v_2(0.5)-v_3(0.25)=4(2)$

or $11v_1-2v_2-v_3=32$

4.13 Cont.

$-v_1(0.5)+v_2(0.5+1)+5(v_1-v_3)=0$

or $9v_1+3v_2-10v_3=0$

$-v_1(0.25)-5(v_1-v_3)+v_3(0.25+1)=0$

or $-21v_1+25_3=0$

$$v_2=\frac{\begin{vmatrix} 11 & 32 & -1 \\ 9 & 0 & -10 \\ -21 & 0 & 25 \end{vmatrix}}{\begin{vmatrix} 11 & -2 & -1 \\ 9 & 3 & -10 \\ -21 & 0 & 25 \end{vmatrix}}=\underline{-0.606V}$$

4.14 Attach a current source of 1A to the given network. The node equations are:

$$v_1\left(\frac{1}{100}+\frac{1}{100}+\frac{1}{100}\right)-v_2\left(\frac{1}{100}\right)-v_3\left(\frac{1}{100}\right)=1$$

$$-v_1\left(\frac{1}{100}\right)+v_2\left(\frac{1}{100}+\frac{1}{100}+\frac{1}{100}\right)-v_3\left(\frac{1}{100}\right)=0$$

$$-v_1\left(\frac{1}{100}\right)-v_2\left(\frac{1}{100}\right)+v_3\left(\frac{1}{100}+\frac{1}{100}+\frac{1}{100}\right)=0$$

$$v_1=\frac{\begin{vmatrix} 1 & -\frac{1}{100} & -\frac{1}{100} \\ 0 & \frac{3}{100} & -\frac{1}{100} \\ 0 & -\frac{1}{100} & \frac{3}{100} \end{vmatrix}}{\begin{vmatrix} \frac{3}{100} & -\frac{1}{100} & -\frac{1}{100} \\ -\frac{1}{100} & \frac{3}{100} & -\frac{1}{100} \\ -\frac{1}{100} & -\frac{1}{100} & \frac{3}{100} \end{vmatrix}}=50V$$

Hence $R_{eq}=50/1=\underline{50\Omega}$

4.15 Connect a current source of 1A to the network

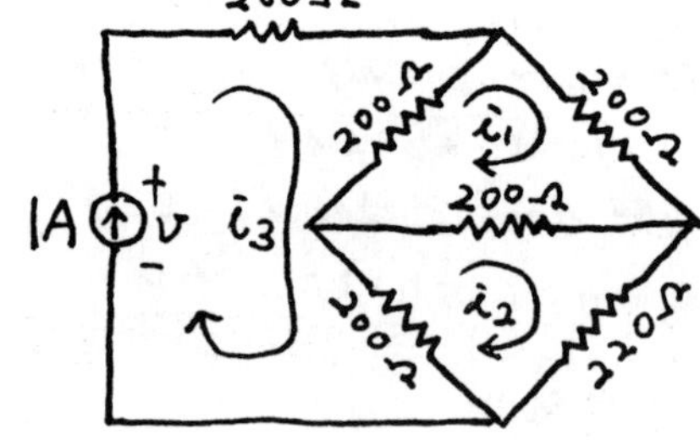

The loop equations are

$$\begin{cases} i_3=1 \\ 600i_1-200i_2=200 \\ -200i_1+620i_2=200 \end{cases} \qquad i_1=\frac{\begin{vmatrix} 200 & -200 \\ 200 & 620 \end{vmatrix}}{\begin{vmatrix} 600 & -200 \\ -200 & 620 \end{vmatrix}}=0.494A$$

$$i_2=\frac{\begin{vmatrix} 600 & 200 \\ -200 & 200 \end{vmatrix}}{\begin{vmatrix} 600 & -200 \\ -200 & 620 \end{vmatrix}}=0.482A$$

Hence $v=1\times200+200\times0.494+220\times0.482=404.84V$

$R_{eq}=v/1=\underline{404.84\Omega}$

4.16 (a) With the current source i killed.

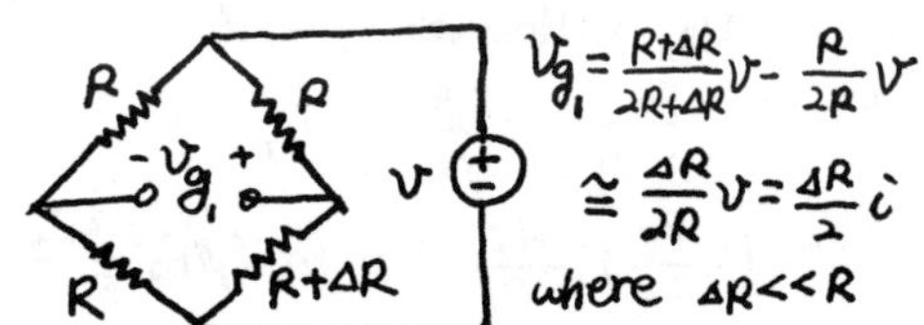

$$v_{g_1} = \frac{R+\Delta R}{2R+\Delta R}v - \frac{R}{2R}v \cong \frac{\Delta R}{2R}v = \frac{\Delta R}{2}i$$

where $\Delta R << R$

With the voltage source v killed

$i(2R+\Delta R) - i_2R - i_3(R+\Delta R) = 0$, and

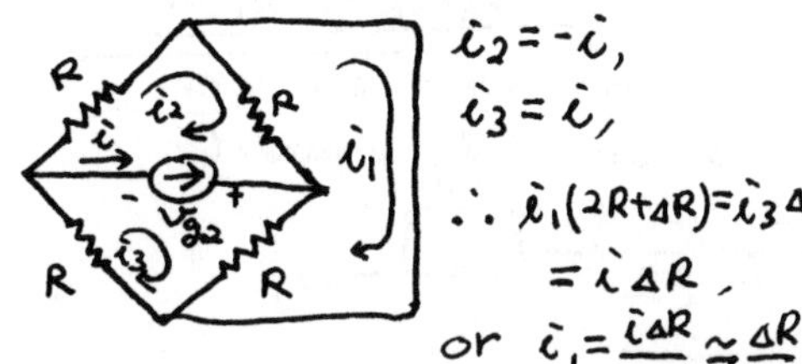

$i_2 = -i$,

$i_3 = i$,

$\therefore i_1(2R+\Delta R) = i_3\Delta R = i\Delta R$,

or $i_1 = \frac{i\Delta R}{2R+\Delta R} \cong \frac{\Delta R}{2R}i$

$$v_{g_2} = (i_3 - i_1)(R+\Delta R) + i_3R = 2i_3R - i_1R + i_3\Delta R - i_1\Delta R$$
$$\cong 2i_3R - i_1R + i_3\Delta R \quad (\because \Delta R << R,\ i_1 = \frac{\Delta R}{2R}i << i)$$
$$= 2iR - \frac{\Delta R}{2R}iR + i\Delta R = 2iR + \frac{1}{2}i\Delta R$$

Hence $v_g = v_{g_1} + v_{g_2} = 2iR + \frac{1}{2}i\Delta R + \frac{1}{2}i\Delta R = 2iR + i\Delta R \cong 2iR = \underline{2v}$

(b) Consider $\Delta R = 0$

With the current source killed, $v_{g_1} = 0$

With voltage source killed,

$$\begin{cases} i_1(2R) - i_2R - i_3R = 0 \\ i_2 = -i \\ i_3 = i \end{cases}$$

$i_1 = 0$, $v_{g_2} = i(2R//2R) = iR = v$.

Hence $v_g = v_{g_1} + v_{g_2} = 0 + v = \underline{v}$

4.17 With the current source killed,

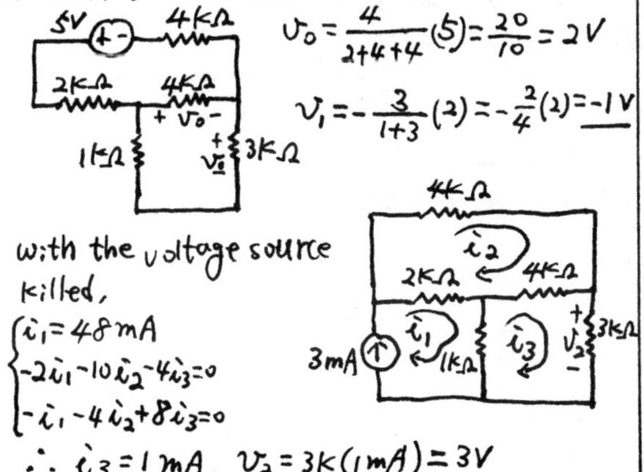

$v_0 = \frac{4}{2+4+4}(5) = \frac{20}{10} = 2V$

$v_1 = -\frac{3}{1+3}(2) = -\frac{3}{4}(2) = \underline{-1V}$

with the voltage source killed,

$$\begin{cases} i_1 = 48\,mA \\ -2i_1 - 10i_2 - 4i_3 = 0 \\ -i_1 - 4i_2 + 8i_3 = 0 \end{cases}$$

$\therefore i_3 = 1\,mA$, $v_2 = 3K(1mA) = 3V$

$v = v_1 + v_2 = -1 + 3 = \underline{2V}$

4.18

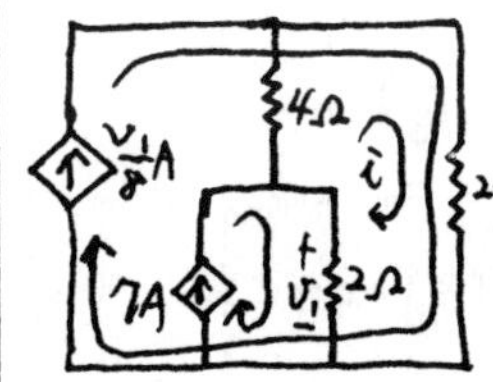

For the mesh with only resistors, KVL yields

$(2+4+20)i - 7(2) + 20(\frac{v_1}{8}) = 0$ since $v_1 = 2(7-i)$

or $i = \frac{14-v_1}{2}$, then

$\frac{26}{2}(14-v_1) + \frac{20}{8}v_1 = 14$ or $\underline{v_1 = 16V}$

4.19

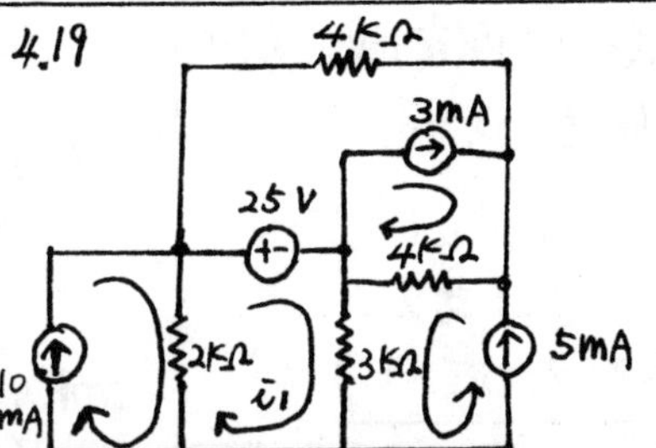

i_1 in mA, KVL for the mesh of i_1

$2(i_1 - 10) + 3(i_1 + 5) = -25$, then $i_1 = -4mA$

$i_{3K\Omega} = i_1 + 5mA = 1\,mA$,

$P_{3K\Omega} = i_{3K\Omega}^2 R = (3K\Omega)(1mA)^2 = \underline{3mW.}$

4.20

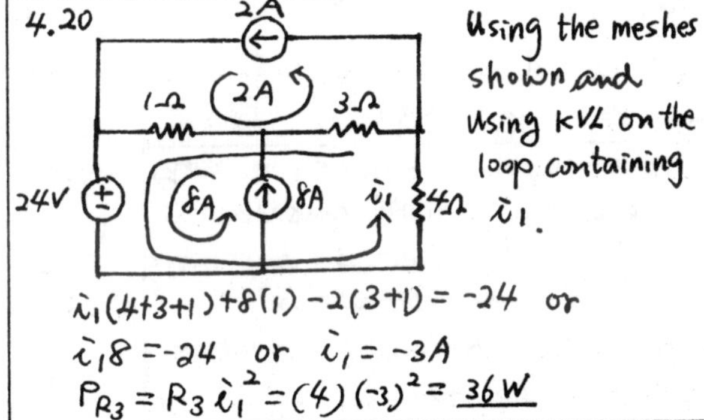

Using the meshes shown and using KVL on the loop containing i_1.

$i_1(4+3+1) + 8(1) - 2(3+1) = -24$ or

$i_1 8 = -24$ or $i_1 = -3A$

$P_{R_3} = R_3 i_1^2 = (4)(-3)^2 = \underline{36W}$

4.21 Let the voltages at inverting and noninverting input terminals of the op-amp be v_1 and v_2 respectively. By virtual short, open and nodal analysis

$$\begin{cases} v_1 = v_2 & -(1) \\ v_1(\frac{1}{R_3} + \frac{1}{R_4}) - v_{in}(\frac{1}{R_3}) - v_{out}(\frac{1}{R_4}) = 0 & -(2) \\ v_2(\frac{1}{R_2} + \frac{1}{R_1}) - v_{in}(\frac{1}{R_2}) = 0 & -(3) \end{cases}$$

From (3) $v_{in} = \frac{(\frac{1}{R_2} + \frac{1}{R_1})}{(\frac{1}{R_2})}v_2 = \frac{(\frac{1}{R_2} + \frac{1}{R_1})}{(\frac{1}{R_2})}v_1$

or $v_1 = \frac{\frac{1}{R_2}}{(\frac{1}{R_1} + \frac{1}{R_2})}v_{in}$.

From (2), $\frac{\frac{1}{R_2}}{\frac{1}{R_1} + \frac{1}{R_2}}v_{in}(\frac{1}{R_3} + \frac{1}{R_4}) - v_{in}(\frac{1}{R_3}) = v_{out}(\frac{1}{R_4})$

$\because R_1 = R_3 = 30K\Omega$, $R_2 = R_4 = 10K\Omega$, $v_{in} = 7V \Rightarrow \underline{v_{out} = \frac{14}{3}V}$

4.22 By virtual short, $v_1 = v_{out}$.

By Nodal analysis

$$\begin{cases} v_1\left(\frac{1}{R_1}+\frac{1}{R_2}+\frac{1}{R_3}\right)-v_{in}\left(\frac{1}{R_1}\right)-v_2\left(\frac{1}{R_3}\right)=0 & (1)\\ -v_1\left(\frac{1}{R_3}\right)+v_2\left(\frac{1}{R_3}+\frac{1}{R_4}\right)-v_3\left(\frac{1}{R_4}\right)=0 & (2)\\ -v_2\left(\frac{1}{R_4}\right)+v_3\left(\frac{1}{R_4}+\frac{1}{R_5}\right)-v_{out}\left(\frac{1}{R_5}\right)=0 & (3)\end{cases}$$

(1) + (2) + (3)

$$v_1\left(\frac{1}{R_1}+\frac{1}{R_2}\right)-v_{in}\left(\frac{1}{R_1}\right)+v_3\left(\frac{1}{R_5}\right)-v_{out}\left(\frac{1}{R_5}\right)=0$$

∵ virtual short, $v_3=0$

$$v_{out}\left(\frac{1}{R_1}+\frac{1}{R_2}\right)-v_{in}\left(\frac{1}{R_1}\right)-v_{out}\left(\frac{1}{R_5}\right)=0$$

$$v_{out}\left(\frac{1}{R_1}+\frac{1}{R_2}-\frac{1}{R_5}\right)=v_{in}\left(\frac{1}{R_1}\right)$$

$$v_{out}=v_{in}\left(\frac{1}{R_1}\right)\Big/\left(\frac{1}{R_1}+\frac{1}{R_2}-\frac{1}{R_5}\right)$$

4.23 Apply KCL at inverting input of first op-amp:

$$\frac{v_g}{1K}+\frac{v_1}{10K}+\frac{v_2}{5K}=0$$

Apply KCL at noninverting input of voltage follower:

$$v_2\left(\frac{1}{1K}+\frac{1}{2K}\right)-v_1\left(\frac{1}{2K}\right)=0$$

$v_2 = -10\ V$, $i = -2mA$

4.24 (a) (1) Superposition principle holds when every device in the network considered is linear; that is, superposition principle applies to linear circuit only.

(2) Both KCL and KVL apply to lumped circuit model only.

(b) An op-amp modelled as a three-terminal device violates KCL. There are other terminals such as the ones connected to power supply, ground, ... etc. are missing in this model. For KCL to be valid, all these terminals have to be considered in an op-amp model.

4.25 (a) The superposition principle holds for linear circuit. Thus the op-amp has to operate in its linear region. This means the output of the op-amp, v_2, is smaller than its saturation region.

4.25 Cont.

(b) With v_{g_2} killed and virtual short, $\frac{R_2}{R_2+R_1}v_{g_1}=\frac{R_A}{R_A+R_f}v_{21}$

$$v_{21}=\left(1+\frac{R_f}{R_A}\right)\left(\frac{R_T}{R_1}\right)v_{g1},\quad R_T=R_1/\!/R_2=\frac{R_1R_2}{R_1+R_2}$$

Similarly, $v_{22}=\left(1+\frac{R_f}{R_A}\right)\left(\frac{R_T}{R_2}\right)v_{g_2}$

Hence $v_2=v_{21}+v_{22}=\left(1+\frac{R_f}{R_A}\right)\left(\frac{R_T}{R_1}v_{g_1}+\frac{R_T}{R_2}v_{g_2}\right)$

4.26 By the Thevenin-Norton equivalent, the given circuit is equal to

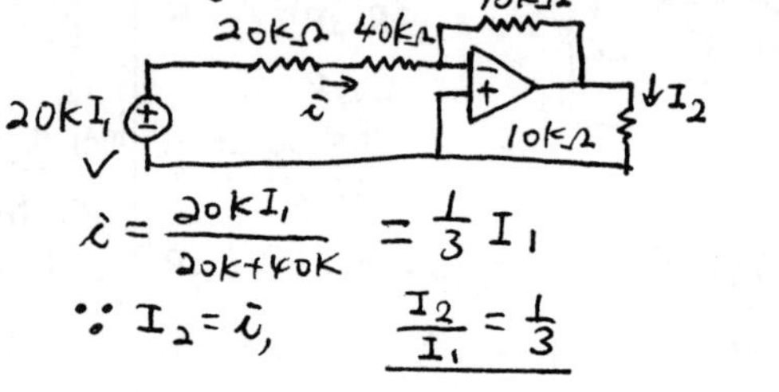

$$i=\frac{20kI_1}{20k+40k}=\frac{1}{3}I_1$$

∵ $I_2=i$, $\quad \frac{I_2}{I_1}=\frac{1}{3}$

4.27 By superposition principle, one can solve the problem in two steps.

(1) By virtual short and virtual open

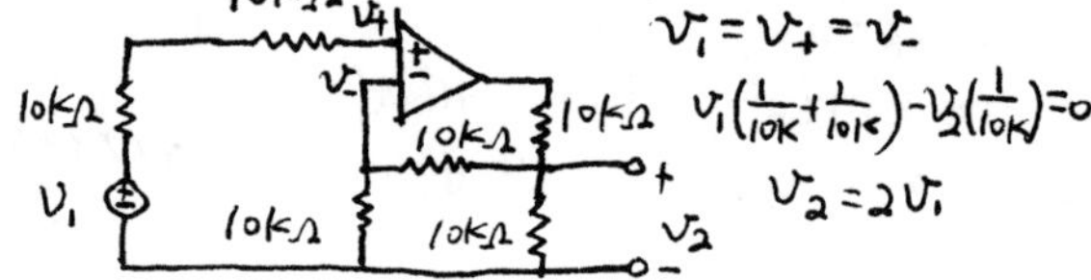

$v_1=v_+=v_-$

$v_1\left(\frac{1}{10K}+\frac{1}{10K}\right)-v_2\left(\frac{1}{10K}\right)=0$

$v_2=2v_1$

(2) By virtual short and virtual open

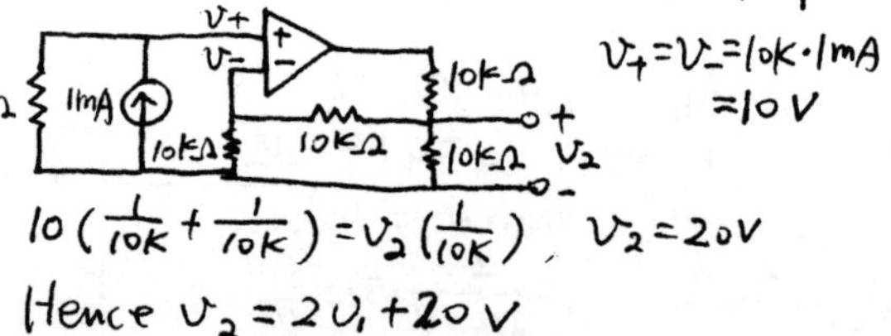

$v_+=v_-=10k\cdot 1mA = 10\ V$

$$10\left(\frac{1}{10K}+\frac{1}{10K}\right)=v_2\left(\frac{1}{10K}\right),\quad v_2=20V$$

Hence $v_2 = 2v_1+20\ V$

4.28 Let the voltages at the input terminals of the op amp are v_- and v_+. By virtual short principle $v_-=v_+=\frac{1}{2}v_{in}$

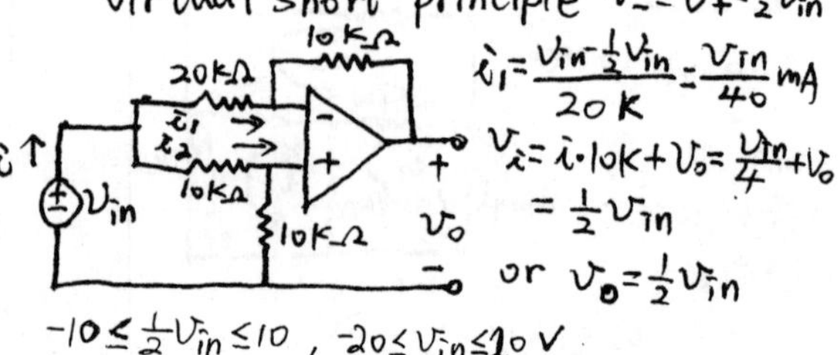

$$i_1=\frac{v_{in}-\frac{1}{2}v_{in}}{20K}=\frac{v_{in}}{40}mA$$

$$v_-=i\cdot 10K+v_0=\frac{v_{in}}{4}+v_0=\frac{1}{2}v_{in}$$

or $v_0=\frac{1}{2}v_{in}$

$-10\le\frac{1}{2}v_{in}\le 10$, $\quad -20\le v_{in}\le 20\ V$

4.29 Let the voltage at the output terminal of the first op amp be v_1 and the voltage at the inverting terminal of the second op amp be v_2.

$$\frac{v_{in}-v_1}{R_3}+\frac{v_{in}}{R_4}=0 \text{ or } v_1=\left(1+\frac{R_3}{R_4}\right)v_{in}$$

$$\frac{0-v_1}{R_5}+\frac{0-v_o}{R_6}=0 \text{ or } v_o=-\frac{R_6}{R_5}v_1$$

$$=-\left(\frac{R_6}{R_5}\right)\left(1+\frac{R_4}{R_3}\right)v_{in}$$

4.30 By virtual short, $v_1=v_-=v_+$, and

$$v\left(\frac{1}{R}+\frac{1}{R}\right)-v_i\left(\frac{1}{R}\right)=0$$

$$v_1\left(\frac{2}{R}\right)=v_i\left(\frac{1}{R}\right), \text{ or } v_i=2v_1$$

Apply KCL at positive input node

$$i_1=\frac{v_1-v_2}{R}+\frac{v_1-v_i}{R}=\frac{v_1-v_2}{R}+\frac{v_1-2v_1}{R}=-\frac{1}{R}v_2$$

Hence the transresistance of the two port network is $r=-R$

4.31 $\frac{0-v_{in}}{R_1}+\frac{0-v_1}{R_2}=0$ or $v_1=-\frac{R_2}{R_1}v_{in}$ —(1)

$$\frac{v_1}{R_2}+\frac{v_1-v_2}{R_4}+\frac{v_1-v_{out}}{R_3}=0$$

or $v_2=-R_4\left[\frac{R_2}{R_1}\left(\frac{1}{R_2}+\frac{1}{R_3}+\frac{1}{R_4}\right)v_{in}+\frac{1}{R_3}v_{out}\right]$ —(2)

$\frac{v_{out}}{R_6}+\frac{v_{out}-v_1}{R_3}+\frac{v_{out}-v_2}{R_5}=0$ or

$$v_{out}\left[\frac{1}{R_3}+\frac{1}{R_5}+\frac{1}{R_6}\right]=\frac{v_1}{R_3}+\frac{v_2}{R_5} \quad —(3)$$

substitute (1) and (2) into (3)

$$v_{out}\left[\left(\frac{1}{R_3}+\frac{1}{R_5}+\frac{1}{R_6}\right)+\frac{R_4}{R_3R_5}\right]=-\left[\frac{R_2}{R_1R_3}+\frac{R_2R_4}{R_1R_5}\left(\frac{1}{R_2}+\frac{1}{R_3}+\frac{1}{R_4}\right)\right]v_{in}$$

The voltage transfer ratio

$$\frac{v_{out}}{v_{in}}=-\frac{\frac{R_2}{R_1R_3}+\frac{R_2R_4}{R_1R_5}\left(\frac{1}{R_2}+\frac{1}{R_3}+\frac{1}{R_4}\right)}{\left(\frac{1}{R_3}+\frac{1}{R_5}+\frac{1}{R_6}\right)+\frac{R_4}{R_3R_5}}$$

4.32 (a) $\begin{cases} v_a(G_1+G_2+G_3+G_4)-v_iG_1-v_2G_4=0 & —(1) \\ -v_aG_3-v_2G_5=0 \text{ or } v_a=-\frac{G_5}{G_3}v_2 & —(2)\end{cases}$

substitute (2) into (1)

$$v_2=-\frac{G_1}{\frac{G_5}{G_3}(G_1+G_2+G_3+G_4)+G_4}v_{in}$$

$$=-\frac{0.25}{\frac{1}{2}\times3.75+2}v_i=0.52\cos(1000t)\text{V}$$

(b) $v_a(G_1+G_2+G_3+G_4)-v_iG_1-v_2G_4=0$ —(1)

$v_2(G_4+G_5)-v_aG_4=0$ or $v_a=\frac{G_4+G_5}{G_4}v_2$ —(2)

4.32 Cont.

substitute (2) into (1)

$$G_2=\frac{G_1}{\frac{(G_4+G_5)}{G_4}(G_1+G_2+G_3+G_4)-G_4}v_i$$

$$=\frac{0.25}{1.25\times3.75-2}8\cos1000t=0.74\cos1000t$$

The answer obtained by (b) is not correct because the current flows into the output terminal of the second op amp is not considered.

4.33.

```
*PROB.4_33
*DETERMINE EQUIVALENT
*RESISTANCE USING PSPICE
*SCHEMATICS NETLIST
R_R1 1 2 200
R_R2 2 3 200
R_R3 2 4 200
R_R4 3 4 200
R_R5 3 0 200
R_R6 4 0 220
I_I1 0 1
.DC I_I1 1 1 1
.PRINT DC V(1)
.END
*************************
****DC TRANSFER CURVES
*************************
  I_I1          V(1)
  1.000E+00     4.048E+02
*THE EQUIVALENT RESISTANCE

 R=V(1)/I_I1
  =404.8/1
  =404.8 OHM
```

4.34

```
*PROB.4_34
*DETERMINE OUTPUT VOLTAGE V(2)
*CIRCUIT DESCRIPTION
R_R1 1 3 0.5
R_R2 2 3 2
R_R3 3 4 4
R_R4 4 0 1
R_R5 2 0 1
V_V1 1 0 DC 4
G_G1 2 4 3 4 5
.END
*SMALL SIGNAL BIAS SOLUTION
NODE VOLTAGE NODE VOLTAGE
(1)  4.0000  (2)  -.6061
(3)  3.0303  (4)  2.5455

*THE OUTPUT VOLTAGE V(2)
 FOUND IS -0.6061V
```

4.35

```
*PROB4_35
*DETERMINE THE CURRENT
*IN THE 5K RESISTOR
R_R1 1 5 1K
R_R2 1 2 10K
R_R3 2 3 2K
R_R4 3 0 1K
R_R5 4 1 5K
R_R6 4 0 4K
X_U1 0 1 2 OPAMP
X_U2 3 4 4 OPAMP
V_V1 5 0 DC 5
.INC C:\PS\OPAMP.LIB
*INCLUDING C:\PS\OPAMP.LIB *
*OP AMP SUBCIRCUIT:
.SUBCKT OPAMP 1 2 3
*NODE 1:+ 2:- 3:output
RIN    1    2    1MEG
E1     4    0    1 2 100K
R0     4    3    30
.ENDS
**** RESUMING PS4174.CIR ***
.DC V_V1 5 5 1
.PRINT DC I(R_R5)
.END
**** DC TRANSFER CURVES
  V_V1         I(R_R5)
  5.000E+00   -2.000E-03

*THE CURRENT IN THE 5K
RESISTOR IS I(R_R5)=-.002A
```

4.36

```
*TRANSRESISTANCE OF
*AN TWO-PORT NETWORK
R_R1 1 2 20K
R_R2 2 3 20K
R_R3 3 4 20K
R_R4 1 6 20K
R_R5 4 6 20K
R_R6 4 5 20K
R_R7 5 0 20K
X_U1 1 3 2 OPAMP
X_U2 6 5 4 OPAMP
I_I1 6 0 DC 1
I_I2 1 0 DC 1
.INC C:\PS\OPAMP.LIB
*INCLUDING C:\PS\OPAMP.LIB ****
*OP AMP SUBCIRCUIT:
.SUBCKT OPAMP 1 2 3
*NODE 1:+ 2:- 3:OUTPUT
RIN   1   2   1MEG
E1    4   0   1 2 100K
R0    4   3   30
.ENDS
**** RESUMING PS4174.CIR ****
.PRINT DC V(2)
.END
******************************
*SMALL SIGNAL BIAS SOLUTION
******************************
NODE VOLTAGE   NODE VOLTAGE
(1)  20.00E+03 (2)  80.00E+03
(3)  20.00E+03 (4) -40.00E+03
(5)-20.00E+03  (6) -20.00E+03
(X_U1.4) 80.18E+03 (X_U2.4)-40.

*THE TRANSRESISTANCE IS
 R=V(1)/I_I1
  =20000/1
  =20000 OHM
```

Chapter 5
Energy - Storage Elements

5.1 Capacitors

5.1 **A 0.1uF capacitor has a current of $3\cos(2000t)$mA. Find its voltage $v(t)$ if $v(0)$=-15V.**

5.2 **Find the charge residing on each plate of a 2uF capacitor that is charged to 50V. If the same charge resides on a 1uF capacitor, what is the voltage?**

5.3 **A 10uF capacitor has a voltage $v(t) = f(t)$V as shown. Find the current i at t=-7, -3, 1, 3, and 7ms.**

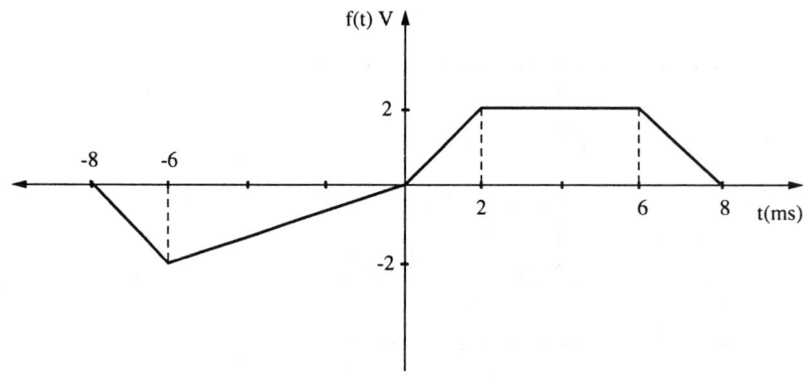

Problem 5.3

5.4 **If $f(t)$ in Prob.5.3 is the current in mA of a 1uF capacitor, find its voltage at t=-6, 0, 2, and 8ms.**

5.5 **A 1F capacitor has $i = 2e^{-2t}$A for $t \geq 0$. If its voltage is determined to be -5V at t=3s, what must the voltage be at t=1s ?**

5.2 Energy stored in capacitors

5.6 **Find the energy stored in the capacitor of problem 5.5 above (a) at t=3s (b) in the limit as $t \to \infty$.**

5.7 **A simple capacitor consisting of two parallel plates is found to have a charge of 3uC on each plate when a voltage of 12V is applied to it, find its capacitance and the energy stored.**

5.3 Series and parallel capacitors

5.8 If the subcircuit has the terminal law $v = \frac{1}{2}\int_0^t i dt$ and i is found to be $9\cos(3t)$ A when v is $sin(3t)$V, find c_1, and c_2.

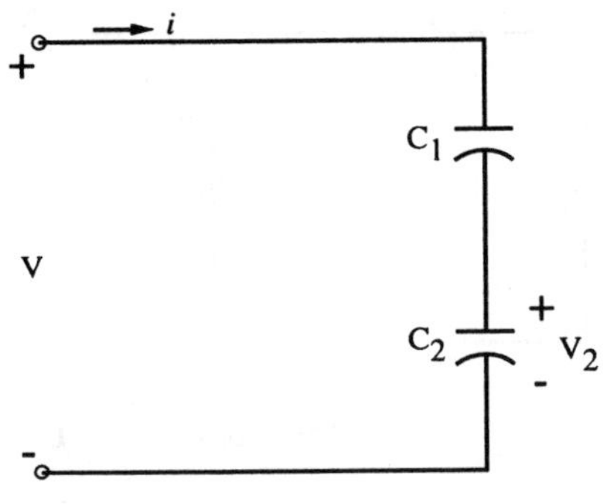

Problem 5.8

5.9 The capacitances shown are all in uF. Find C_{eq} at terminals a-b.

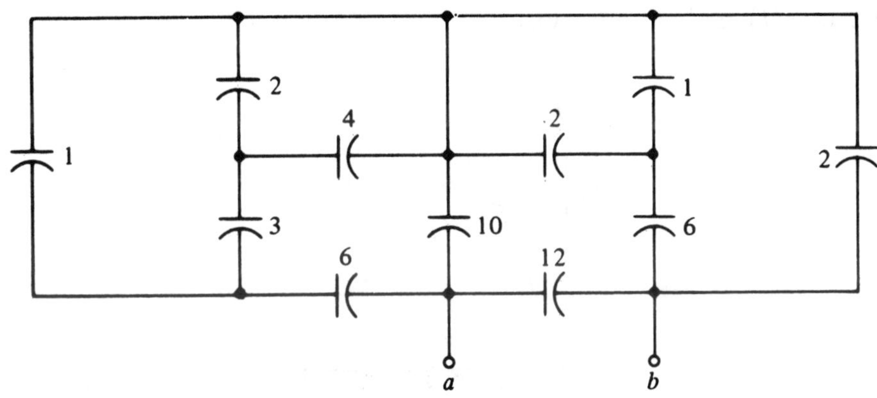

Problem 5.9

5.10 Find the time t at which the voltage source in the circuit shown delivered maximum power to the rest of the system. What will this maximum power be? Assume the switch is closed at time t=0 and the capacitor is discharged at $t=0^-$.

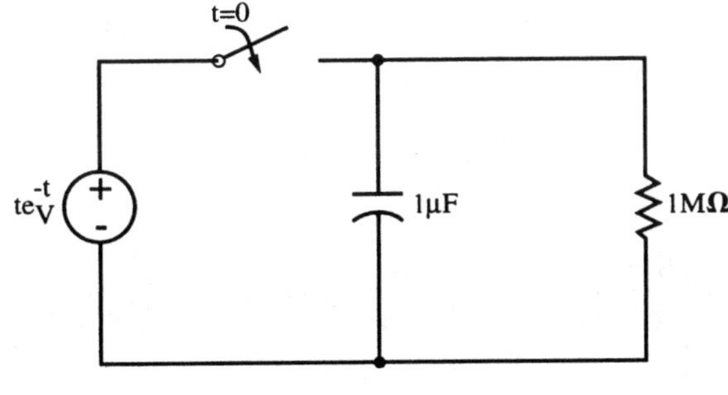

Problem 5.10

5.11 Find $i_1(0^-)$, $i_1(0^+)$, $i_2(0^-)$, and $i_2(0^+)$, if the switch is opened at t=0, and $v_1(0^-)$=18V and $v_2(0^-)$=6V.

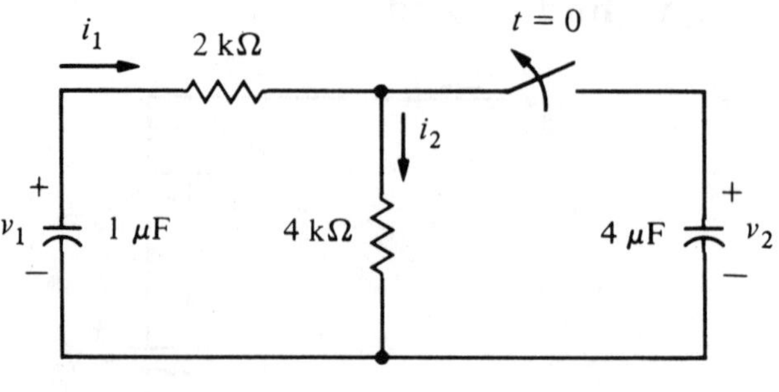

Problem 5.11

5.4 Inductors

5.12 Find the flux linkage of a 50mH inductors at 10ms, 40ms and 80ms, if the current is $f(t)$ mA as shown.

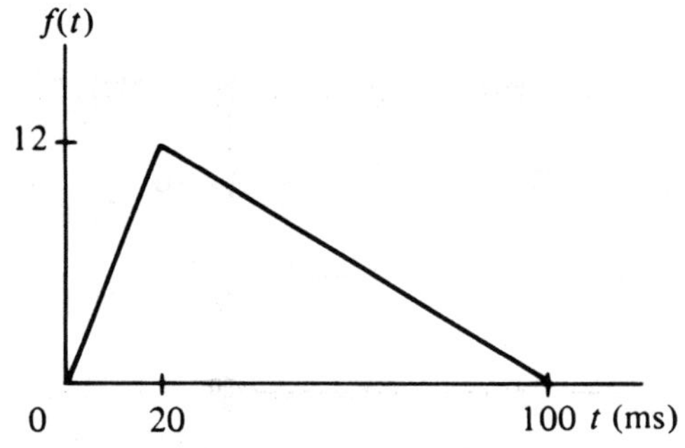

Problem 5.12

5.13 Find the voltage across a 10H inductor at t=10ms, 40ms, and 80ms if its current is $f(t)$ mA, given in Prob.5.12.

5.14 Find the current i through a 120H inductor at t=20ms, t=60ms and t=100ms if its voltage is $f(t)$ V, given in Prob.5.12. and $i(0)$=2mA.

5.5 Energy stored in inductors

5.15 The voltage of 0.25H inductor is $v = 6\cos(2t)$V and the current at t=0 is 0. Find the energy stored in the inductor at $\frac{3\pi}{4}$s and the power delivered to the inductor at $\frac{\pi}{8}$s.

5.16 If $i(0)$=2A is the initial current in a 1H inductor, find the current for $0 \leq t \leq 30$ms for the voltage waveform shown. Assume passive sign convention is satisfied.

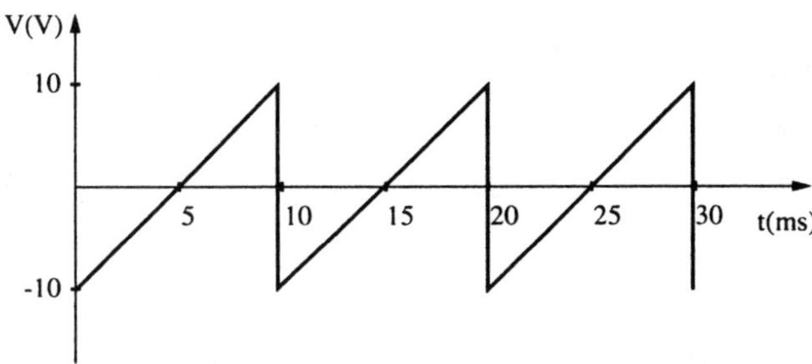

Problem 5.16

5.6 Series and parallel inductors

5.17 Determine L_{eq}.

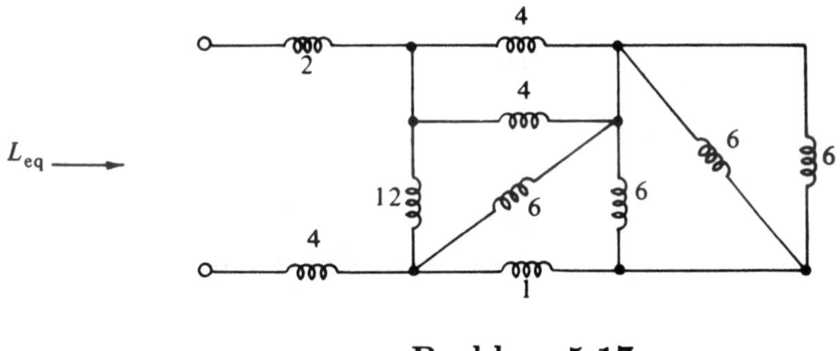

Problem 5.17

5.18 Construct a 11.11uF capacitor out of 4.7uF capacitors only. Show the detail of your design. How many 4.7uF capacitors do you need?

5.19 Construct a 12.35mH inductor out of 10mH inductors only. Show the detail of your design. How many 10mH inductirs do you need ?

5.7 Practical capacitors and inductors

5.20 Practical capacitors can be modelled as a resistor in parallel with an ideal capacitor. The leakage resistance R_c is inversely proportional to the capacitance C. Therefore the product of the leakage resistance and capacitance

R_cC is often given by manufacturers to specify the capacitor loss. A Teflon and an electrolytic capacitors are specified to have resistance-capacitance products of $10^6\Omega$ and $10^3\Omega$ respectively. (a) Assume the two capacitors have the same capacitances. Which one dissipates more power when both capacitors are applied with the same voltage (b) Find the leakage resistance of the electrolytic capacitor for the following capacitance (1) 4.7uF (2) 0.001uF.

5.8 Singular circuits

5.21 Find $v_1(0^+)$, $v_2(0^+)$, $i(0^+)$ and $i(0^-)$ if the switch is closed at t=0 and $v_1(0^-)$=14V and $v_2(0^-)$=6V.

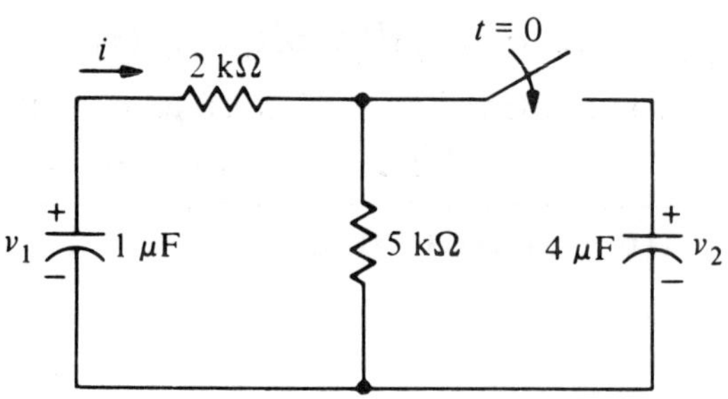

Problem 5.21

5.22 If $i_1(0^-)$=1A and the switch is opened at t=0, find $i_1(0^+)$, $i_2(0^-)$, $i_2(0^+)$, $v(0^-)$, and $v(0^+)$.

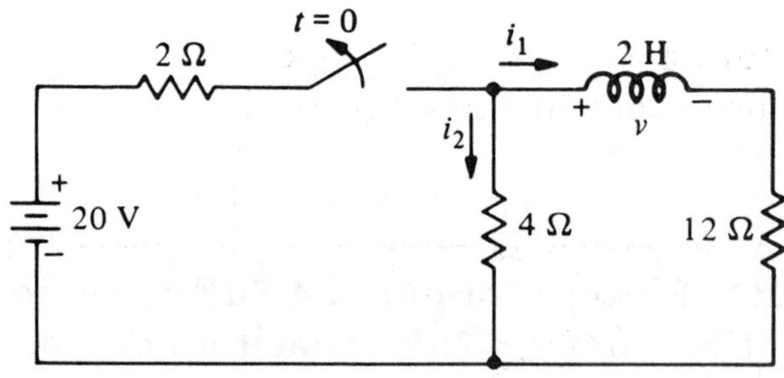

Problem 5.22

5.23 If $v_1(0^-)$=9V, $i_1(0^-)$=3A, and the switch is opened at t=0, find $v_1(0^+)$, $i_1(0^+)$, $i_2(0^-)$, $i_2(0^+)$, $v_2(0^-)$, and $v_2(0^+)$.

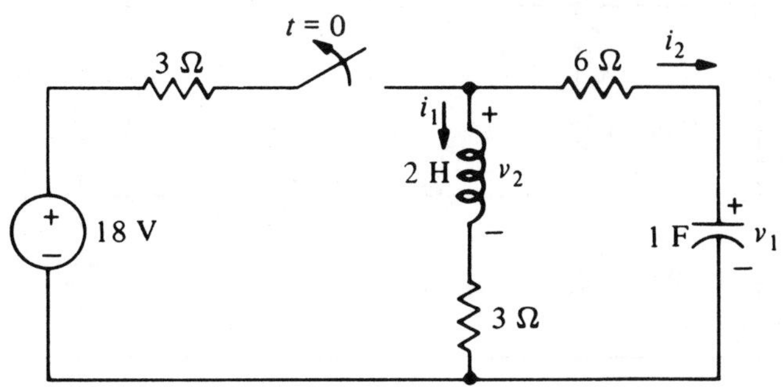

Problem 5.23

5.24 In circuit shown, switch S_1 is closed at t=0s and switch S_2 is closed at t=4s. Find $v(0^+)$, $v(3^+)$, $i(0^+)$, and $i(3^+)$. Assume $v(0^-)$=0V.

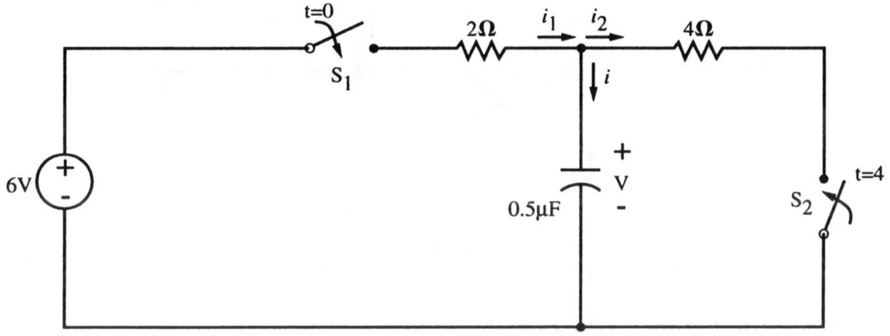

Problem 5.24

5.25 Find the equivalent capacitance C_{eq} at dc steady state. Assume the parasitic capacitance of inductors are negligible (ideal inductors).

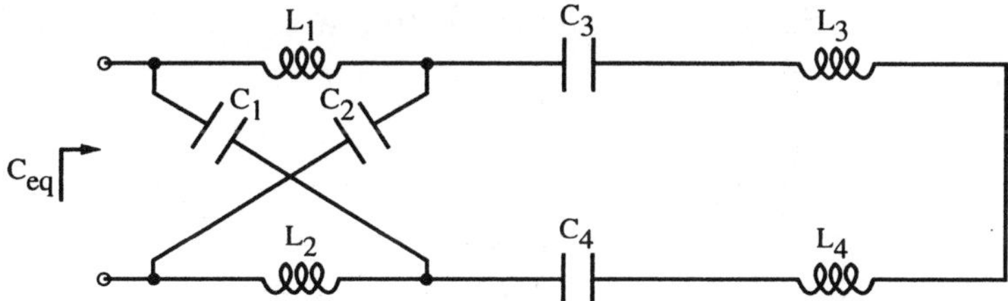

Problem 5.25

5.26 Find the dc steady state voltage v.

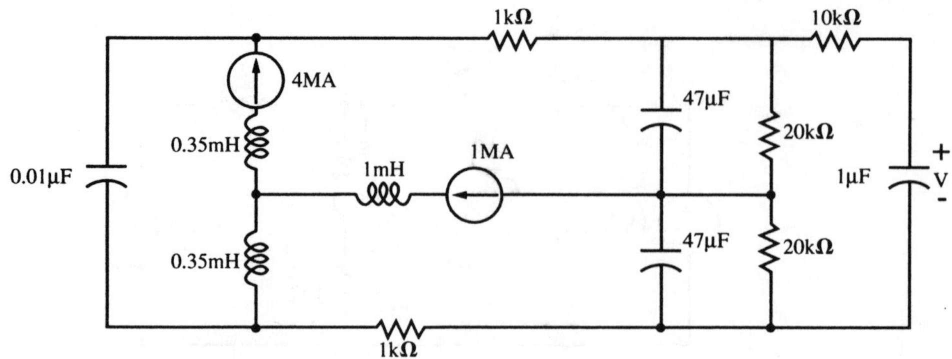

Problem 5.26

5.27 If the circuit shown is in dc steady state at $t = 0^-$, find $i(t)$ for $t = 0^-$ and $t \geq 0$.

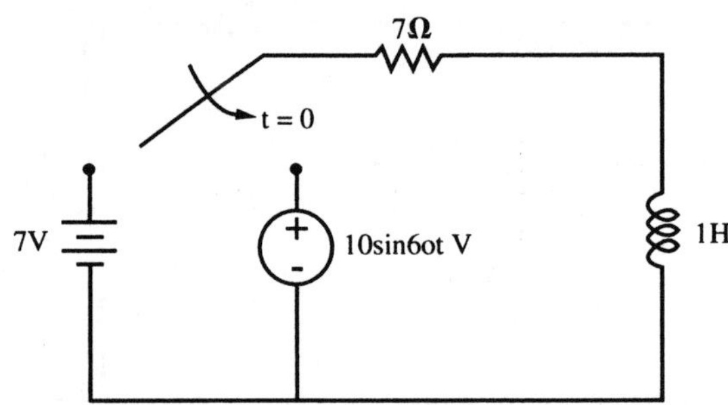

Problem 5.27

5.28 Let $C = \frac{1}{3}$F, $R_1 = R_2 = 3\Omega$ and $v = 9$V. If the current in R_2 at $t = 0^-$ is 1A directed downward, find at $t = 0^-$ and at $t = 0^+$ (a) the charge on the capacitor, (b) the current in R_1 directed to the right, and (c) the current in C directed downward.

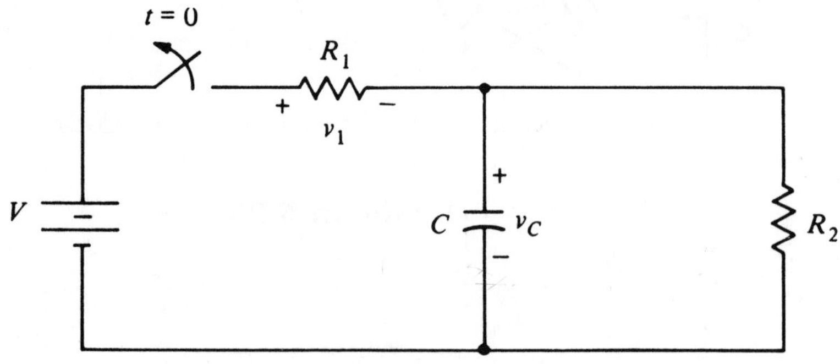

Problem 5.28

5.29 Find i, v_1 and v_2 in dc steady state.

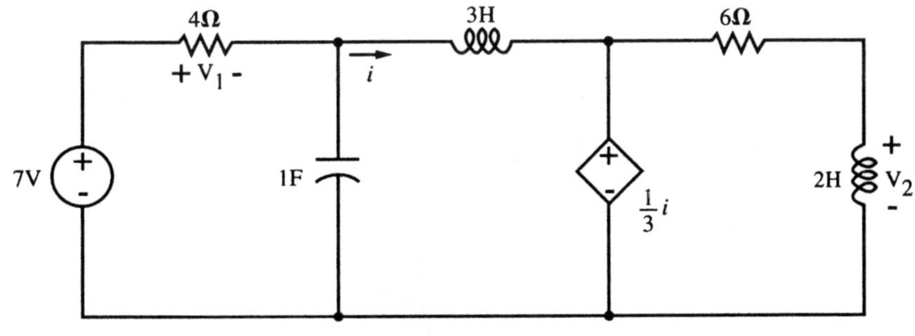

Problem 5.29

5.30 Finde the dc steady state current i_1 and i_2.

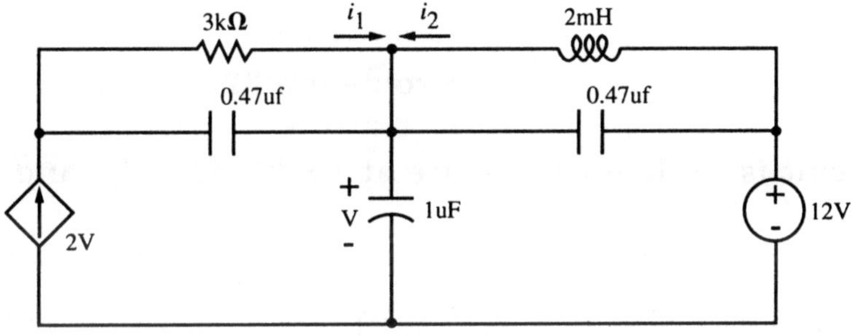

Problem 5.30

5.31 If the network current is in dc steady state at $t = 0^-$, find v_1 and v_2 at $t = 0^-$ and $t = 0^+$.

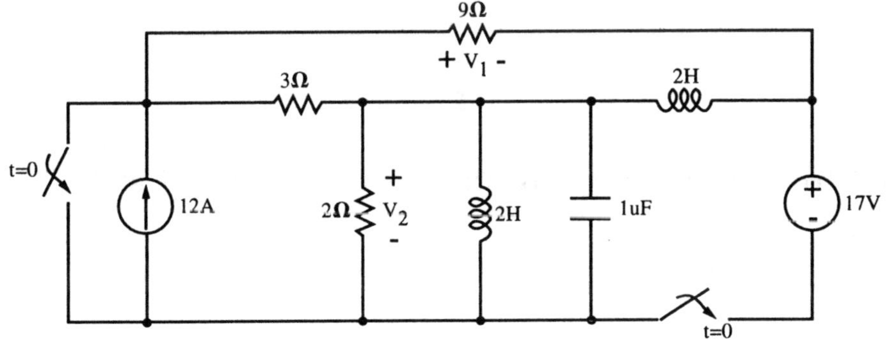

Problem 5.31

5.32 If the current is in dc steady state at $t = 0^-$, find v_1 and v_2 at $t = 0^-$ and $t = 0^+$.

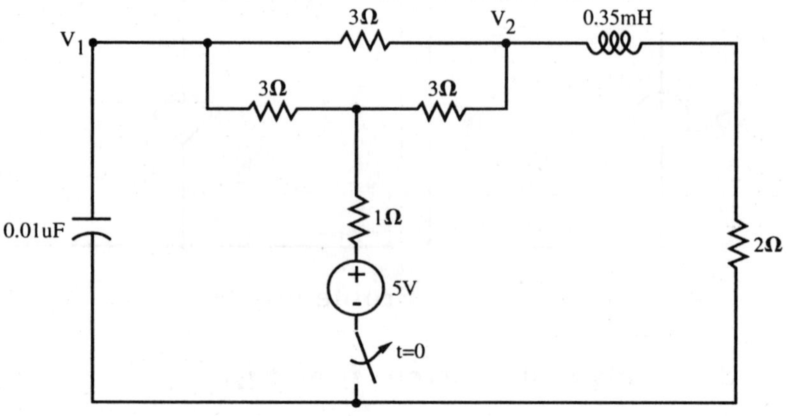

Problem 5.32

5.33 If the circuit is in dc steady state at $t = 0^-$, find $\frac{dv_1}{dt}$ and $\frac{dv_2}{dt}$ at $t = 0^+$.

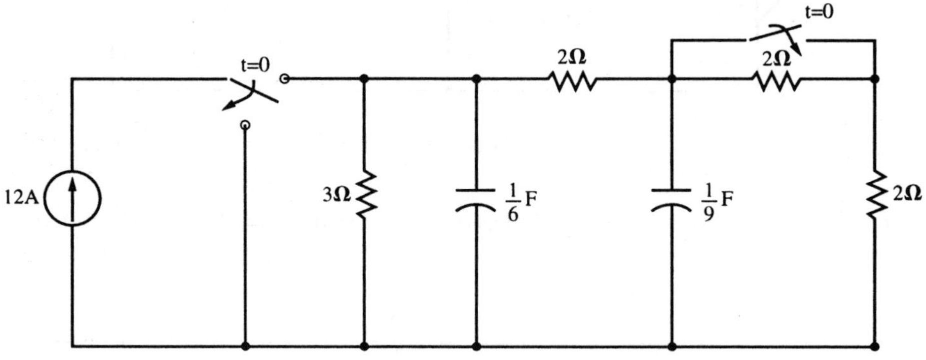

Problem 5.33

5.34 If the circuit is in dc steady state at $t = 0^-$, find $\frac{dv}{dt}$ at $t = 0^+$.

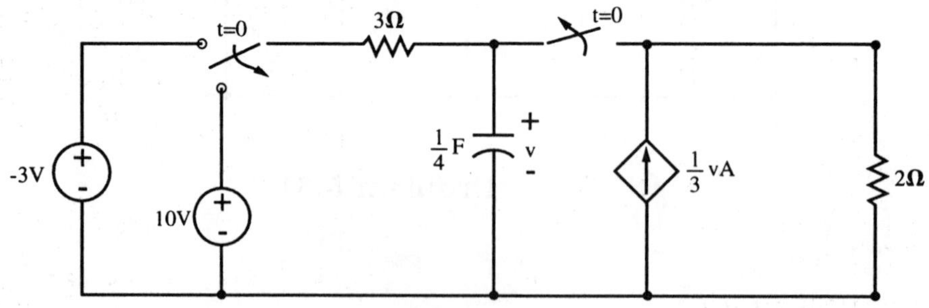

Problem 5.34

5.1 $v = \frac{1}{C}\int_0^t i\,dt + v(0)$

$= \frac{1}{10^{-7}}\int_0^t [3\cos 2000t]\times 10^{-3}\,dt - 15$

$= \underline{15\sin(2000t) - 15 \text{ V}}$

5.2 $q = Cv = (2\times 10^{-6})(50) = \underline{100\mu C}$

on each plate (±)

$v = q/C = 100\times 10^{-6}/10^{-6} = \underline{100 \text{ V}}$

5.3 $i = C\frac{dv}{dt}$; $\frac{dv}{dt}$ is the slope of graph.

$t = -7\text{ms}: i = (10^{-5})\frac{0-2}{[-6-(-8)]10^{-3}} = \underline{-10\text{mA}}$

$t = -3\text{ms}: i = (10^{-5})\frac{0+2}{[0+6]10^{-3}} = \underline{3.33 \text{ mA}}$

$t = 1\text{ms}: i = (10^{-5})\frac{2-0}{(2-0)10^{-3}} = \underline{10\text{mA}}$

$t = 3\text{mS}: i = (10^{-5})(0) = \underline{0 \text{ mA}}$

$t = 7\text{ms}: i = (10^{-5})\frac{0-2}{[8-6]10^{-3}} = \underline{-10\text{mA}}$

5.4 $v = \frac{1}{C}\int_{-\infty}^t i\,dt + v(t_0)$;

area under the graph is the integral and $v(t_0) = 0$ at less than -8 ms

$v(-6\text{ms}) = \frac{1}{10^{-6}}[\frac{1}{2}(2\text{ms})(-2\text{mA})] + 0 = \underline{-2\text{V}}$

$v(0) = \frac{1}{10^{-6}}[\frac{1}{2}(6\text{ms})(-2\text{mA})] - 2 = \underline{8\text{V}}$

$v(2\text{ms}) = \frac{1}{10^{-6}}[\frac{1}{2}(2\text{ms})(2\text{mA})] - 8 = \underline{-6\text{V}}$

$v(8\text{ms}) = \frac{1}{10^{-6}}[(2)(4) + \frac{1}{2}(2)(2)] - 6 = \underline{4\text{V}}$

5.5 $v(3) = \frac{1}{C}\int_1^3 i\,dt + v(1)$

$-5 = \frac{1}{1}\int_1^3 2e^{-2t}\,dt + v(1)$

$= -e^{-2t}\big|_1^3 + v(1)$

$-5 = -e^{-6} + e^{-2} + v(1)$

or $\underline{v(1) = e^{-6} - e^{-2} - 5\text{V}}$

5.6 (a) Energy stored in the capacitor

$W_C(t) = \frac{1}{2}Cv^2(t) = \frac{1}{2}(1)(-5)^2 = \underline{12.5 \text{ J}}$

(b) $W_C(\infty) = \frac{1}{2}Cv^2(\infty)$

$v(\infty) = \lim_{t\to\infty} v(t) = \lim_{t\to\infty}\left[\frac{1}{C}\int_3^\infty i\,dt + v(3)\right]$

$= \lim_{t\to\infty}\left[\int_3^t 2e^{-2t}\,dt - 5\right] = e^{-6} - 5$

Hence $W_C(\infty) = \frac{1}{2}Cv^2(\infty) = \frac{1}{2}(1)(e^{-6}-5)^2$

$= \underline{\frac{1}{2}(e^{-6}-5)^2 \text{ J}}$

5.7 $q = CV$

$3\times 10^{-6} = C\times 12$, $C = 0.25\times 10^{-6} = \underline{0.25 \text{ uF}}$

Energy stored in the capacitor is

$W_C = \frac{1}{2}CV^2 = \frac{1}{2}(\frac{1}{4})(12)^2 = \underline{18 \text{ J}}$

5.8 $C = \frac{C_1C_2}{C_1+C_2} = 2F$

and $i = C_2\frac{dv_2}{dt}$, $9\cos 3t = C_2\frac{d(\sin 3t)}{dt}$,

Hence $C_2 = 3F$

$\frac{3C_1}{C_1+3} = 2$, $\underline{C_1 = 6F}$

5.9

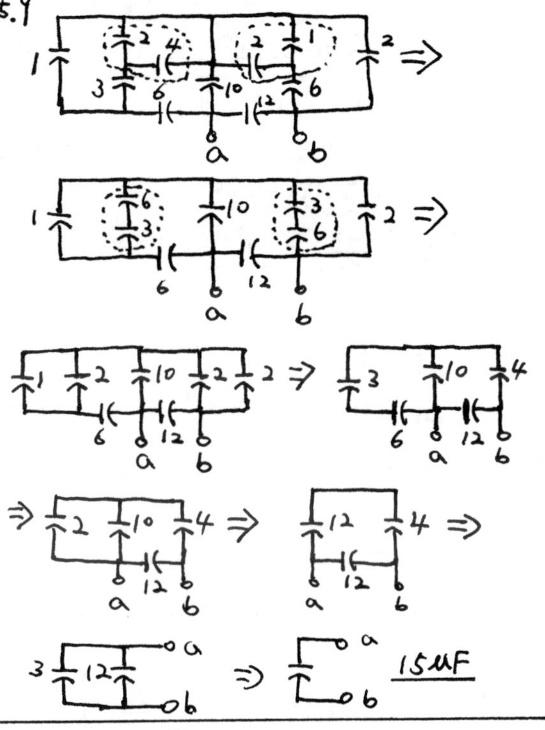

5.10 $p = p_C + p_R$

$p_C = v_C(t)\,i_C(t)$ and

$i_C(t) = C\frac{dv}{dt} = 10^{-6}\times\frac{d[te^{-t}]}{dt} = 10^{-6}[e^{-t} - te^{-t}]$

$p = 10^{-6}\times te^{-t}[e^{-t} - te^{-t}] + te^{-t}(\frac{te^{-t}}{10^{-6}})$

$= 10^{-6}[te^{-2t} - t^2e^{-2t} + t^2e^{-2t}]$

$= 10^{-6}te^{-2t}$

$\frac{dp}{dt} = 0$, $10^{-6}[e^{-2t} - 2te^{-2t}] = 0$

$10^{-6}\times e^{-2t}(1-2t) = 0$, $\underline{t = \frac{1}{2}\text{s}}$

$p_{max} = 10^{-6}(\frac{1}{2})(e^{-2\times\frac{1}{2}}) = \underline{5\times 10^{-5}e^{-1} \text{ W}}$

5.11 $v_1(0^+) = v_1(0^-) = 18V$

$v_2(0^+) = v_2(0^-) = 6V$

$i_2(0^-) = \frac{v_2(0^-)}{4k\Omega} = \underline{1.5\,mA}$

$i_1(0^-) = \frac{v_1(0^-) - v_2(0^-)}{2k\Omega} = \underline{6\,mA}$

At $t = 0^+$, the switch is open and

$i_1(0^+) = i_2(0^+)$

$\therefore\ i_1(0^+) = i_2(0^+) = \frac{v_1(0^+)}{(2+4)k\Omega} = \underline{3\,mA}$

5.12 $\lambda = Li$

$\lambda(10ms) = (50\times10^{-3})(\frac{12}{2}\,mA) = \underline{300\,\mu Wb}$

$\lambda(40ms) = (0.05)[\frac{3}{4}(12mA)] = \underline{450\mu Wb}$

$\lambda(80ms) = (0.05)(\frac{12}{4}\,mA) = \underline{150\mu Wb}$

5.13 $v = L\,di/dt$, di/dt equals slope

$v(10ms) = (10)\frac{12}{20} = \underline{6V}$

$v(40ms) = (10)(\frac{-12}{100-20}) = \underline{-1.5V}$

$v(80ms) = (10)(\frac{-12}{80}) = \underline{-1.5V}$

5.14 $i = \frac{1}{L}\int_0^t v\,dt + i_0$

$i(20\,ms) = \frac{1}{120}[\frac{1}{2}(20ms)12] + 2mA = \underline{3mA}$

$i(60\,ms) = \frac{1}{120}[\frac{1}{2}(12 + \frac{12}{2})(40ms)] + 3\,mA$

$= \underline{6\,mA}$

$i(100ms) = \frac{1}{120}[\frac{1}{2}(12)(80\,ms)] + 3mA$

$= \underline{7\,mA}$

5.15 $i(3\pi/4) = \frac{1}{0.25}\int_0^{3\pi/4} 6\cos 2t\,dt + 0$

$= 12(\sin\frac{3\pi}{2}) = -12A$

$W(3\pi/4) = \frac{1}{2}Li^2 = \frac{1}{2}(0.25)(12^2) = \underline{18J}$

$i(\pi/8) = 12(\sin\pi/4) = \frac{12}{\sqrt{2}}\,A$

$v(\pi/8) = 6\cos(\pi/4) = \frac{6}{\sqrt{2}}\,V$

$p(\pi/8) = vi = (\frac{12}{\sqrt{2}})(\frac{6}{\sqrt{2}}) = \underline{36W}$

5.16 $v(t) = 2t - 10$ for $0 \le t \le 10ms$

$i = \int_0^t v(t)dt + i(0)$

$= \int_0^t (2t-10)dt + 2$

$= (t^2 - 10t)\Big|_0^t + 2 = t^2 - 10t + 2$

$i = \int_{10}^t v(t)dt + i(10) = (t-10)^2 - 10(t-10) + 2$ for $10 \le t \le 20ms$

Similarly, $i = (t-20)^2 - 10(t-20) + 2$ for $20 \le t \le 30\,ms$

5.16 Cont.

Hence $i(t) = t^2 - 10t + 2$, $\quad 0 \le t \le 10ms$

$= (t-10)^2 - 10(t-10) + 2$, $\quad 10 \le t \le 20ms$

$= \underline{(t-20)^2 - 10(t-20) + 2}$, $\quad 20 \le t \le 30ms$

5.17

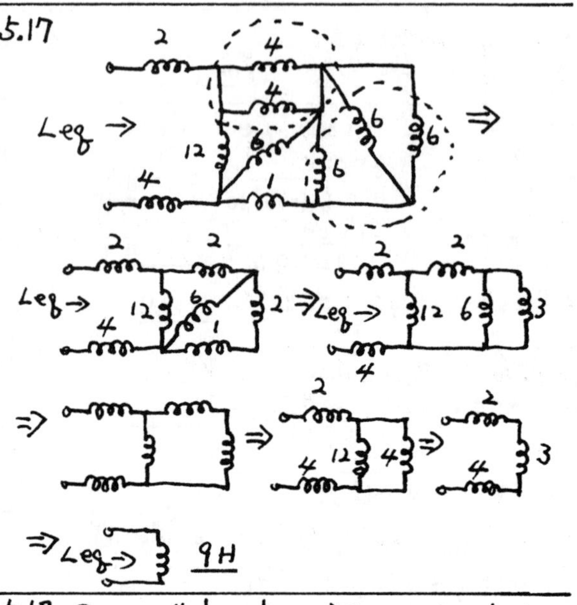

5.18 By parallel and series combination one can construct capacitors of arbitrary values of the form $\frac{P}{q}\times 47\,\mu F$, where P, q are integers

Suppose we choose the building block of the type $2^n \times 47\ \mu F$

$11.11 = a_1\times2^n\times47 + a_2\times2^{n-1}\times47 + a_3\times2^{n-2}\times47 + \cdots$

$= 47(a_1\times2^n + a_2\times2^{n-1} + a_3\times2^{n-2} + \cdots)$

$a_1\times2^n + a_2\times2^{n-1} + a_3\times2^{n-1} + \cdots = 2.3638$

$\because$ the polynomial can approach any real number as close as possible, one can construct a $11.11\,\mu F$ capacitor with desired accuracy by properly choosing the set of coefficients $a_1, a_2, \cdots a_n$.

$a_1\times2^1 + a_2\times a^0 + a_3\times2^{-1} + a_4\times2^{-2} + \cdots = 2.36$

By long division, we find

$a_1 = 1,\ a_2 = 0,\ a_3 = 0,\ a_4 = 1,\ a_5 = 0,\ a_6 = 1$ and $a_7 = 1$ with remainder 0.01625

5.18 Cont.

Hence one can construct a 11.11 μF capacitor by the following combination of 4.7 μF capacitors.

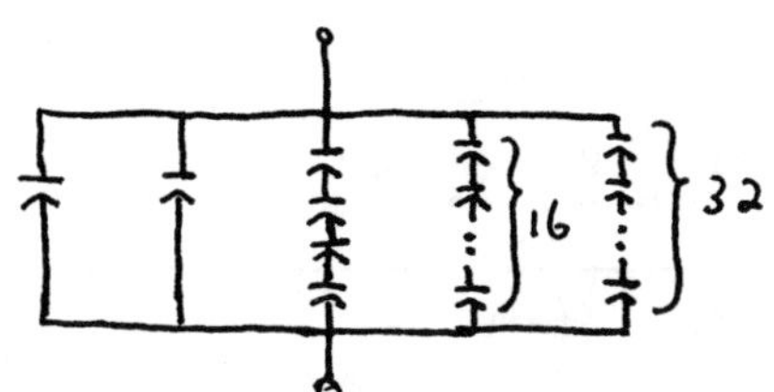

It takes $32+16+4+2=\underline{54}$ 4.7 μF capacitors to construct a 11.11 μF one with the error of $4.7\times 0.01625 = 0.08\ \mu F$

5.19 Similar to the solution of <5.18>

$12.35 = 10(a_1 \times 2^n + a_2 \times 2^{n-1} + a_3 \times 2^{n-2} + \cdots)$

$1.235 = a_1 \times 2^n + a_2 \times 2^{n-1} + a_3 \times 2^{n-1} + \cdots$

By long division, we find the coefficients: $a_1 = 1$, $a_2 = a_3 = 0$, $a_4 = a_5 = a_6 = 1$

The combination of 10 mH inductor is:

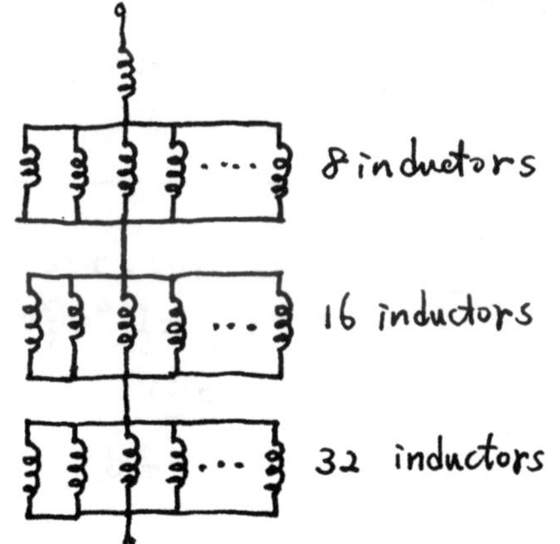

Total of $(1+8+16+32) = \underline{57}$ inductors of 10mH are used.

The error is $10 \times 0.01625 \cong 0.16$ mH

5.20 (a) Assume both capacitors have equal capacitance c and a voltage v is applied.

$R_{Tef} = \frac{10^6}{c}\ \Omega$, $R_{electro} = \frac{10^3}{c}\ \Omega$

$P_{Tef} = \frac{1}{2}\frac{v^2}{R_{Tef}}$, $P_{electro} = \frac{1}{2}\frac{v^2}{R_{electro}}$

$P_{electro} > P_{Tef}$.

5.20 Cont.

(b) $R_{4.7\mu F} = \frac{10^3}{4.7\times10^{-6}} = \underline{2.13\times10^8\ \Omega}$

$R_{0.001\mu F} = \frac{10^3}{1\times10^{-9}} = \underline{1\times10^{12}\ \Omega}$

5.21 $v_1(0^+) = v_1(0^-) = \underline{14V}$

$v_2(0^+) = v_2(0^-) = \underline{6V}$

$i_1(0^-) = \frac{v_1(0^-)}{(2+5)k\Omega} = \underline{2mA}$

$i_1(0^+) = \frac{v_1(0^+) - v_2(0^+)}{2k\Omega} = \frac{14-6}{2k\Omega} = \underline{4mA}$

5.22

2Ω
v_1
2H
i_1
i_2
+ v −
20V
4Ω
12Ω
$t=0^-$

By KCL, $i_1(0^-) + i_2(0^-) + \frac{v_1(0^-) - 20}{2} = 0$

or $2i_2(0^-) + v_1(0^-) = 18$, Since

$v_1(0^-) = i_2(0^-)4$, $i_2(0^-) = \underline{3A}$

$v(0^-) = v_1(0^-) - i_1(0^-)(12) = 12 - 12 = \underline{0V}$

At $t = 0^+$, $i_1(0^+) = i_1(0^-) = \underline{1A}$

$i_2(0^+) = -i_1(0^+) = \underline{-1A}$

5.23

3Ω
v
6Ω
i_2
i_1
v_2
18V
1F
v_1
3Ω
$t=0^-$

At $t=0^-$
$v_1(0^-) = 9V$,
$i_1(0^-) = 3A$,

By KCL, $\frac{v(0^-) - 18}{3} + i_1(0^-) + \frac{v(0^-) - v_1(0^-)}{6} = 0$

or $3v(0^-) = 27$ or $v(0^-) = 9V$

$i_2(0^-) = \frac{v(0^-) - v_1(0^-)}{6} = \frac{9-9}{6} = \underline{0A}$

$v_2(0^-) = v(0^-) - i_1(0^-)3 = 9 - 9 = \underline{0V}$

6Ω
i_2
i_1
2H
v_2
v_1
3Ω
$t=0^+$

At $t=0^+$,
$i(0^+) = i_1(0^-) = \underline{3A}$
$v_1(0^+) = v_1(0^-) = 9V$
$i_2(0^+) = -i_1(0^+) = \underline{-3A}$

By KCL

$v_2(0^+) = 6i_2(0^+) + v_1(0^+) + 3i_2(0^+)$
$= \underline{-18V}$

5.24 By continuity principle of capacitor voltage

$v(0^+) = v(0^-) = \underline{0V}$

$i(0^+) = \frac{6-0}{2} = \underline{3A}$

$v(3^+) = v(3^-) = \underline{6V}$

$i_2(3^+) = \frac{6}{4} = 1.5A$

By KCL $\frac{6-6}{2} = i(3^+) + i_2(3^+) = i(3^+) + 1.5A$

$i(3^+) = \underline{-1.5A}$

5.25 At DC steady state, all inductors are short circuited.

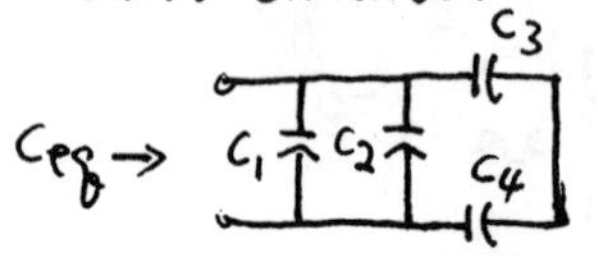

$C_{eq} = C_1 + C_2 + \frac{C_3 C_4}{C_3 + C_4}$

5.26 By superposition

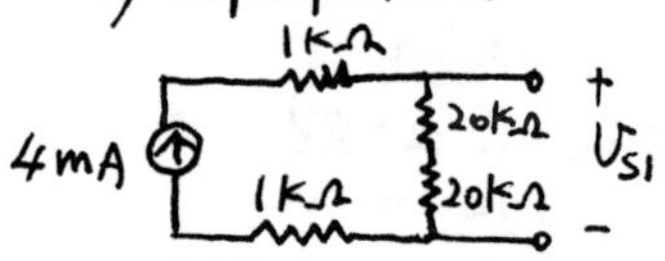

$V_{S1} = (4mA)(40k\Omega) = 160V$

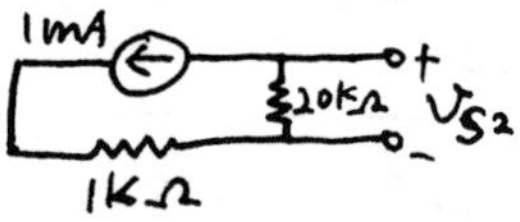

$V_{S2} = (-1mA)(20k\Omega) = -20V$

Hence $v = V_{S1} + V_{S2} = \underline{140V}$

5.27 At $t=0^-$

$i(0^-) = \frac{7}{7} = \underline{1A}$

At $t=0^+$

$10\sin 60t = 7i + 1\times\frac{di}{dt} = 7i + \frac{di}{dt}$

$(10\sin 60t - 7i)dt = di$

$\int (10\sin 60t - 7i)dt = \int di + 1$

$-\frac{1}{6}\cos 60t - 1 = 7ti + i$

$i(t) = \frac{-\frac{1}{6}\cos 60t - 1}{7t+1}\ A$

5.28 $v_C(0^-) = IR_2 = 1(3) = 3V = v_C(0^+)$

(a) $q(0^-) = q(0^+) = Cv(0^-) = \frac{1}{3}(3) = \underline{1C}$

(b) $i_{R_1}(0^-) = \frac{V_1(0^-)}{R_1} = \frac{V - V_C(0^-)}{R_1} = \frac{9-3}{3} = \underline{2A}$

$i_{R_1}(0^+) = \underline{0}$ since the circuit is open.

(c) $i_C(0^-) = i_{R_1}(0^-) - \frac{V_C(0^-)}{R_2} = 2 - \frac{3}{3} = \underline{1A}$

$i_C(0^+) = i_{R_1}(0^+) - \frac{V_C(0^+)}{R_2} = 0 - \frac{3}{3} = \underline{-1A}$

5.29 At DC steady state

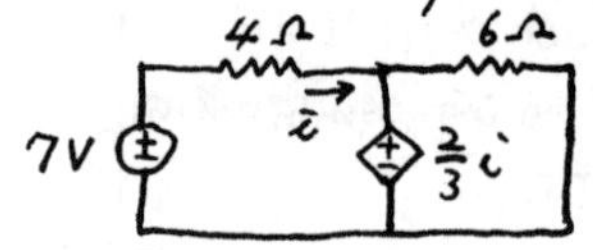

$\frac{7 - \frac{1}{3}i}{4} = i$ or $7 - \frac{i}{3} = 4i$

$\frac{13i}{3} = 7$ or $i = \frac{21}{13}A$

$v_1 = 4i = \frac{84}{13}V$, $v_2 = \underline{0V}$

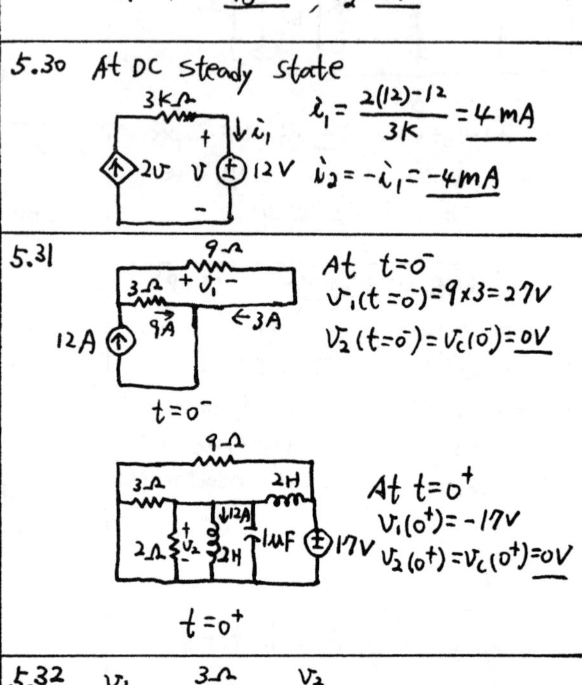

5.30 At DC steady state

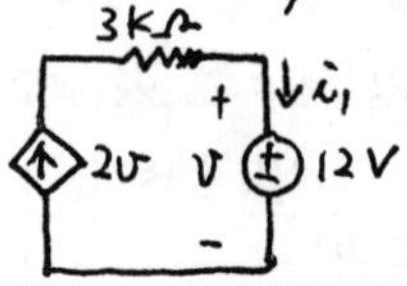

$i_1 = \frac{2(12)-12}{3K} = \underline{4mA}$

$i_2 = -i_1 = \underline{-4mA}$

5.31

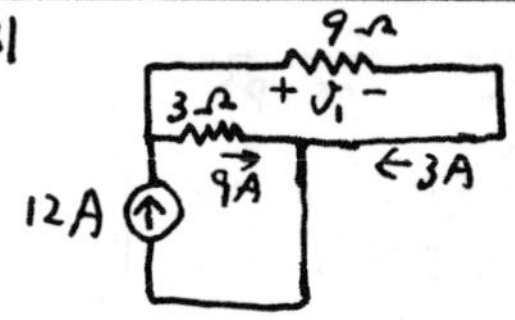

$t=0^-$

At $t=0^-$

$v_1(t=0^-) = 9\times 3 = 27V$

$V_2(t=0^-) = V_C(0^-) = \underline{0V}$

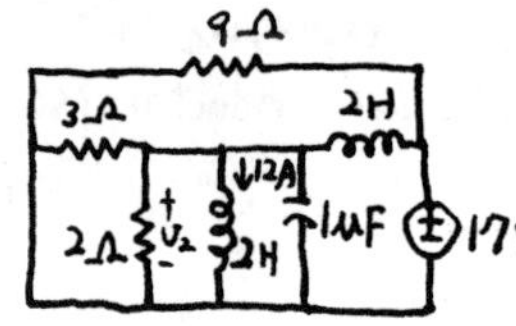

$t=0^+$

At $t=0^+$

$V_1(0^+) = -17V$

$V_2(0^+) = V_C(0^+) = \underline{0V}$

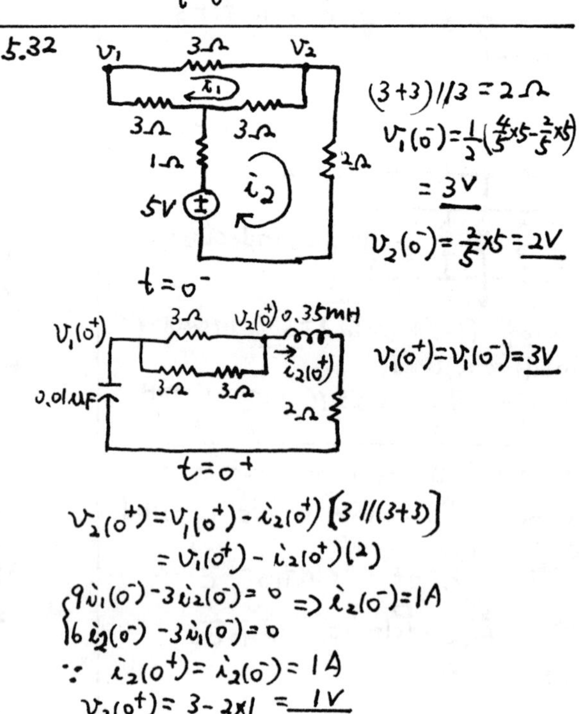

5.32

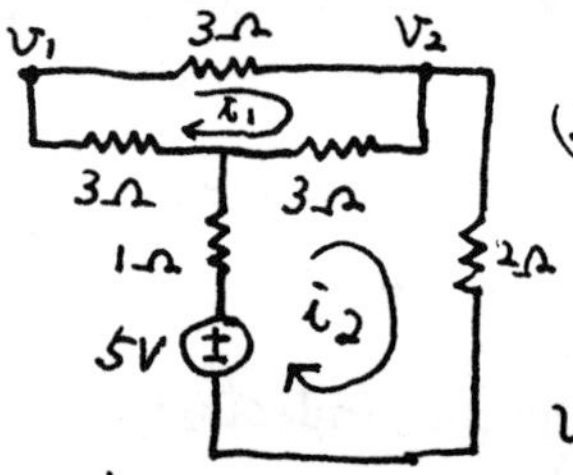

$t=0^-$

$(3+3)//3 = 2\Omega$

$V_1(0^-) = \frac{1}{2}(\frac{4}{5}\times 5 - \frac{2}{5}\times 5) = \underline{3V}$

$V_2(0^-) = \frac{2}{5}\times 5 = \underline{2V}$

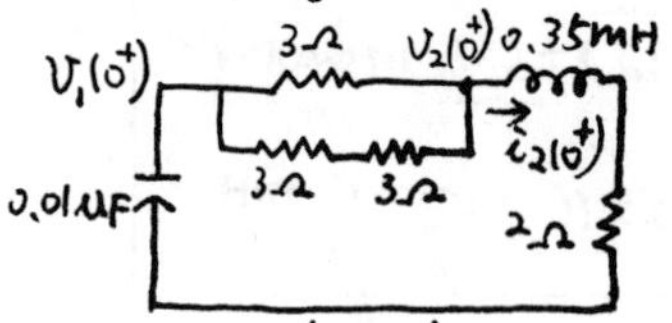

$t=0^+$

$V_1(0^+) = V_1(0^-) = \underline{3V}$

$V_2(0^+) = V_1(0^+) - i_2(0^+)[3//(3+3)]$

$= V_1(0^+) - i_2(0^+)(2)$

$\begin{cases} 9i_1(0^-) - 3i_2(0^-) = 0 \\ 6i_2(0^-) - 3i_1(0^-) = 0 \end{cases} \Rightarrow i_2(0^-) = 1A$

$\therefore i_2(0^+) = i_2(0^-) = 1A$

$V_2(0^+) = 3 - 2\times 1 = \underline{1V}$

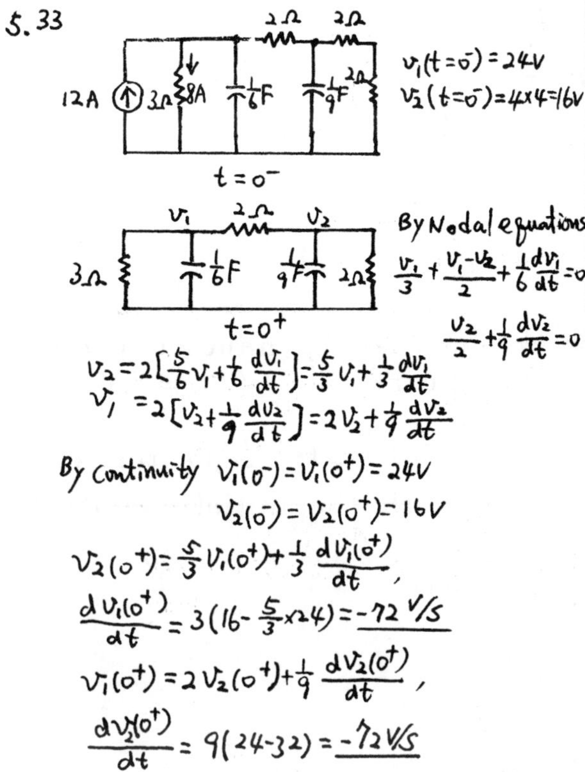

5.33

$v_1(t=0^-)=24V$

$v_2(t=0^-)=4\times4=16V$

By Nodal equations

$\frac{v_1}{3}+\frac{v_1-v_2}{2}+\frac{1}{6}\frac{dv_1}{dt}=0$

$\frac{v_2}{2}+\frac{1}{9}\frac{dv_2}{dt}=0$

$v_2=2\left[\frac{5}{6}v_1+\frac{1}{6}\frac{dv_1}{dt}\right]=\frac{5}{3}v_1+\frac{1}{3}\frac{dv_1}{dt}$

$v_1=2\left[v_2+\frac{1}{9}\frac{dv_2}{dt}\right]=2v_2+\frac{1}{9}\frac{dv_2}{dt}$

By continuity $v_1(0^-)=v_1(0^+)=24V$

$v_2(0^-)=v_2(0^+)=16V$

$v_2(0^+)=\frac{5}{3}v_1(0^+)+\frac{1}{3}\frac{dv_1(0^+)}{dt}$,

$\frac{dv_1(0^+)}{dt}=3(16-\frac{5}{3}\times24)=\underline{-72\ V/s}$

$v_1(0^+)=2v_2(0^+)+\frac{1}{9}\frac{dv_2(0^+)}{dt}$,

$\frac{dv_2(0^+)}{dt}=9(24-32)=\underline{-72\ V/s}$

5.34

At $t=0^-$

$\frac{v+3}{3}+\frac{1}{4}\frac{dv}{dt}-\frac{1}{3}v+\frac{v}{2}=0$

$\frac{1}{4}\frac{dv}{dt}+v\left[\frac{1}{3}-\frac{1}{3}+\frac{1}{2}\right]=-1$ or $\frac{1}{4}\frac{dv}{dt}+\frac{1}{2}v=-1$

Because at $t=0^-$, the circuit is in steady state, $\frac{1}{2}v=-1$, $v(0^-)=-2V$

at $t=0^+$

$\frac{1}{4}\frac{dv}{dt}+\frac{v-10}{3}=0$ or $\frac{dv}{dt}=4(\frac{10}{3}-v)$

$\frac{dv(0^+)}{dt}=4(\frac{10}{3}-v(0^+))=4\left[\frac{10}{3}-v(0^-)\right]$

$=4\left[\frac{10}{3}+2\right]=\underline{\frac{64}{3}\ V/s}$

Chapter 6
First-order Circuits

6.1 Simple RC and RL circuits

6.1 **Find a switched first-order inductive network such that the inductor current for** t **>0 satisfies** $0.003\frac{di}{dt} + 8i = 2$**A with the initial condition** $i_L(0^-) = \frac{9}{4}$**A.**

6.2 **Find a switched first-order capacitive network such that the capacitor voltage** V_c **for** t **>0 satisfies** $0.004\frac{dv_c}{dt} + 7v_c = 14$**V with the initial condition** $v_c(0^-) = -3$**V.**

6.2 Time constants

6.3 **Find the time constant** τ **of the network.**

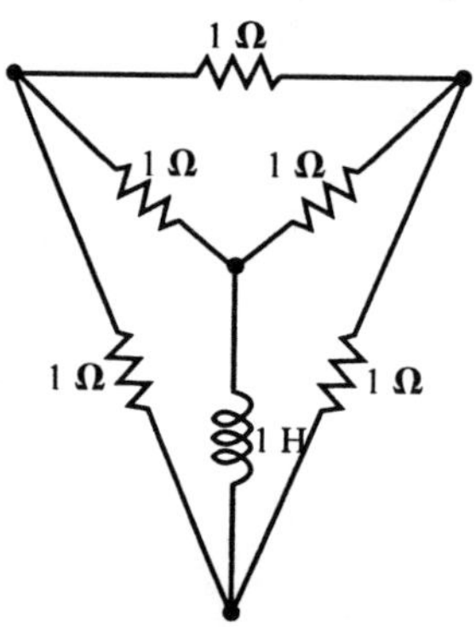

Problem 6.3

6.4 **Find the time constant of the circuit shown. Determine i_t, t >16ms if $i(16ms^{-}) = \frac{10}{e}$A.**

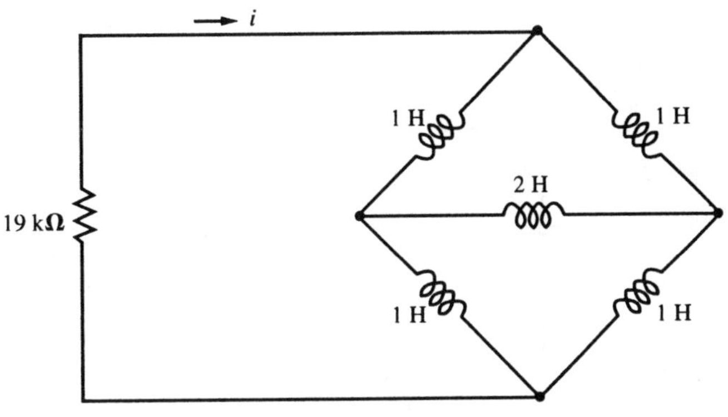

Problem 6.4

6.3 General first order circuits without sources

6.5 **Find $v(t)$ and $i(t)$ for t >0 if $v(0)$=5V.**

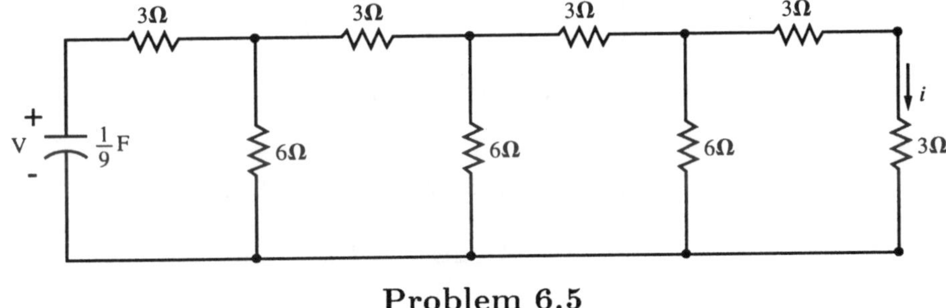

Problem 6.5

6.6 **If initially there is 40μ coulombs of charge stored on each plate of the capacitor at $t = 0^{-}$. Determine the value of R such that at t=10ms, $v_c = \frac{20}{e}$V.**

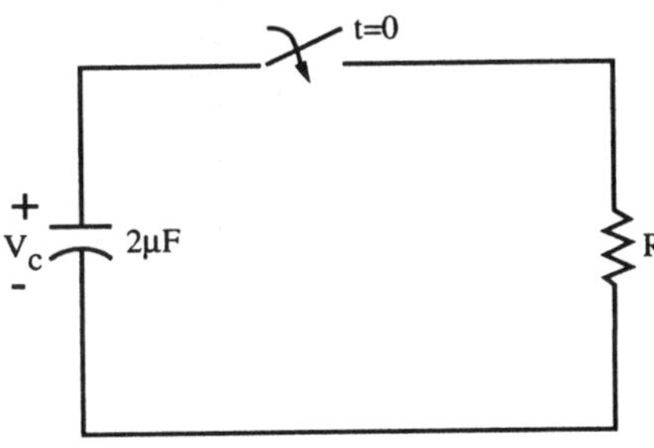

Problem 6.6

6.4 Circuits with dc sources

6.7 The switch in the network opens at t=1s, at which time v=10V. For t >1, find $v(t)$ and $W_c(t)$.

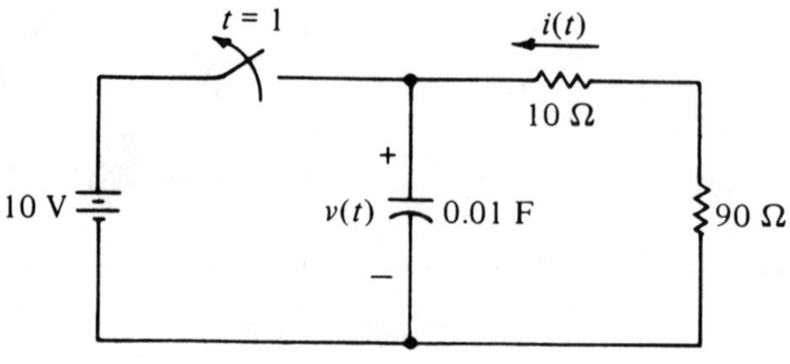

Problem 6.7

6.8 Find v for t >0 if $i(0)$=2A.

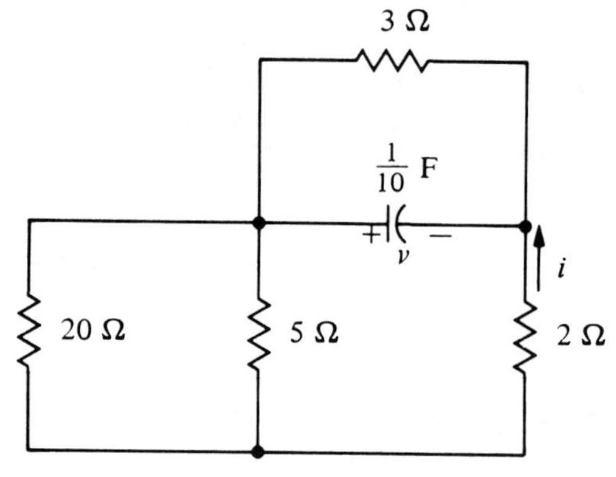

Problem 6.8

6.9 Find v for t >0 if $v(0)$=4V.

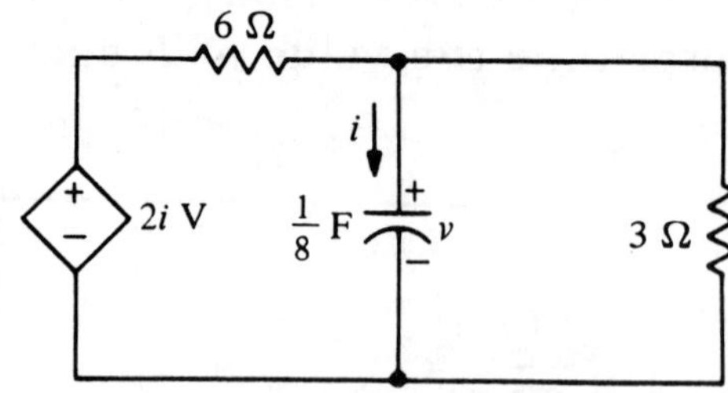

Problem 6.9

6.10 Find i for $t > 0$ if the circuit is in steady state at $t = 0^-$.

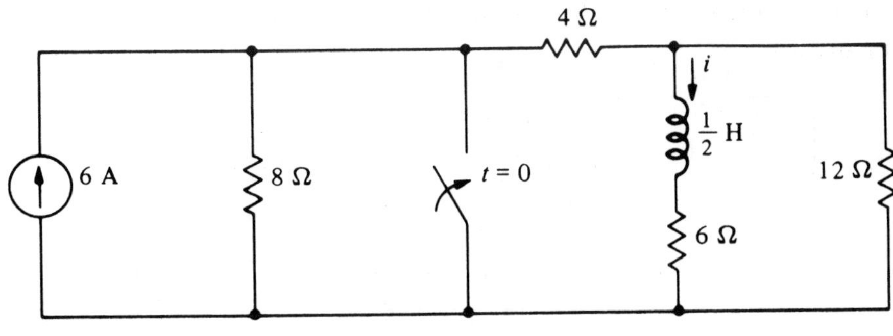

Problem 6.10

6.11 Find i for $t > 0$ if the circuit is in steady state at $t = 0^-$.

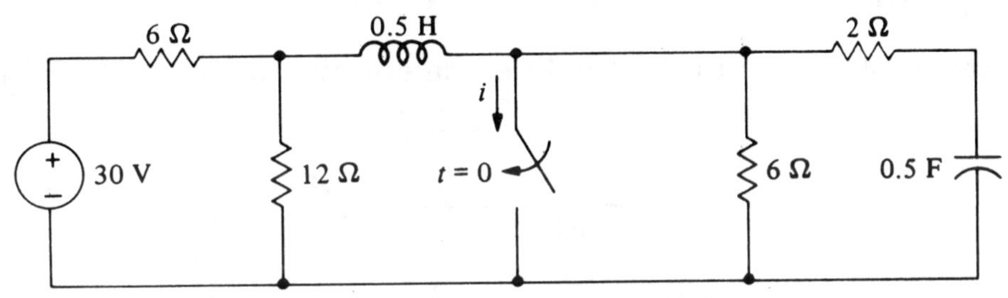

Problem 6.11

6.12 The circuit is in steady state at $t = 0^-$. Find i for $t > 0$ if the switch is moved from position 1 to position 2 at $t = 0$.

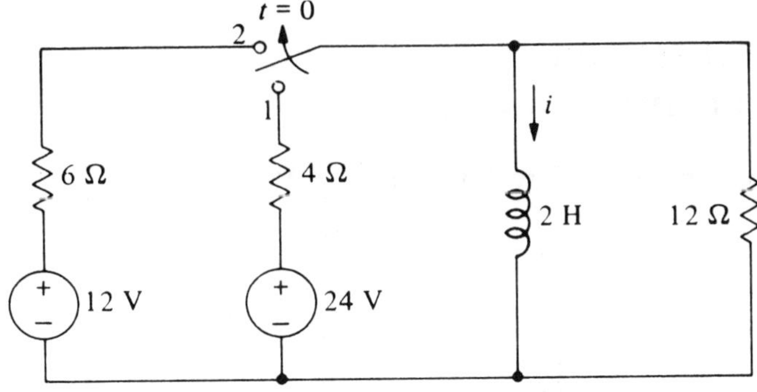

Problem 6.12

6.13 Find v on the capacitor for $t > 0$ of the circuit shown if the circuit is in steady state at $t = 0^-$.

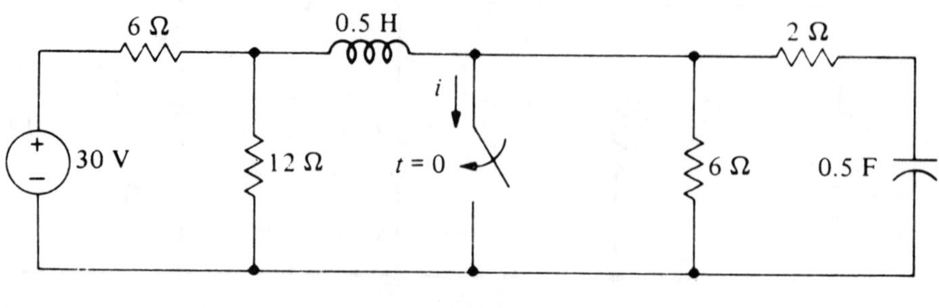

Problem 6.13

6.5 Superposition in first order circuits

6.14 The circuit shown is in *dc* steady state condition at $t = 0^-$. Find v for $t > 0$.

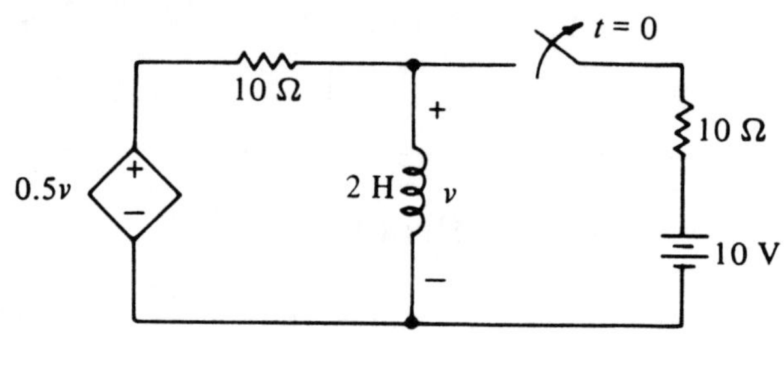

Problem 6.14

6.6 The unit step function

6.15 Find the step response i with $v_g = u(t)$V.

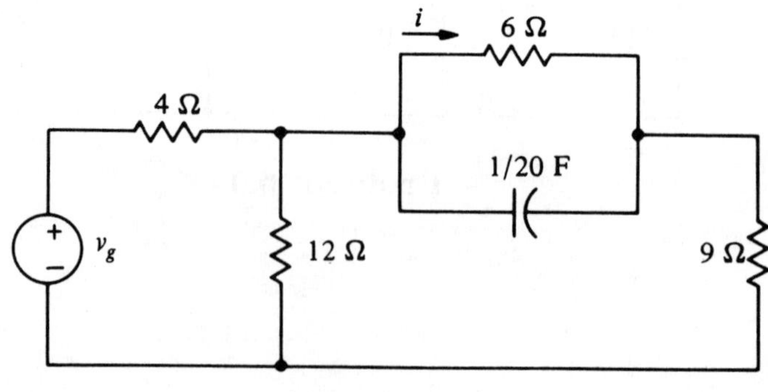

Problem 6.15

6.16 Find v for $t > 0$.

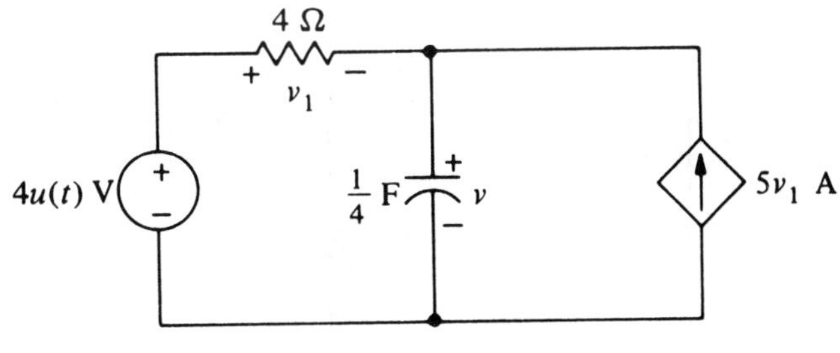

Problem 6.16

6.7 The step and pulse response

6.17 Find the voltage v_1 and v_2 in terms of R_1, R_2, C_1, C_2 and v when $t \to \infty$.

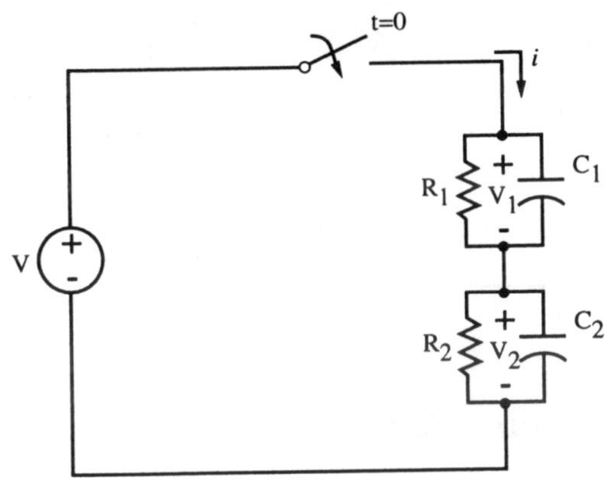

Problem 6.17

6.18 Find v for $t > 0$. Assume the circuit is in *dc* steady state at $t = 0^-$.

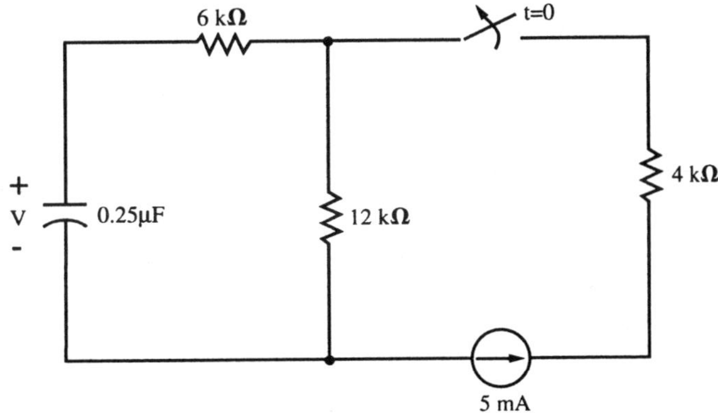

Problem 6.18

6.19 Find the indicated voltage v for $t > 0$.

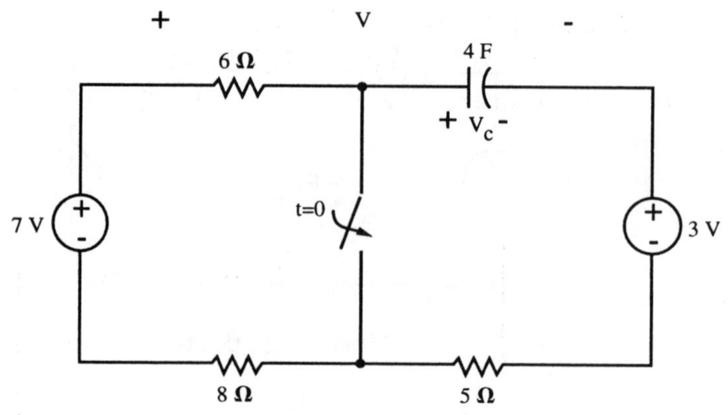

Problem 6.19

6.20 Find the capacitor voltage v_c for the network shown.

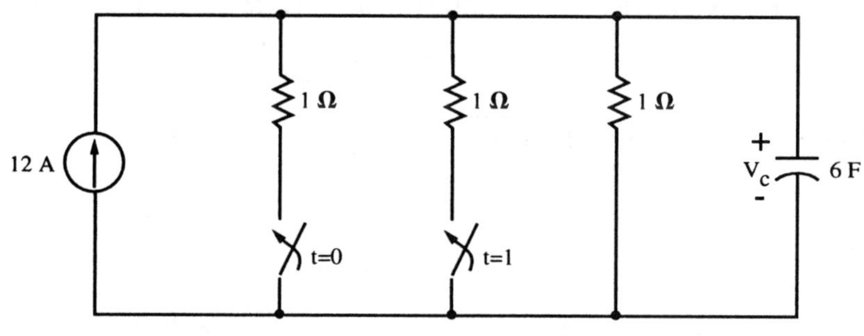

Problem 6.20

6.21 Find v for $t > 4$. Assume dc steady state at $t = 4^-$.

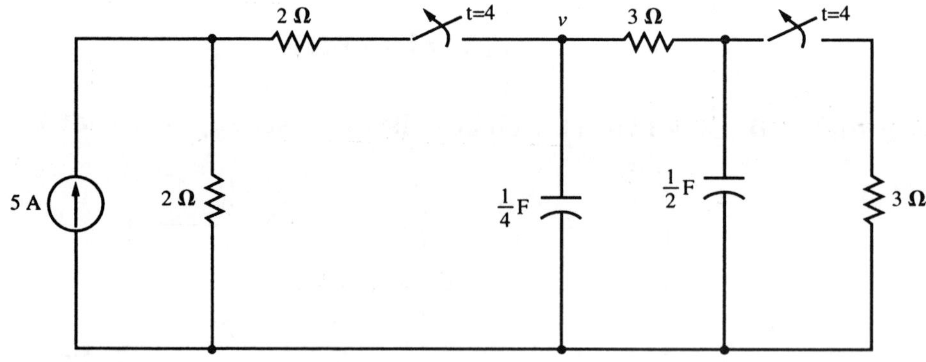

Problem 6.21

6.22 The switch operation is set to toggle between position A and B for $t > 0$ such that

sw $\rightarrow$ A for $2n \leq t < 2n + 1$ sec, n=0,1,2,...

sw $\rightarrow$ B for $2n + 1 \leq t < 2n + 2$ sec, n=0,1,2,...

Sketch the voltage across the 1Ω resistor for $0 \leq t \leq 5$. Assume $v_c(0^-) = 10$V, $i_l(0^-) = 10$A.

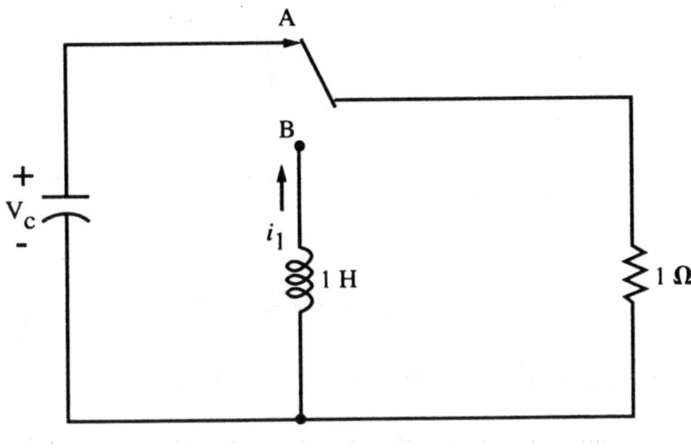

Problem 6.22

6.23 Find V for $t > 0$ if $v_1(0)=0$ and $v_g = 5u(t)$V.

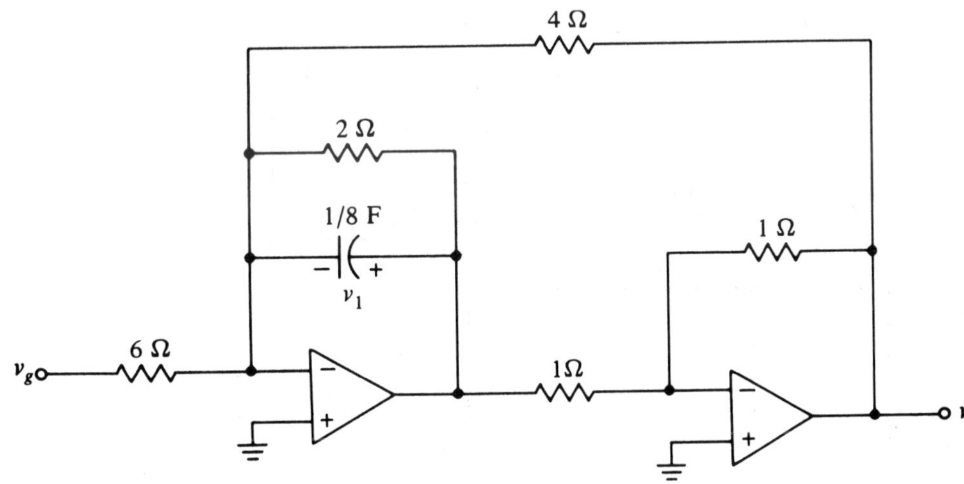

Problem 6.23

6.24 **Find** v **if** $v_g = 3e^{-3(t+4)}u(t+4)$**V and there is no initial energy stored in the capacitor.**

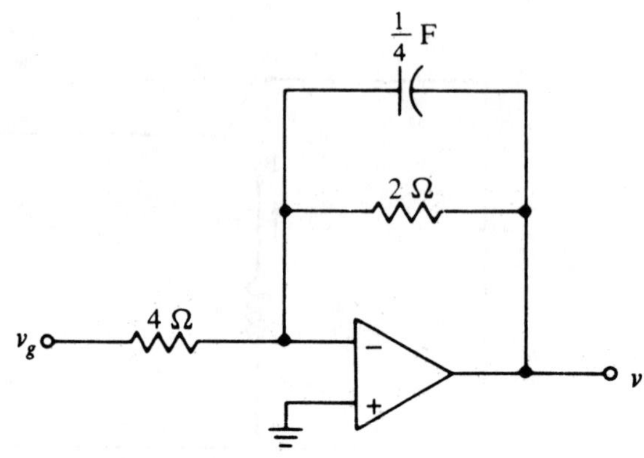

Problem 6.24

6.25 **Find** v_c **and** v_{out} **in terms of** v_{in} **for** t **>0. Assume** $v_g = v_{in}u(t)$**V.**

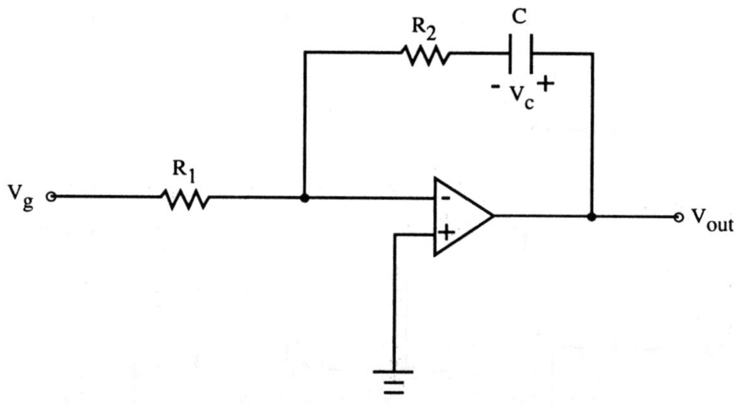

Problem 6.25

6.26 Find $v_2(t)$ for $t > 0$ with $v_c(0^-) = 0$.

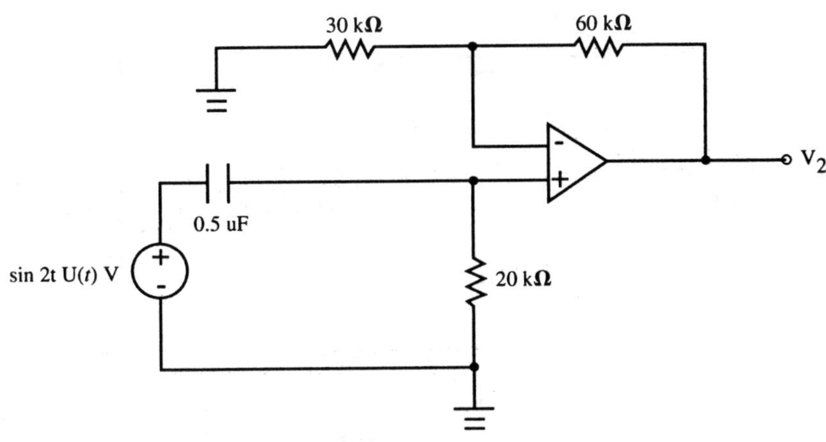

Problem 6.26

6.27 Sketch the voltage given by

$$v(t) = u(-t-3) + 5u(t+3) - 2u(t+2) - 4u(t) + (t-2)u(t-2)$$

6.28 Express $i(t)$ in terms of step and ramp function.

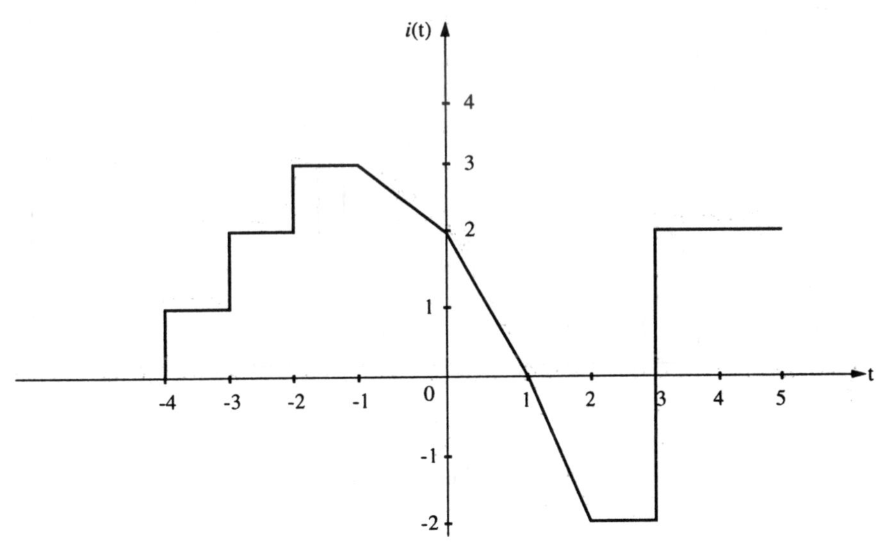

Problem 6.28

6.29 Find v for $t > 0$. Assume steady state at $t = 0^-$.

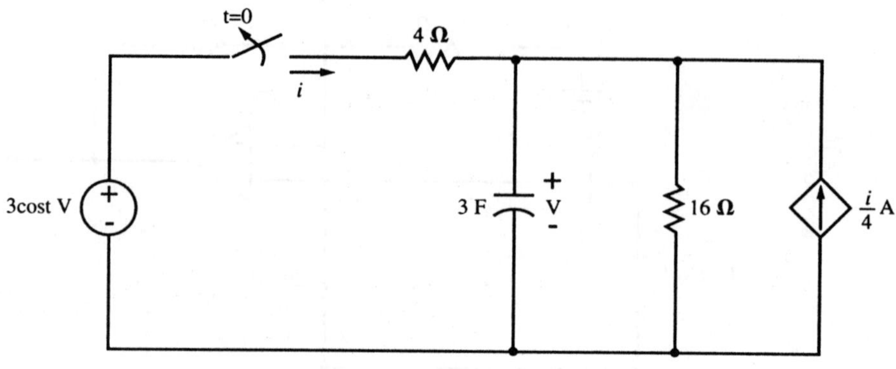

Problem 6.29

6.30 Find i_0 for $t > 0$. Assume steady state at $t = 0^-$.

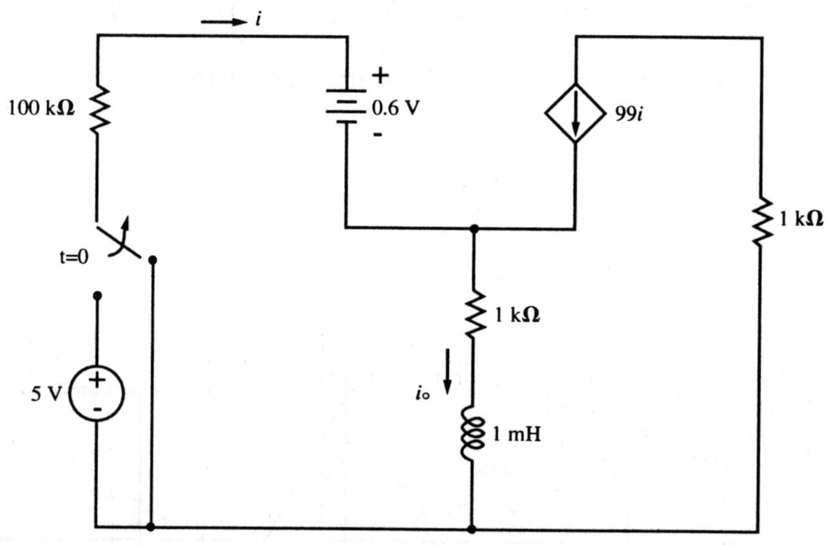

Problem 6.30

6.8 SPICE and the transient response

6.31 Solve Prob.6.23 using SPICE. Plot the output v from t=0 to t=10s.

6.32 Use SPICE to determine the output current of the circuit shown below to a pulse input

$$\nu_g = 1 \quad 0 \leq t \leq 1 \ ms$$
$$= 0 \quad elsewhere$$

Plot the response from t=0 to t=20ms.

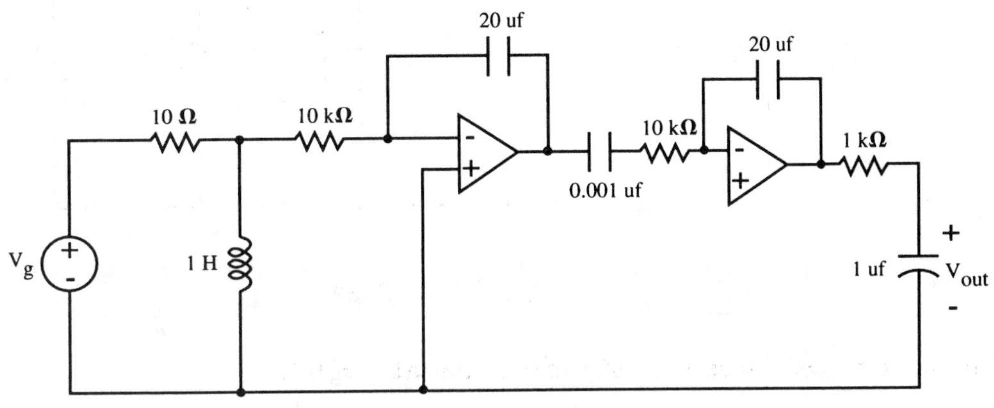

Problem 6.32

6.33 Use SPICE to solve Prob.6.30. Plot the response from t=0 to t=10ms.

6.34 Use SPICE to determine the output voltage of the circuit shown below with R=10KΩ, c= 0.001μF, $v_g = 10\sin(10wt)$V. Plot the response from t=0 to t=10ms.

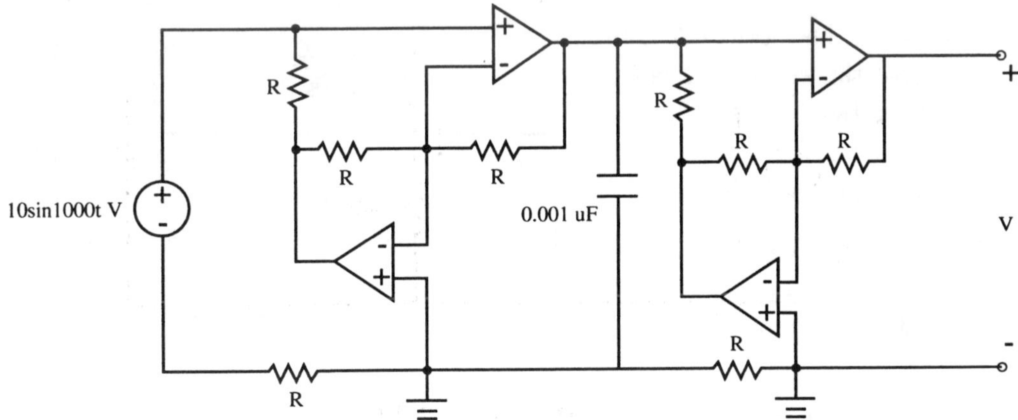

Problem 6.34

6.1 The characteristic equation is

$3s + 8000 = 0$, or $s = -\frac{8000}{3}$

and the forced response $i_f = \frac{2}{8} = \frac{1}{4}$ A

$i_L = \frac{1}{4} + Ke^{-\frac{8000}{3}t}$

$i_L(0) = \frac{1}{4} + K = \frac{9}{4}$, $K = 2$

$\therefore\ i_L = \frac{1}{4} + 2e^{-(8000/3)t}$

Any inductor-resistor network which satisfies $\frac{L}{R} = \frac{3}{8000}$ is a possible solution.

One can choose $L = 30$ mH and $R = 800$ kΩ

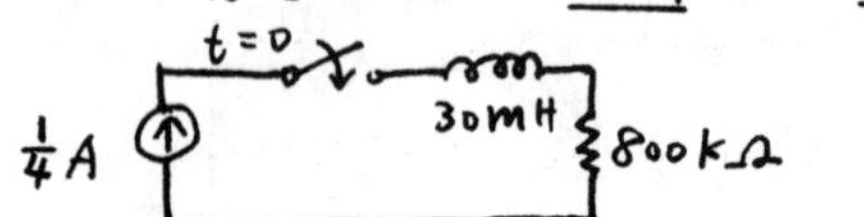

6.2 The characteristic equation is

$0.004s + 7 = 0$, $s = -\frac{7}{0.004} = -\frac{7000}{4}$

and forced response $v_C = \frac{14}{7} = 2$V

$v_C = 2 + Ke^{-\frac{7000}{4}t}$

$v_C(0^-) = 0$, $2 + K(1) = -3$, $K = -5$

$\therefore\ v_C = 2 - 5e^{-\frac{7000}{4}t}$

Any capacitor-resistor network which satisfies $RC = \frac{4}{7000}$ is a possible solution. Let $C = \frac{1}{7}\mu F$, $R = 4$KΩ

t=0

2V $\frac{1}{7}\mu F$ v_C 4KΩ

6.3 To determine the time constant τ, one need to find the equivalent resistance R_{eq} seen by the inductor.

To find R_{eq} we will force a current i to flow into the circuit.

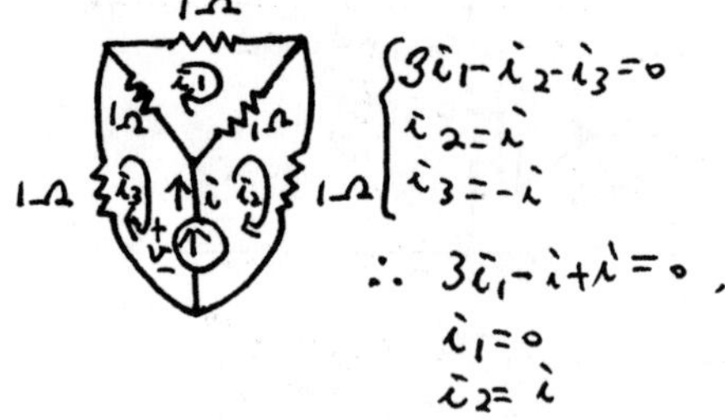

$$\begin{cases} 3i_1 - i_2 - i_3 = 0 \\ i_2 = i \\ i_3 = -i \end{cases}$$

$\therefore\ 3i_1 - i + i = 0$,

$i_1 = 0$

$i_2 = i$

By KVL around right loop

$v = i + i = 2i$, $R_{eq} = \frac{v}{i} = 2\Omega$

equivalent circuit is: 2Ω 1H

Time constant $\tau = \frac{L}{R} = \frac{1}{2} = 0.5$s

6.4 To determine the time constant τ, one need to find the equivalent inductance L_{eq} seen by the resistor

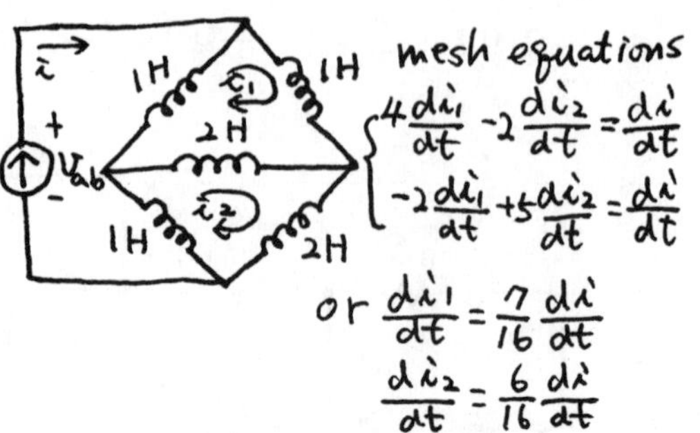

mesh equations

$$\begin{cases} 4\frac{di_1}{dt} - 2\frac{di_2}{dt} = \frac{di}{dt} \\ -2\frac{di_1}{dt} + 5\frac{di_2}{dt} = \frac{di}{dt} \end{cases}$$

or $\frac{di_1}{dt} = \frac{7}{16}\frac{di}{dt}$

$\frac{di_2}{dt} = \frac{6}{16}\frac{di}{dt}$

By KVL around the outer loop

$V_{ab} = 1\frac{di_1}{dt} + 2\frac{di_2}{dt} = \frac{19}{16}\frac{di}{dt}$

$L_{eq} = \frac{19}{16}$H,

$\tau = \frac{L}{R} = \frac{19/16}{19\times10^3} = \frac{1}{16\times10^3} = \frac{1}{16}$ ms

$i(t) = Ke^{-16000t}$ and $i(16\text{ms}) = Ke^{-1} = \frac{10}{e}$

$\therefore\ K = 10$, Hence $i(t) = 10e^{-16000t}$ A

6.5 The equivalent circuit is

$\tau = RC = \frac{6}{9} = \frac{2}{3}$s

+ v − $\frac{1}{9}$F 6Ω

$v = Ke^{-\frac{3}{2}t}$, $v(0) = 5 = K$

$\therefore\ v = 5e^{-\frac{3}{2}t}$ V

$i = \frac{v_R}{3}$ and $v_R = \frac{1}{16}v = \frac{5}{16}e^{-(3/2)t}$

Hence $i = \frac{v_R}{3} = \frac{5}{48}e^{-\frac{3}{2}t}$ A

6.6 $q = Cv_C(0^-)$, $40\times10^{-6} = 2\times10^{-6}\times v_C(0^-)$

$v_C(0^-) = 20$V,

$\tau = RC = 2\times10^{-6}R$, $v_C(t) = 20\times e^{-\frac{t}{20\times10^{-6}R}}$ V

at $t = 10$ms

$v_C(t) = 20\times e^{-\frac{t}{2\times10^{-6}R}} = 20e^{-\frac{10^{-2}}{2\times10^{-6}R}}$

$= 20\times e^{-\frac{5\times10^3}{R}} = \frac{20}{e}$

$R = 5\times10^3\,\Omega = 5$ KΩ

6.7 $v(t) = v(1)e^{-(t-1)/RC}$, where

$RC = (0.01)(10+90) = 1$s, $v(1) = 10$V,

$v(t) = 10e^{1-t}$ V

$w_C(t) = \frac{1}{2}Cv^2 = \frac{1}{2}(0.01)(10e^{1-t})^2$

$= 0.5e^{2(1-t)}$ J

6.8 Replacing the 20Ω and 5Ω parallel combination by 4Ω, we see that

$v(0) = (4+2)\,i(0) = 12V$

R_{eq} = resistance seen by capacitor $= (4+2)(3)/(4+2+3) = 2\Omega$

$1/\tau = 1/R_{eq}C = \frac{1}{2(1/10)} = 5s^{-1}$

$\therefore v = 12e^{-5t}\ V$

6.9 KCL gives $\frac{1}{8}\frac{dv}{dt} + \frac{v}{3} + \frac{v-2i}{6} = 0$,

$i = \frac{1}{8}\frac{dv}{dt}$, $\therefore \frac{dv}{dt} + 6v = 0$ and

$\frac{dv}{v} = -6dt$, $\ln v = -6t + \ln k$

or $v = ke^{-6t}$, $v(0) = 4 = k$, $v = 4e^{-6t}\ V$

6.10 At $t=0^-$, the inductor is a short circuit. Thus for the circuit shown current division gives

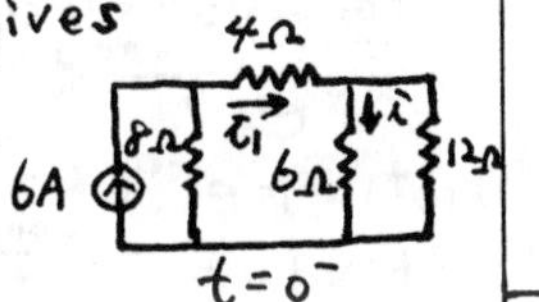

4Ω, 8Ω, i_1, 6Ω, i, 12Ω, 6A, $t=0^-$

$i_1(0^-) = \frac{8(6)}{4+8+\frac{6(12)}{6+12}} = 3A$,

$i(0^-) = \frac{12}{6+12}(i_1) = 2A$,

for $t>0$, $R_{eq} = 6 + \frac{12(4)}{4+12} = 9\Omega$

$\frac{1}{\tau} = \frac{R_{eq}}{L} = \frac{9}{1/2} = 18s^{-1}$, $i = 2e^{-18t}A$

6.11

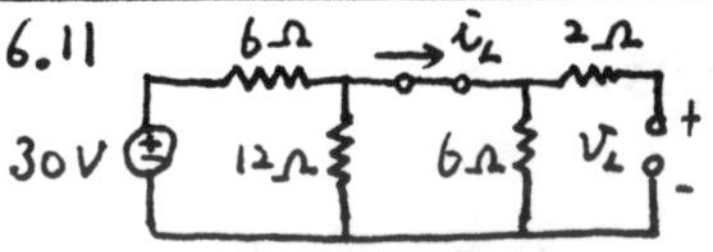

6Ω, i_L, 2Ω, 30V, 12Ω, 6Ω, v_c +, -

At $t=0^-$, by voltage division

$v_c(0^-) = \frac{6(12)/(6+12)}{6+4}(30) = 12V$, then

$i_L(0^-) = \frac{v_c(0^-)}{6} = 2A$. At $t=0^+$, the closing of the switch seperates the circuit into two independent circuits due to the short-circuit.

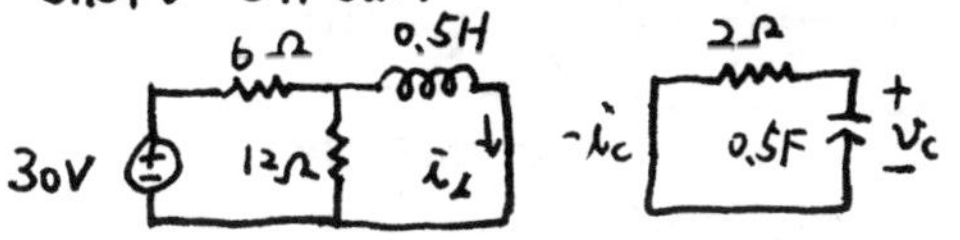

6Ω, 0.5H, 30V, 12Ω, i_L ; $-i_c$, 2Ω, 0.5F, v_c +, -

For $t>0$, $i = i_L - i_c$ as shown above.

In the capacitor circuit,

$-i_c = -0.5\frac{dv_c}{dt} = -0.5\,v_c(0^+)\frac{d}{dt}(e^{-\frac{t}{RC}})$

$= (0.5)(12)(\frac{1}{2(0.5)})e^{-t} = 6e^{-t}A$

In the inductor circuit $i_L = i_{Ln} + i_{Lf}$ where $i_{Lf} = 30/6 = 5A$ (inductor is a short-circuit).

6.11 Cont.

With the source dead the inductor sees the resistance $R_{eq} = \frac{6(12)}{6+12} = 4\Omega$.

Then $\frac{R_{eq}}{L} = \frac{4}{1/2} = 8$ and $i_L = 5 + Ke^{-8t}$,

$i_L(0^+) = i_L(0^-) = 2 = 5 + K$, $K = -3$ and

$i_L = 5 - 3e^{-8t}A$

$\therefore i = 5 - 3e^{-8t} + 6e^{-t}A$ $(i_L - i_c = i)$

6.12 At $t=0^-$, the inductor is a short circuit, thus $i(0^-) = 24/4 = 6A$.

For $t>0$, KCL gives for v_L

$\frac{v_L - 12}{6} + i + \frac{v_L}{12} = 0$ or $3v_L + 12i = 24$

Setting $v_L = 2\frac{di}{dt}$ gives $\frac{di}{dt} + 2i = 4$

Seperation of variables yields

$\frac{di}{i-2} + 2dt = 0$ and $\ln(i-2) = -2t + \ln k$

$\therefore i - 2 = ke^{-2t}$; Since $i(0^-) = i(0^+)$

$i(0^+) - 2 = k$, $k = 6 - 2 = 4$

then $i = 4e^{-2t} + 2A$ for $t>0$

6.13

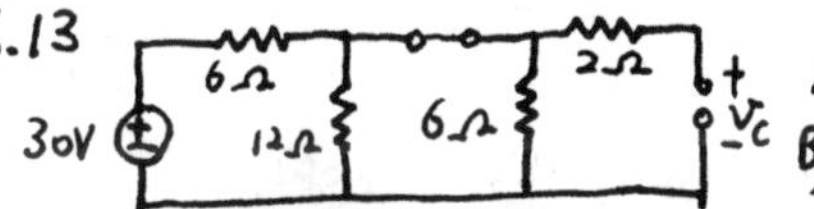

6Ω, 2Ω, 30V, 12Ω, 6Ω, v_c +

At $t=0^-$, By voltage div

$v_c(0^-) = \frac{\frac{(6)(12)}{6+12}}{6+4} = 12V$

At $t>0$, $v_c(0^+) = v_c(0^-)$,

2Ω, v_c +

$\frac{1}{\tau} = \frac{1}{R_{eq}C} = \frac{1}{2(\frac{1}{2})} = 1s^{-1}$

Since $v_{cf} = 0$ then $v_c = Ae^{-t}$

$v_c(0^+) = 12$, $A = 12V$, $v_c = 12e^{-t}V$

6.14 i_L = inductor current downward

$v(0^-) = 0$ (short circuit)

$i_L(0^-) = \frac{10V}{10\Omega} = 1A$, for $t>0$, KVL,

$v + 10i_L - \frac{v}{2} = 0$, $v = 2\frac{di_L}{dt}$. Then

$\frac{di_L}{dt} + 10i_L = 0$ $\therefore i_L(t) = i_L(0^+)e^{-10t} = e^{-10t}A$

Then $v = 2\frac{d}{dt}(e^{-10t}) = -20e^{-10t}V$

6.15 For $t<0$, $v_g = 0$ and therefore the capacitor voltage is zero. Hence $i(0^+) = 0$. With the source dead the resistance seen by the capacitor is:

6.15 Cont.

$$R_{eq} = \frac{6\left[9+\frac{4(12)}{4+12}\right]}{6+9+3} = 4\,\Omega$$

$R_{eq}C = 4\left(\frac{1}{20}\right) = \frac{1}{5}\,s$

Now $i = i_n + i_f = A_1 e^{-5t} + i_{f_1}$

where i_f is the dc steady-state current. With the capacitor open circuited the voltage across the $12\,\Omega$ resistor is $(6+9)i_f = 15\,i_f$.

By voltage division,

$$15\,i_f = \frac{15(12)/(15+12)}{4+(15)(12)/(15+12)}(1),\quad i_f = \frac{1}{24}\,A$$

$\therefore i = A_1 e^{-5t} + \frac{1}{24},\ t>0$

$i(0^+) = 0 = A_1 + \frac{1}{24},\quad A_1 = -\frac{1}{24}$

$\therefore i = \frac{1}{24}(1-e^{-5t})u(t)\,A$

6.16 For $t<0$, the voltage source is zero and therefore the capacitor voltage is also zero. Since there is only one independent source, KCL gives for $t>0$,

$$\frac{1}{4}\frac{dv}{dt} + \frac{v-4}{4} - 5v_1 = 0$$

$v_1 = 4 - v$, therefore, $\frac{dv}{dt} + 21v = 84$

$v = Ae^{-21t} + 4,\quad v(0) = 0 = A+4,\ A = -4$

$\therefore v = 4(1-e^{-21t})u(t)\,V$

6.17 By Nodal equations

$$\begin{cases} C_1\frac{dv_1}{dt} + \frac{v_1}{R_1} = C_2\frac{dv_2}{dt} + \frac{v_2}{R_2} \\ v_1 + v_2 = v \end{cases}$$

$v_1 = v - v_2,\quad \frac{dv_1}{dt} = -\frac{dv_2}{dt}$ ($\because v$ is a constant)

$$-C_1\frac{dv_2}{dt} + \frac{v-v_2}{R_1} = C_2\frac{dv_2}{dt} + \frac{v_2}{R_2},$$

or $(C_1+C_2)\frac{dv_2}{dt} + v_2\left(\frac{1}{R_1}+\frac{1}{R_2}\right) = \frac{v}{R_1}$

forced response $v_2 = \frac{\frac{v}{R_1}}{\frac{1}{R_1}+\frac{1}{R_2}} = \frac{R_2 v}{R_1+R_2}$

and $(C_1+C_2)s + \left(\frac{1}{R_1}+\frac{1}{R_2}\right) = 0,\quad s = -\frac{\frac{1}{R_1}+\frac{1}{R_2}}{C_1+C_2}$

$$v_2 = \frac{R_2 v}{R_1+R_2} + K_2 e^{-\frac{\frac{1}{R_1}+\frac{1}{R_2}}{C_1+C_2}t}$$

Similarly, $v_1 = \frac{R_1 v}{R_1+R_2} + K_1 e^{-\frac{\frac{1}{R_1}+\frac{1}{R_2}}{C_1+C_2}t}$

when $t\to\infty$, $K_1 e^{-\frac{\frac{1}{R_1}+\frac{1}{R_2}}{C_1+C_2}t} \to 0$

and $K_2 e^{-\frac{\frac{1}{R_1}+\frac{1}{R_2}}{C_1+C_2}t} \to 0$

6.17 Cont.

regardless the initial conditions of capacitors C_1 and C_2.

Hence $v_1 = \frac{R_1}{R_1+R_2}v\ V$

$v_2 = \frac{R_2}{R_1+R_2}v\ V$

It is found that this "voltage divider" is determined by R_1 and R_2 but not C_1 and C_2.

6.18 $v(0^-) = 5\times10^{-3}\times12\times10^{3} = 60\,V$

$v(0^-) = v(0^+) = 60\,V$

at $t>0$

$RC = 0.25\times10^{-6}\times18\times10^{3} = 4.5\times10^{-3} = \frac{9}{2}10^{-3}$

$v = K\times e^{-\frac{2000}{9}t}$

$v(0^+) = Ke^{-0} = K = 60\,V$

$v(t) = 60e^{-\frac{2000}{9}t}\,V$

6.19 $v_c(0^-) = 7-3 = 4V$, at $t=0^+$, the switch is closed

The natural response due to initial capacitor voltage

$v_N = v_c = 4e^{-\frac{1}{20}t}$

Kill the 7V voltage source,

$v_c = -3V$

$i = 0$

$\therefore v_{S1} = -3V$

Kill the 3V voltage source

$v_c = 0V$

$i = \frac{7}{6+8} = \frac{1}{2}A$

$v_{S2} = \frac{1}{2}\times6 = 3V$

By superposition, $v = v_N + v_{S1} + v_{S2}$

$= 3-3+4e^{-\frac{1}{20}t} = 4e^{-\frac{t}{20}}\,V$

6.20 $v_c(t=0^-) = 12\times 1 = 12V$

for $0 \le t < 1$, $\tau = RC = \frac{1}{2}\times 6 = 3s$

$v_c(t) = 6 + ke^{-\frac{t}{3}}$

$v_c(t=0^-) = v_c(t=0^+) = 6+k = 12$, $k=6$

$v_c(t) = 6+6e^{-\frac{t}{3}}$, for $0\le t<1$

At $t=1$, $v_c(t=1) = 6+6e^{-\frac{1}{3}}$

12A, $\frac{1}{3}\Omega$, 6F, $+\ v_c\ -$ ($t>1$); 12A, 1Ω, 1Ω, v_c, 6F ($0\le t<1$)

$\tau = RC = \frac{1}{3}\times 6 = 2s$

$v_c(t) = 4 + Ke^{-\frac{t-1}{2}}$

$v_c(t=1^-) = v_c(t=1^+) = 4+K = 6+6e^{-\frac{1}{3}}$,

$K = 2+6e^{-\frac{1}{3}}$, $v_c(t) = 4+(2+6e^{-\frac{1}{3}})e^{-\frac{t-1}{2}}$

$= 4+2e^{-\frac{t-1}{2}} + e^{-\frac{3t-1}{6}}$

Hence $v_c(t) = 6+6e^{-\frac{t}{3}}$ for $0\le t<1$

$= 4+2e^{-\frac{t-1}{2}} + e^{-\frac{3t-1}{6}}$ for $t\ge 1$

6.21 At $t=4^-$, because of DC steady state

5A, 2Ω, 2Ω, 3Ω, 3Ω, v, v_1, i_1, i_2

$i_1 = \frac{8}{8+2}\times 5 = 4A$

$i_2 = \frac{2}{8+2}\times 5 = 1A$

$v(4^-) = 6\cdot 1 = 6V$

$v_1(4^-) = 3\cdot 1 = 3V$

At $t=4^+$

v, 3Ω, v_1, $\frac{1}{4}F$, $\frac{1}{2}F$

$\begin{cases} \frac{1}{4}\frac{dv}{dt} + \frac{v-v_1}{3} = 0 \text{ or } v_1 = \frac{3}{4}\frac{dv}{dt} + v \\ \frac{v_1-v}{3} + \frac{1}{2}\frac{dv_1}{dt} = 0 \text{ or } 2v_1 - v + \frac{3dv_1}{dt} = 0 \end{cases}$

$\frac{dv_1}{dt} = \frac{3}{4}\frac{d^2v}{dt} + \frac{dv}{dt}$,

$3\left[\frac{3}{4}\frac{d^2v}{dt^2} + \frac{dv}{dt}\right] + 2\left[\frac{3}{4}\frac{dv}{dt} + v\right] - 2v = 0$,

$\frac{d^2v}{dt^2} + 2\frac{dv}{dt} = 0$, $\frac{dv}{dt}\left(\frac{dv}{dt}+2\right) = 0$,

$\frac{dv}{dt} = 0$ or $\frac{dv}{dt} + 2 = 0$,

$v = A_1 e^{-2(t-4)} + A_2$, $v(4^-) = v(4^+) = 6 = A_1 + A_2$

$v_1 = \frac{3}{4}\frac{dv}{dt} + v = \frac{3}{4}(-2)A_1 e^{-2(t-4)} + A_1 e^{-2(t-4)} + A_2$

$= -\frac{1}{2}A_1 e^{-2(t-4)} + A_2$,

$v_1(4^-) = v_1(4^+) = -\frac{1}{2}A_1 + A_2 = 3$,

$\begin{cases} A_1 + A_2 = 6 \\ -\frac{1}{2}A_1 + A_2 = 3 \end{cases}$, $A_1 = 2$, $A_2 = 4$, $v(t) = 2e^{-2(t-4)} + 4V$

6.22 The time constant of the circuit when the S.W. is at position A is $\tau_c = RC = 1S$, $v_c(t) = v(t_n)e^{-t}$, $2n \le t < 2n+1$, and $n = 0,1,2,\cdots$

Similarly, the time constant of the circuit for the SW at position B is $\tau_I = \frac{L}{R} = 1S$, $i_I = i(t_n)e^{-t}$, $2n+1 \le t < 2n+2$, and $n = 0,1,2,\cdots$

$v_R = i_R \cdot 1 = i(t_n)\times e^{-t}$

With $v_c(0^-) = 10V$, $i_I(0^-) = 10A$

$v_c(0^+) = 10V$, $i_I(0^+) = 10A$

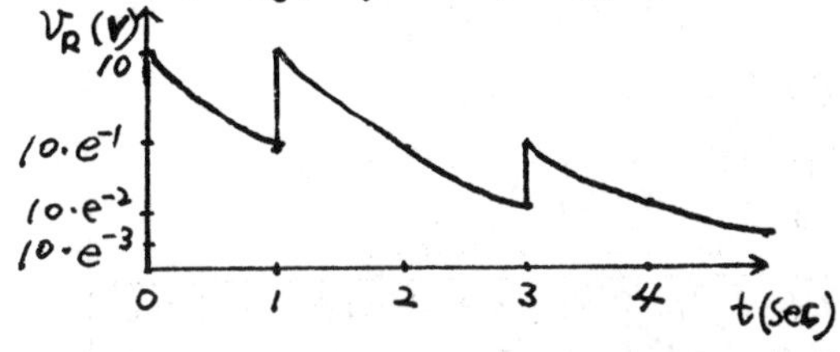

6.23 For $t>0$ KCL at the inverting op amp input terminals yields

$\frac{5}{6} + \frac{1}{8}\frac{dv_1}{dt} + \frac{v_1}{2} + \frac{v}{4} = 0$, $v_1 + v = 0$

Note that v_1 is the output voltage of the first op amp. Eliminating v results in $\frac{dv_1}{dt} + 2v_1 = -\frac{20}{3}$

$\therefore v = v_n + v_f = Ae^{-2t} - \frac{10}{3}$

$v_1(0^+) = v_1(0^-) = 0$, $A = \frac{10}{3}$

$\therefore v = \frac{10}{3}(e^{-2t} - 1)V$, $t>0$

6.24 KCL gives at the inverting op amp terminals, $\frac{1}{4}\frac{dv}{dt} + \frac{v}{2} + \frac{v_g}{4} = 0$ or

$\frac{dv}{dt} + 2v = -3e^{-3(t+4)}u(t+4)$

for $t > -4$,

$ve^{2t} = A + \int(-3e^{-3(t+4)})(e^{2t})dt$

$= A + 3e^{-(t+12)}$

$v = Ae^{-2t} + 3e^{-3(t+4)}$

$v(-4) = Ae^{8} + 3 = 0$, $A = -3e^{-8}$

$\therefore v = [3e^{-3(t+4)} - 3e^{-2(t+4)}]u(t+4)$ V

6.25 Nodal analysis gives at the inverting op amp terminals, $C\frac{dv}{dt} + \frac{v_{in}}{R_1} = 0$,

$\frac{dv_c}{dt} = -\frac{v_{in}}{R_1 C}$, $v_c = -\frac{1}{R_1 C}\int_{0^+}^{t} v_{in}\,dt + v_c(0^+)$

6.25 Cont.

$v_c(0^+) = v_c(0^-) = 0$ since the excitation is a step function, $v_c(t) = -\frac{1}{R_1C}\int_{0^+}^{t} v_{in}\,dt$

$$V_{out} = v_c + i_c R_2$$
$$= -\frac{1}{R_1C}\int_{0^+}^{t} v_{in}\,dt + C\frac{dv_c}{dt}R_2$$
$$= -\frac{1}{R_1C}\int_{0^+}^{t} v_{in}\,dt + C\left(-\frac{v_{in}}{R_1C}\right)\cdot R_2$$
$$= -\frac{1}{R_1C}\int_{0^+}^{t} v_{in}\,dt - \frac{R_2}{R_1}v_{in}\ \text{V}$$

6.26

$$\begin{cases} 0.5\dfrac{d(v_1 - \sin 2t)}{dt} + \dfrac{v_1}{20\times10^3} = 0, \text{ or } \dfrac{dv_1}{dt} + 10^{-4}v_1 = 2\cos 2t \\ v_1 = \dfrac{30K}{(30+90)K}v_2 = \dfrac{v_2}{3}, \text{ or } v_2 = 3v_1 \end{cases}$$

$v_1 = Ae^{-0.0001t} + 2\times10^4\cos 2t$

$v_c(0^-) = 0$, $v_1(0^-) = A + 2\times10^4\cos 2t = 0$,

$A = -2\times10^4\cos 2t$

$v_1(t) = 2\times10^4\cos 2t\,(1 - e^{-0.0001t})$

$v_2(t) = 6\times10^4\cos 2t\,(1 - e^{-0.0001t})$ V

6.27

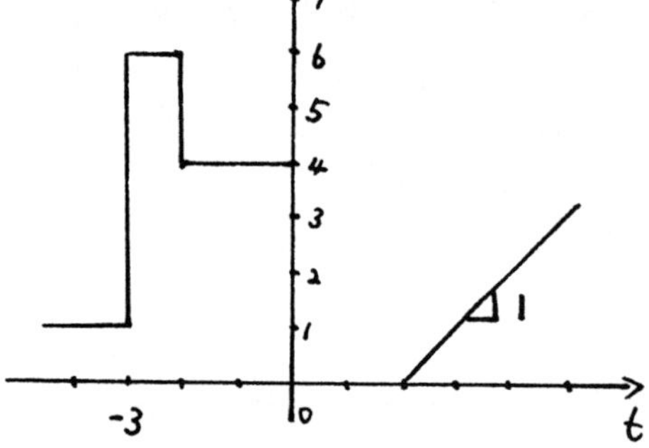

6.28 $i(t) = u(t+4) + u(t+3) + u(t+2) - (t+1)u(t+1) - 2tu(t) + 2(t-2)u(t-2) + 4u(t-3)$

6.29 Steady state at $t = 0^-$

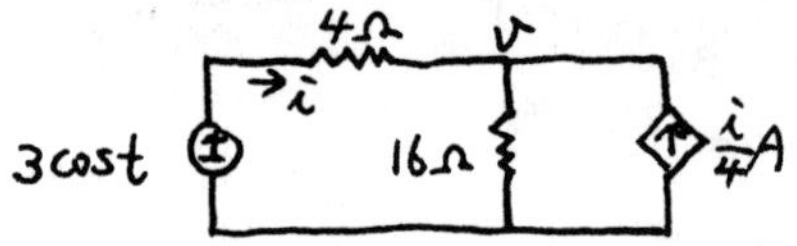

$$\frac{v - 3\cos t}{4} + \frac{v}{16} - \frac{1}{4}\left(\frac{v - \cos 3t}{4}\right) = 0$$
$$\frac{3}{4}\left(\frac{v - 3\cos 3t}{4}\right) + \frac{v}{16} = 0$$
$\frac{1}{4}v = \frac{9}{16}\cos 3t$, or $v = \frac{9}{4}\cos 3t$, $v(0^-) = \frac{9}{4}$ V

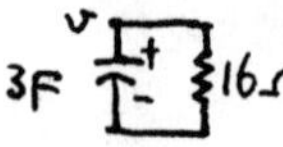

$3\frac{dv}{dt} + \frac{v}{16} = 0$, $v = Ae^{-\frac{1}{48}t}$

$v(0^+) = A = \frac{9}{4}$, $v(t) = \frac{9}{4}e^{-\frac{t}{48}}$ V

6.30 steady state at $t = 0^-$

$10^5 i + 0.6 + (i + 99i)10^3 = 5$, $2\times10^5 i = 4.4$

or $i = 2.2\times10^{-5}$

$= 22\times10^{-6}$ A

$i(0^-) = 22\times10^{-6}$ A

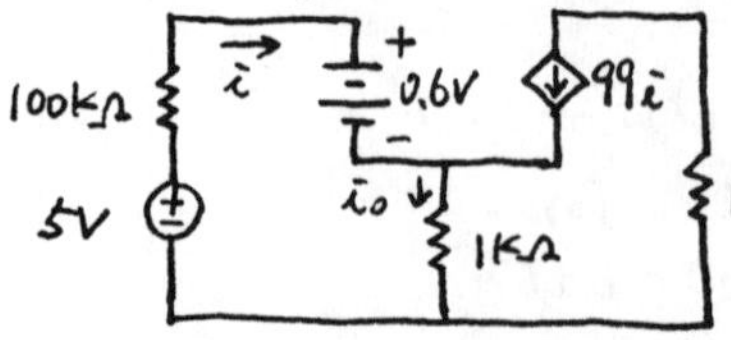

for $t > 0$, $10^5 i + 0.6 + (i + 99i)10^3 + 10^{-3}\frac{d(100i)}{dt} = 0$

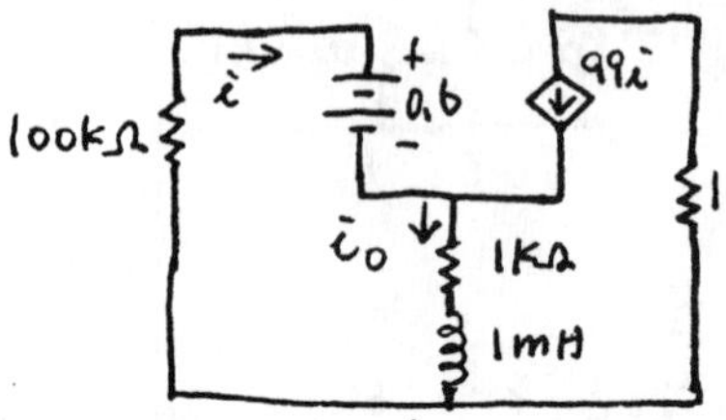

$0.1\frac{di}{dt} + 2\cdot10^5 i + 0.6 = 0$

$\frac{di}{dt} + 2\cdot10^6 i = 0$,

$2\cdot10^6 i_f = 6$,

$i_f = \frac{6}{2\times10^6} = -3\times10^{-6}$

$i = A_1 e^{-2\times10^6 t} - 3\times10^{-6}$

$i(0^+) = i(0^-) = A_1 - 3\times10^{-6} = 22\times10^{-6}$ or

$A_1 = 25\times10^{-6}$

$i(t) = 25\times10^{-6}e^{-2\times10^6 t} - 3\times10^{-6}$

Hence $i_o(t) = 100\,i(t)$

$= 25\times10^{-4}e^{-2\times10^6 t} - 3\times10^{-4}$ A

$= 2.5\,e^{-2\times10^6 t} - 0.3$ mA

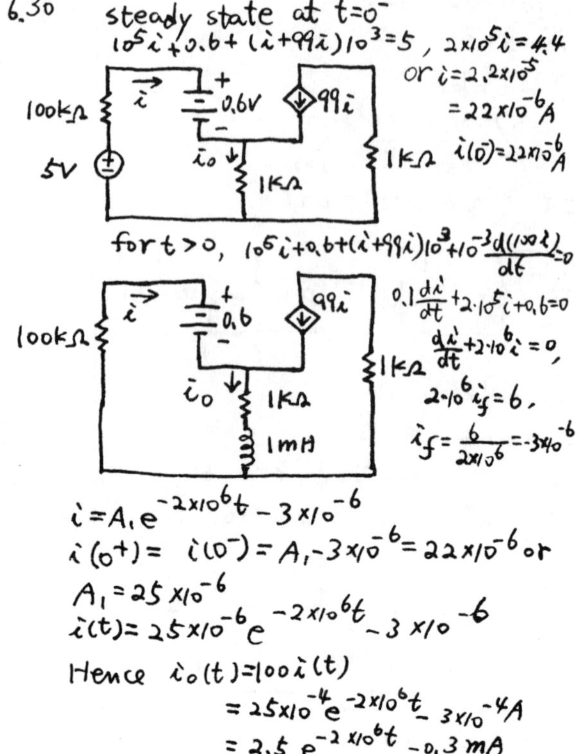

6.31

```
*PROB.6_31
*TRANSIENT AND STEADY STATE
*RESPONSES OF AN AMPLIFIER
R_R1       2 1 1
C_C1       3 2 0.125
R_R2       3 2 2
R_R3       3 4 4
R_R4       1 4 1
R_R5       5 3 6
V_VG       5 0 DC 1
X_U3       0 3 2 OPAMP
X_U4       0 1 4 OPAMP
R_OUT      0 4 1
.INC       C:\PS\OPAMP.LIB
*INCLUDING C:\PS\OPAMP.LIB *
*OP AMP SUBCIRCUIT:
.SUBCKT OPAMP 1 2 3
*NODE 1:+ 2:- 3:OUTPUT
RIN    1    2    1MEG
E1     4    0    1 2 100K
R0     4    3    30
.ENDS
**** RESUMING PS4174.CIR ***
*Output statements:
.DC        V_VG  -5  +5   1
.TRAN      100MS 10000MS UIC
*OUTPUT IS V(4)
.END

*PLOT OF V(4) IS SHOWN IN
THE FOLLOWING FIGURE:
```

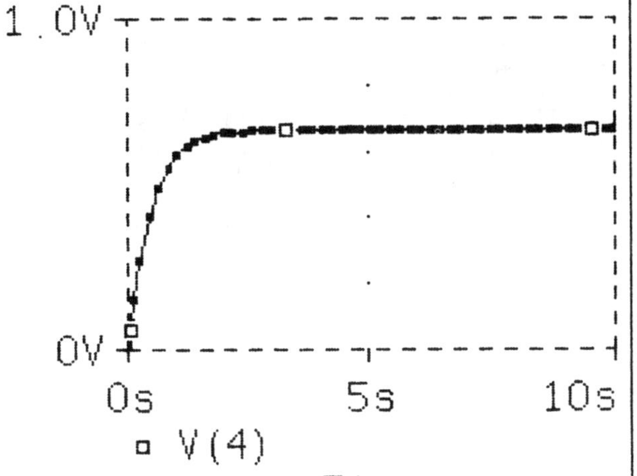

6.32

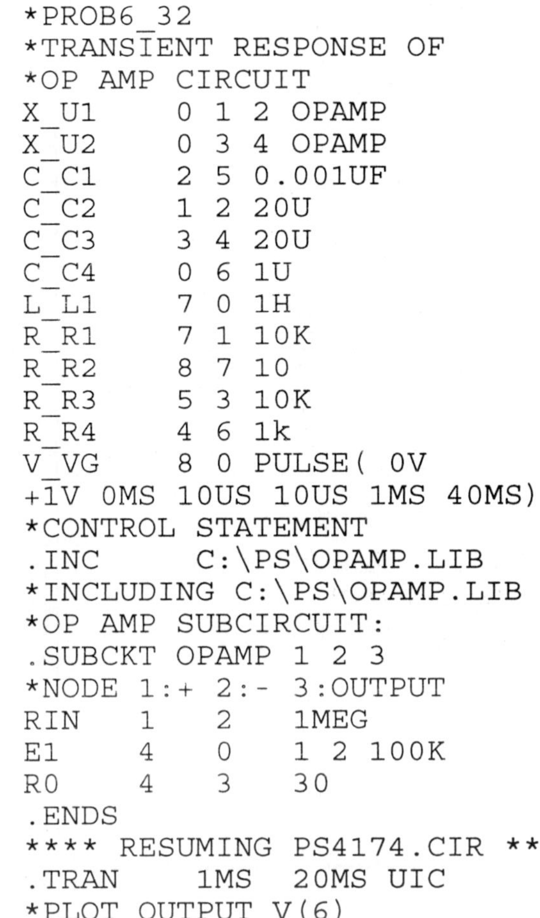

```
*PROB6_32
*TRANSIENT RESPONSE OF
*OP AMP CIRCUIT
X_U1     0 1 2 OPAMP
X_U2     0 3 4 OPAMP
C_C1     2 5 0.001UF
C_C2     1 2 20U
C_C3     3 4 20U
C_C4     0 6 1U
L_L1     7 0 1H
R_R1     7 1 10K
R_R2     8 7 10
R_R3     5 3 10K
R_R4     4 6 1k
V_VG     8 0 PULSE( 0V
+1V 0MS 10US 10US 1MS 40MS)
*CONTROL STATEMENT
.INC      C:\PS\OPAMP.LIB
*INCLUDING C:\PS\OPAMP.LIB
*OP AMP SUBCIRCUIT:
.SUBCKT OPAMP 1 2 3
*NODE 1:+ 2:- 3:OUTPUT
RIN    1    2    1MEG
E1     4    0    1 2 100K
R0     4    3    30
.ENDS
**** RESUMING PS4174.CIR **
.TRAN     1MS  20MS UIC
*PLOT OUTPUT V(6)
.END
```

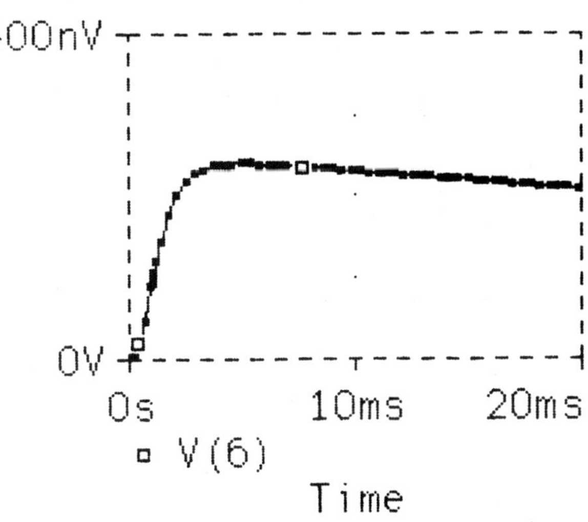

6.33

```
* Schematics Netlist *
V_V1      1 2 dc 0.6
R_R1      1 4 100k
R_R2      5 2 1k
R_R3      0 6 1k
L_L1      0 5 1mH
+IC=22*10e-6
F_F1      6 2 VF_F1 99
VF_F1     3 1 0V
V_V2      4 0 PULSE
+(5 0 0 0 0 10Us 3s)
*CONTROL STATEMENT
.INC      C:\PS\OPAMP.LIB
.TRAN     1MS 10MS
.END
```

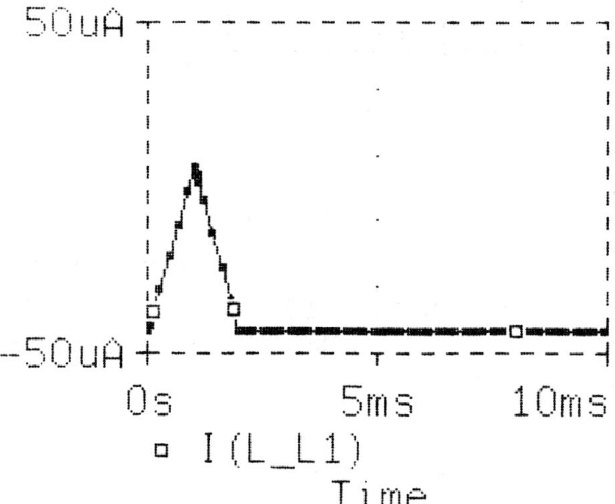

6.34

```
*PROB.6_34
*TRANSIENT RESPONSE OF
*OP AMP CIRCUIT
R_R1      2 1 10K
R_R2      3 2 10K
R_R3      4 3 10K
R_R4      5 4 10K
R_R5      6 5 10K
R_R6      8 6 10K
R_R7      7 0 10K
R_R8      0 0 10K
C_C1      0 4 1U
V_V1      8 7
+SIN (0 10 1000 0 0 0)
X_U1     8 5 4 OPAMP
X_U2     0 5 6 OPAMP
X_U3     0 2 3 OPAMP
X_U4     4 2 1 OPAMP
*CONTROL STATEMENT
.INC     C:\PS\OPAMP.LIB
*INCLUDING C:\PS\OPAMP.LIB
*OP AMP SUBCIRCUIT:
.SUBCKT OPAMP 1 2 3
*NODE 1:+ 2:- 3:OUTPUT
RIN    1    2    1MEG
E1     4    0    1 2 100K
R0     4    3    30
.ENDS
**** RESUMING PS4174.CIR **
.TRAN  500US 10MS  UIC
.END
```

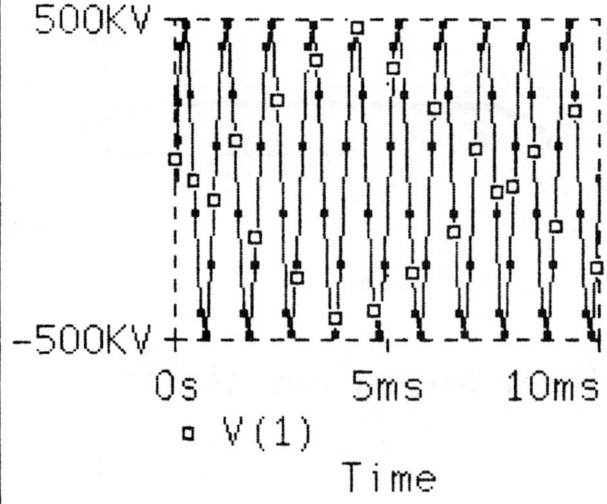

Chapter 7
Second-Older Circuit

7.1 Circuits with two storage elements

7.1 Write describing equations for the indicated variables. Are these second order circuits?

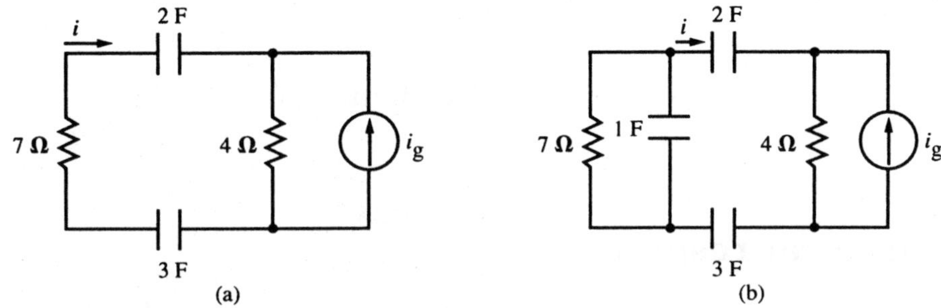

Problem 7.1

7.2 An electromechanical device called "Piezoelectric crystal" is commonly found in electronic design to provide timing reference. Such device can be characterized by a pure electrical equivalent circuit as shown. Write the describing equation for the indicated variable in the crystal's equivalent circuit.

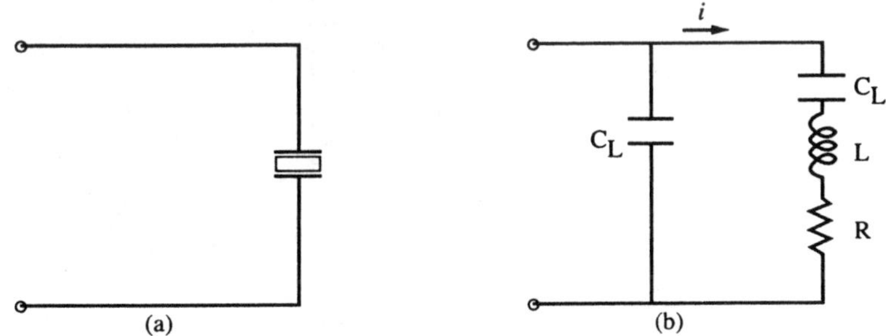

Problem 7.2

7.2 Second-order equations

7.3 Write a second-order differential equation satisfied by v_0 with R_1=2KΩ, R_2=3KΩ, R_3=9KΩ, C=0.001uF and L=1mH.

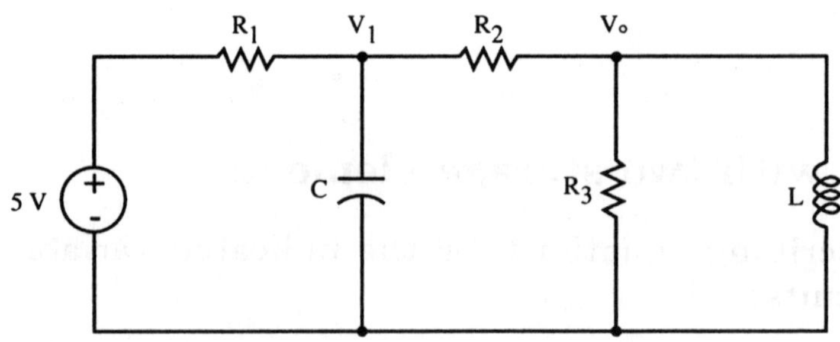

Problem 7.3

7.3 The natural response

7.4 Find the natural frequency ω_0 and damping ratio of the circuit in Prob.7.3.

7.5 Write a second-order differential equation satisfied by v_1 with the circuit shown.

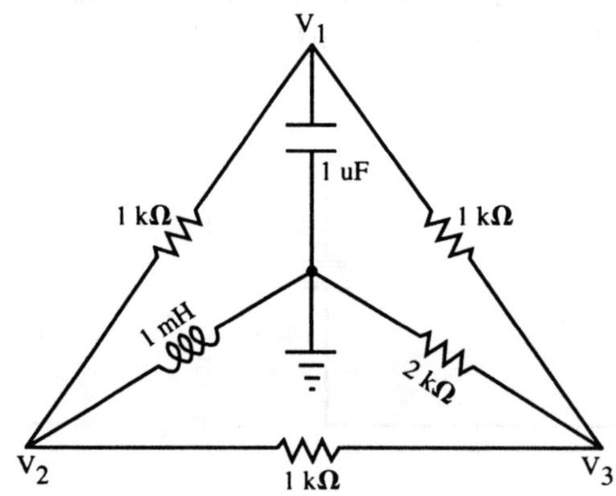

Problem 7.5

7.6 Find the describing equation of v. Is this a second-order differential equation? What function does this circuit perform?

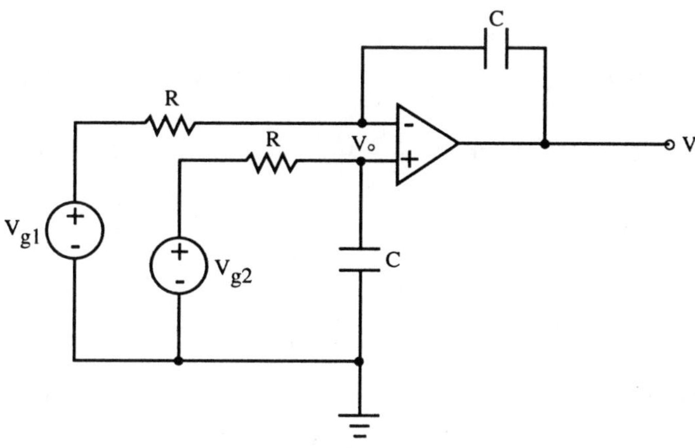

Problem 7.6

7.7 Find the describing equation for v.

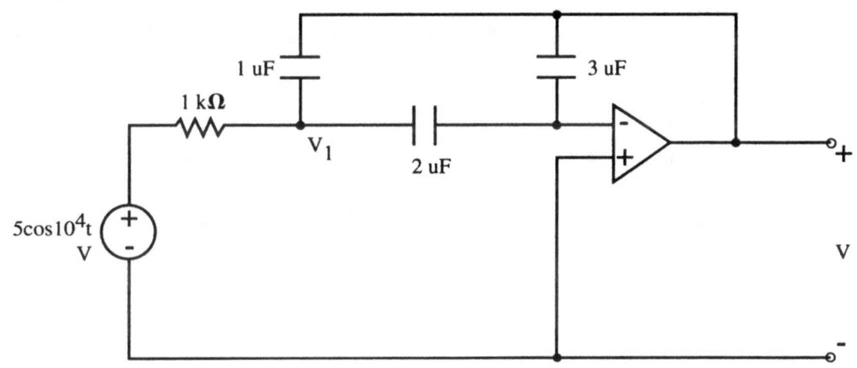

Problem 7.7

7.8 Find the describing equation for v.

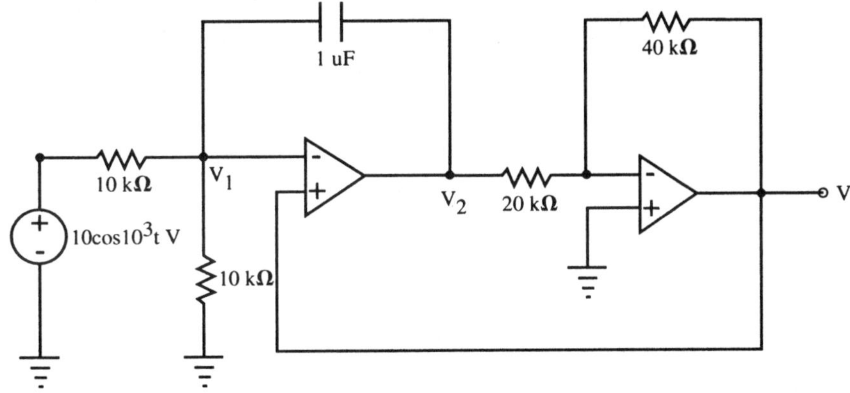

Problem 7.8

7.9 Verify that $x = A\cosh(2x) + B\sinh(2x)$ is a solution of the the differential equation $\frac{d^2x}{dt} - 4x = 0$.

7.10 Find a differential equation of the form $\frac{d^2x}{dt^2} + A_1\frac{dx}{dt} + A_2x = 0$, such that $x_1 = e^{1000-(\pi-1)t}$ and $x_2 = e^{-1000-(\pi+1)t}$ are general solutions.

7.11 Find the natural response of $\frac{d^2x}{dt^2} + a_1\frac{dx}{dt} + a_0x = f(t)$, if a_1=18, a_0=81, $x(0)$=3 and $\frac{dx}{dt}\mid_0$=-27.

7.4 The forced response

7.12 Find the forced solution of $\frac{d^2x}{dt^2} + 6\frac{dx}{dt} + 9x = f(t)$, with $f(t) = 4e^{-2t} + t$.

7.13 Find the forced response if $\frac{d^2x}{dt^2} + 4\frac{dx}{dt} + 3x = f(t)$, where $f(t) = 2e^{-3t} - e^{-t}$.

7.5 The total response

7.14 Find the complete solution of the equation $\frac{d^2x}{dt^2} + 4\frac{dx}{dt} + 3x = -\sin 3t$, with the initial conditions $x(0) = \frac{dx(0)}{dt} = 0$.

7.15 Find the complete response if $\frac{d^2x}{dt^2} + 9x = \sin(3t)$, and $x(0) = \frac{dx(0)}{dt} = 0$.

7.16 Find the complete response if $\frac{d^2x}{dt^2} + 2\sqrt{2}\frac{dx}{dt} + 2x = f(t)$, where $f(t) = e^{-\sqrt{2}t}$ and $x(0) = 1$, $\frac{dx}{dt}\mid_0 = 1 - \sqrt{2}$.

7.6 The unit step response

7.17 Find i for $t > 0$ if the circuit is in steady state at $t = 0^-$.

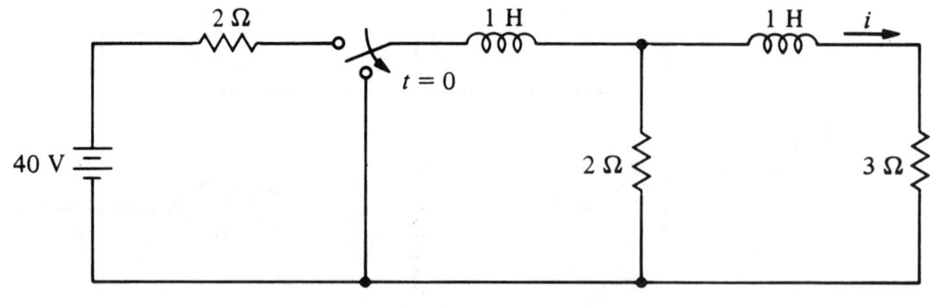

Problem 7.17

7.18 Find v for $t > 0$ if the circuit is in steady state at $t = 0^-$.

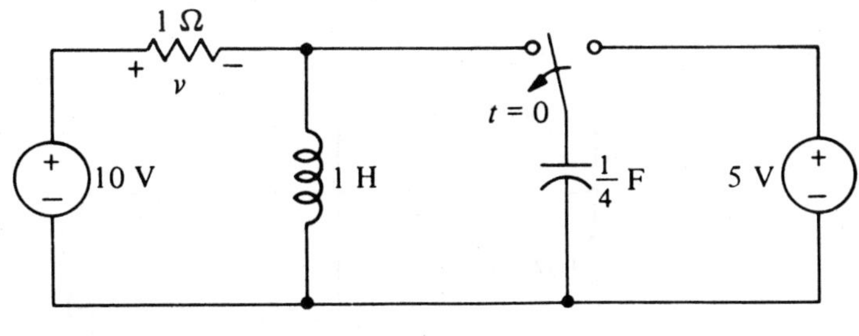

Problem 7.18

7.19 Find i for $t > 0$ if the circuit is in steady state at $t = 0^-$.

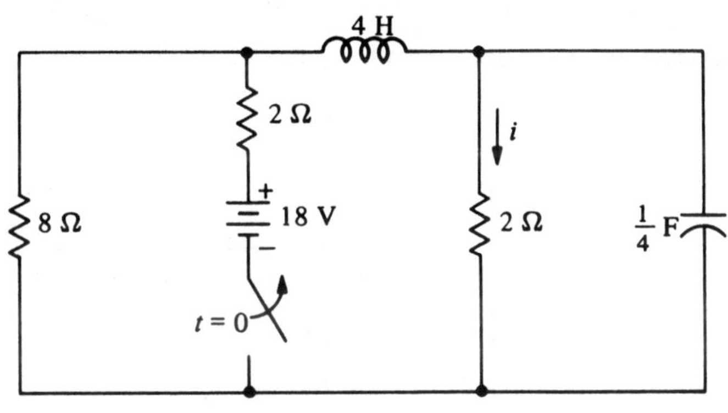

Problem 7.19

7.20 Find v for $t > 0$ if (a)$i_g = 2u(t)$A, and (b)$i_g = 2e^{-t}u(t)$A.

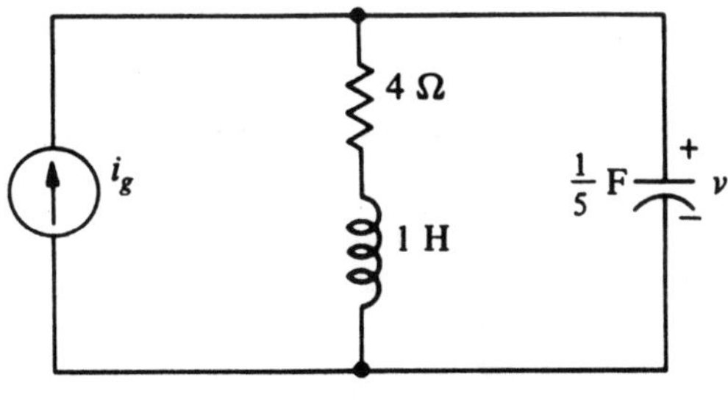

Problem 7.20

7.21 Find i for $t > 0$ if the circuit is in steady state at $t = 0^-$.

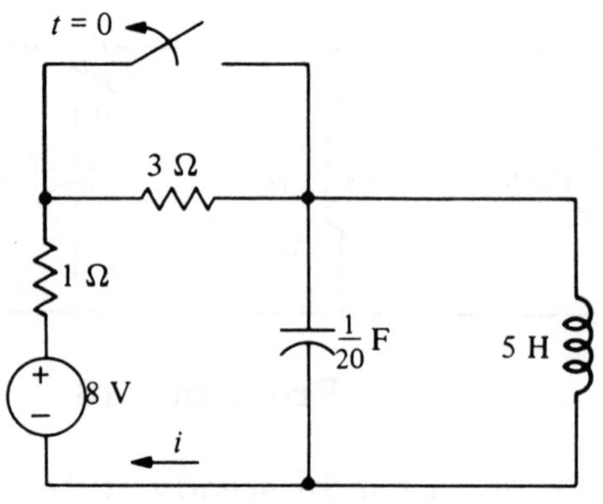

Problem 7.21

7.22 Solve Prob.7.21 if the 8V source is replaced by a source of $36e^{-4t}$V with the same polarity, and there is no initial stored energy.

7.23 Find the natural response v with v_0=1V and $i(0)$=-2A.

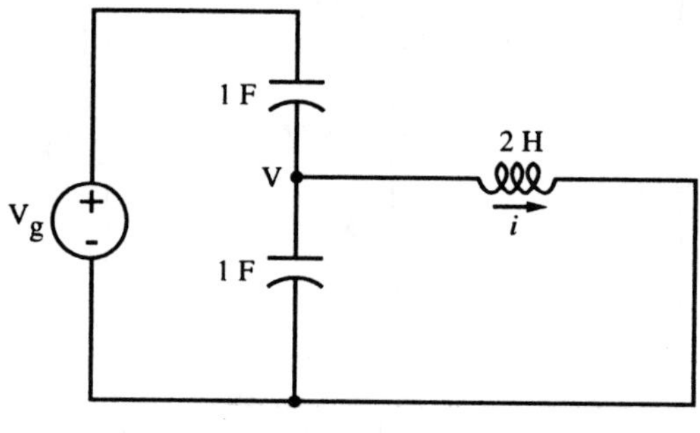

Problem 7.23

7.24 Find the forced response of Prob.7.23 with $v_g = \sin \frac{t}{2}$V.

7.25 Find v, $t > 0$, if $v(0)$=4V and $i(0)$=2A.

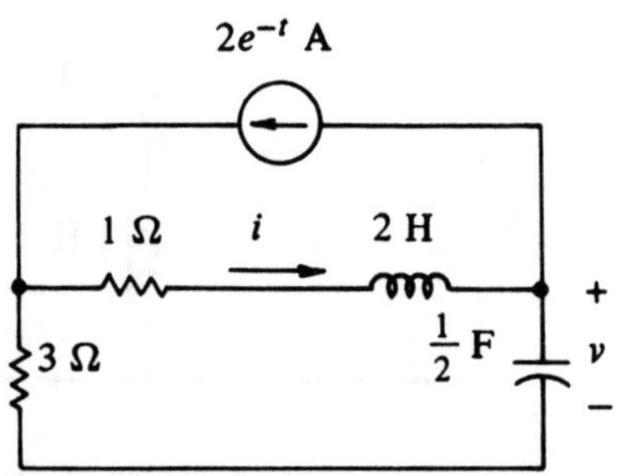

Problem 7.25

7.26 Find i for $t > 0$ if $L = \frac{8}{3}$H.

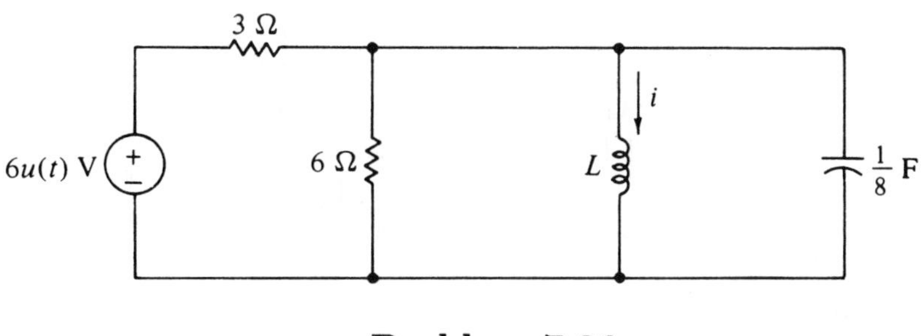

Problem 7.26

7.27 Find the forced solution $i_f(t)$.

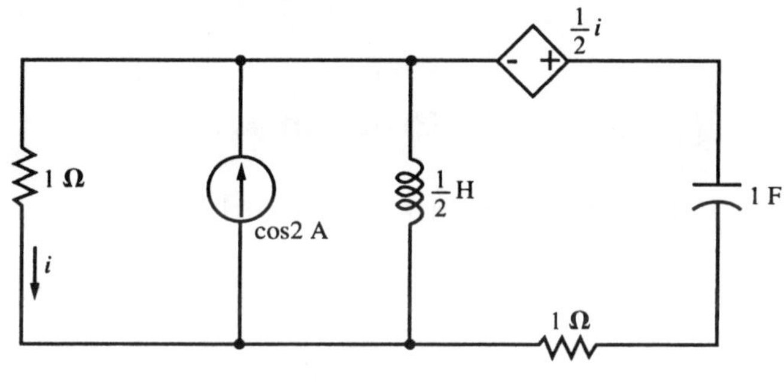

Problem 7.27

7.28 Find i for $t > 0$. Assume no initial stored energy.

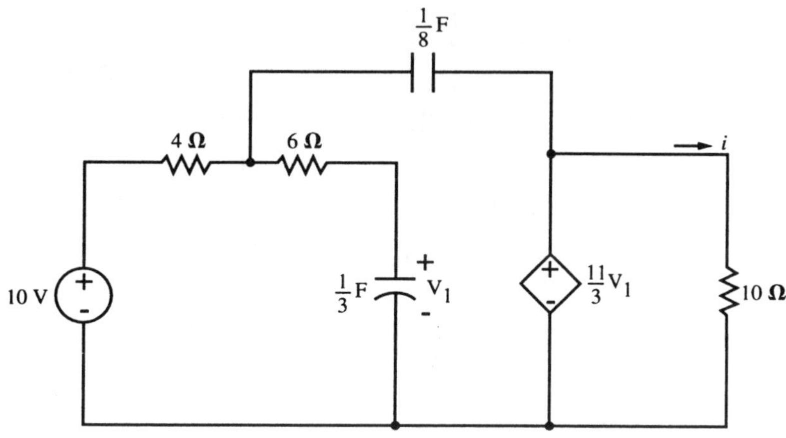

Problem 7.28

7.29 Find the voltage v for the circuit shown.

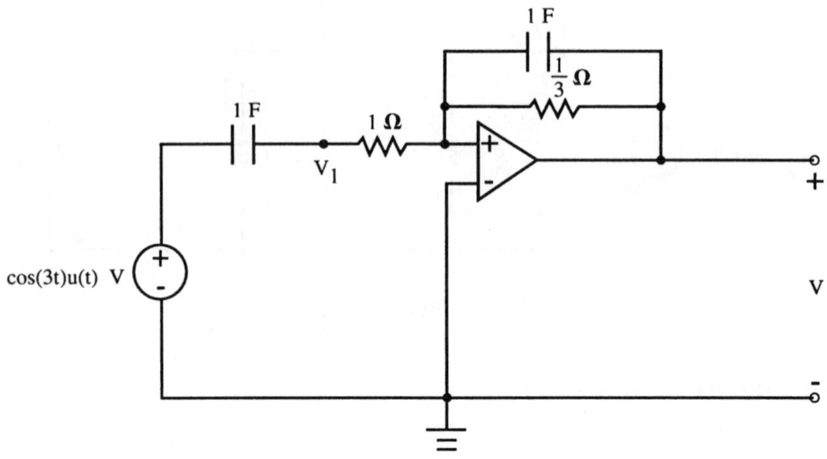

Problem 7.29

7.30 Find v for $t > 0$ if there is no initial stored energy.

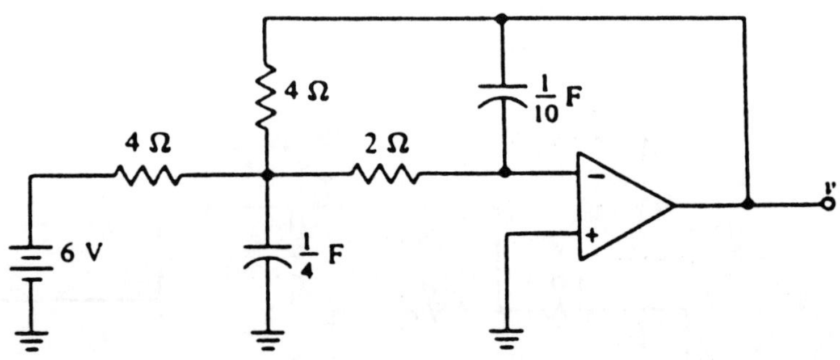

Problem 7.30

7.31 Find v for the circuit shown below with initial conditions $v(0)$=**0V**, $v_1(0)$=**5V**.

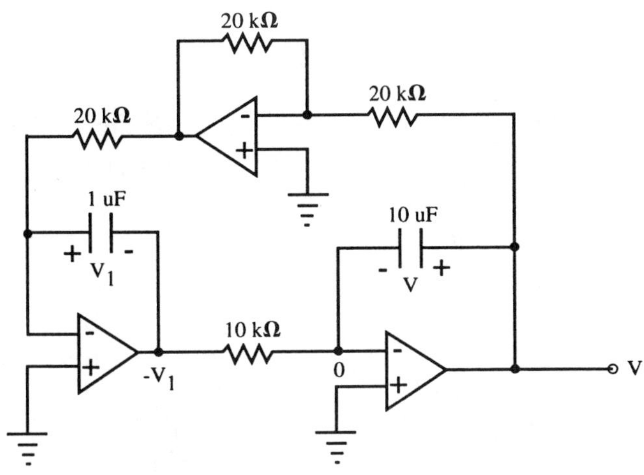

Problem 7.31

7.32 Solve Prob.7.28 using SPICE.

7.33 Use SPICE to determine the output voltage v of the circuit shown below. Use the improved op amp model with R_{in}=1MΩ, R_{out}=30Ω and open loop gain of 10^5.

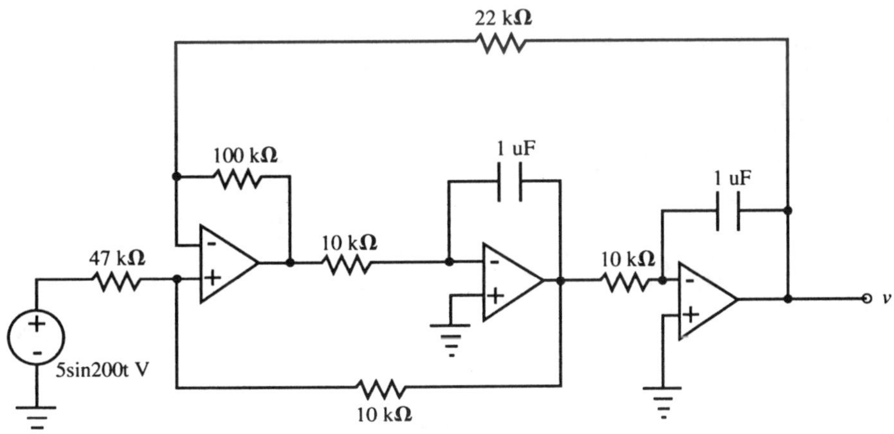

Problem 7.33

7.34 Use SPICE to solve Prob.7.31.

7.35 Higher-order differential equation may be solved by the same procedure of guessing a trial from solution as one does in second-order case. Find the natural response, the force response and the total response of the equation $\frac{d^3x}{dt} + 7\frac{d^2x}{dt} + 14\frac{dx}{dt} + 8x = 24$.

7.1 (a) Apply KVL around the left loop

$$\frac{1}{2}\int_{t_0}^{t_1} i(\tau)d\tau + (i+i_g)\cdot 4 + \frac{1}{3}\int_{t_0}^{t_1} i(\tau)d\tau + v_{c2}(t_0) + 7i = 0$$

Differentiate both side of the equation

$$\frac{1}{2}i + 4\frac{di}{dt} + \frac{1}{3}i + 7\frac{di}{dt} = -4\frac{di_g}{dt}$$

$$11\frac{di}{dt} + \frac{5}{6}i = -4\frac{di_g}{dt}$$

It is a first order differential equation.

(b) Apply KVL around the left most loop

$$7i_0 + \frac{1}{1}\int_{t_0}^{t_1}(i_0 - i)d\tau + v_{c_1}(t_0) = 0$$

$$\text{or } 7\frac{di_0}{dt} = i_0 - i \quad \text{—}\langle 1\rangle$$

KVL around the middle loop gives

$$v_{c1}(t_0) + \frac{1}{2}\int_{t_0}^{t_1} i\,dt + 4(i+i_g) + \frac{1}{3}\int_{t_0}^{t_1} i\,d\tau + v_{c2}(t_0) + 1\int_{t_0}^{t_1}(i - i_0)d\tau + v_{c3}(t_0) = 0,$$

$$\frac{1}{2}i + \frac{4di}{dt} + 4\frac{di_g}{dt} + \frac{1}{3}i + i - i_0 = 0,$$

$$i_0 = \frac{11}{6}i + 4\frac{di}{dt} + 4\frac{di_g}{dt},$$

differentiate both side of the equation

$$\frac{di_0}{dt} = \frac{11}{6}\frac{di}{dt} + 4\frac{d^2i}{dt^2} + 4\frac{d^2i_g}{dt^2} \quad \text{—}\langle 2\rangle$$

substitute ⟨2⟩ into ⟨1⟩

$$\frac{11}{6}i + 4\frac{di}{dt} + 4\frac{di_g}{dt} - \frac{77}{6}\frac{di}{dt} + 28\frac{d^2i}{dt^2} + 28\frac{d^2i_g}{dt^2} = i$$

$$28\frac{d^2i}{dt^2} - \frac{53}{6}\frac{di}{dt} + \frac{11}{6}i = -28\frac{d^2i_g}{dt^2} - 4\frac{di_g}{dt}$$

It is a second order circuit.

7.2 KVL gives

$$\frac{1}{C_2}\int_{t_0}^{t} i\,d\tau + V_{c2}(t_0) + L\frac{di}{dt} + Ri + \frac{1}{C_1}\int_{t_0}^{t_1} i\,d\tau + V_{c1}(t_0) = 0$$

Differentiate both sides of the equation

$$\left(\frac{1}{C_1} + \frac{1}{C_2}\right)i + L\frac{d^2i}{dt^2} + R\frac{di}{dt} = 0$$

$$\frac{d^2i}{dt^2} + \frac{R}{L}\frac{di}{dt} + \left(\frac{1}{C_1} + \frac{1}{C_2}\right)\frac{1}{L}i = 0$$

7.3 Nodal analysis gives

$$\frac{v_1 - 5}{R_1} + C\frac{dv_1}{dt} + \frac{v_1 - v_0}{R_2} = 0$$

$$\text{or } \frac{dv_1}{dt} + v_1\left(\frac{1}{R_1C} + \frac{1}{R_2C}\right) - K_1 = \frac{v_0}{R_2C} \quad \text{—}\langle 1\rangle$$

7.3 Cont.

$$\frac{R_2+R_3}{R_3} + \frac{v_0}{3} + \frac{1}{L}\int_{t_0}^{t_1} v_0\,d\tau + K_2 = 0 \text{ or}$$

$$\frac{dv_1}{dt} = \frac{dv_0}{dt}\left(\frac{R_2+R_3}{R_3}\right) + \frac{R_2}{L}v_0 \quad \text{—}\langle 2\rangle$$

substitute ⟨2⟩ into ⟨1⟩

$$\frac{R_2+R_3}{R_3}\frac{dv_0}{dt} + \frac{R_2}{L}v_0 + \left(\frac{1}{R_1C} + \frac{1}{R_2C}\right)\left[\frac{R_2+R_3}{R_3}v_0 + \frac{R_2}{L}\int_{t_0}^{t} v_0\,d\tau + K_2\right] - K_1 = \frac{v_0}{R_2C}, \text{ or}$$

$$\frac{d^2v_0}{dt} + \left[\frac{R_2R_3}{(R_2+R_3)L} + \frac{R_1+R_2}{R_1R_2C} - \frac{R_3}{R_2C(R_2+R_3)}\right]\frac{dv}{dt} + \frac{(R_1+R_2)R_3}{R_1C(R_2+R_3)L}v_0 = 0$$

with $R_1 = 2K\Omega$, $R_2 = 3K\Omega$, $R_3 = 9K\Omega$, $C = 0.001\mu F$ and $L = 1mH$

$$\frac{R_2R_3}{(R_2+R_3)L} + \frac{R_1+R_2}{R_1R_2C} - \frac{R_3}{R_2C(R_2+R_3)} = \frac{17}{6}\times 10^6, \text{ and}$$

$$\frac{(R_1+R_2)R_3}{R_1C(R_2+R_3)L} = \frac{15}{8}\times 10^{12}$$

Hence the differential equation is:

$$\frac{d^2v_0}{dt^2} + \frac{17}{6}\times 10^6 \times \frac{dv_0}{dt} + \frac{15}{8}\times 10^{12} = 0$$

7.4

$$\because \frac{d^2v_0}{dt^2} + \frac{17}{6}\times 10^6\frac{dv_0}{dt} + \frac{15}{8}\times 10^{12} = 0$$

$$w_0 = \sqrt{\frac{15}{8}\times 10^{12}} = 1.37\times 10^6\ Hz$$

$$\zeta = \frac{17}{6}\times 10^6 \Big/ 2w_0 = 1.03$$

The damping ratio is 1.03

7.5 Let $R = 1K\Omega$, $C = 1\mu F$, $L = 1mH$

$$\begin{cases} \frac{V_1 - V_2}{R} + \frac{V_1 - V_3}{R} + C\frac{dV_1}{dt} = 0 & \langle 1\rangle \\ \frac{V_2 - V_1}{R} + \frac{V_2 - V_3}{R} + \frac{1}{L}\int V_2\,dt = 0 & \langle 2\rangle \\ \frac{V_3 - V_2}{R} + \frac{V_3 - V_1}{R} + \frac{V_3}{2R} = 0 \text{ or } V_3 = \frac{2}{5}V_1 + \frac{2}{5}V_2 & \langle 3\rangle \end{cases}$$

substitute ⟨3⟩ into ⟨1⟩ and ⟨2⟩

$$\frac{8}{5}V_1 - \frac{7}{5}V_2 + RC\frac{dV_1}{dt} = 0 \text{ or } V_2 = \frac{8}{7}V_1 + \frac{5}{7}RC\frac{dV_1}{dt} \quad \langle 4\rangle$$

$$\frac{8}{5}V_2 - \frac{7}{5}V_1 + \frac{1}{L}\int V_2\,dt = 0 \quad \langle 5\rangle$$

Substitute V_2 into ⟨5⟩

$$\frac{8}{5}\left(\frac{8}{7}V_1 + \frac{5}{7}RC\frac{dV_1}{dt}\right) - \frac{7}{5}V_1 + \frac{1}{L}\int\left(\frac{8}{7}V_1 + \frac{5}{7}RC\frac{dV_1}{dt}\right)d\tau = 0$$

Differentiate both sides of equation

$$\frac{d^2V_1}{dt^2} + \left(\frac{3}{8} + \frac{5}{8}\frac{RC}{L}\right)\frac{1}{RC}\frac{dV_1}{dt} + \frac{1}{RLC}V_1 = 0$$

$$\because \left(\frac{3}{8} + \frac{5}{8}\frac{RC}{L}\right)\frac{1}{RC} = 10^3,\ \frac{1}{RLC} = 10^6$$

The equation is $\frac{d^2V_1}{dt^2} + 10^3\frac{dV_1}{dt} + 10^6 V_1 = 0$

7.6

$$\begin{cases} \dfrac{v_o - v_{g_1}}{R} + C\dfrac{d(v_o - v)}{dt} = 0 & \langle 1\rangle \\ \dfrac{v_o - v_{g_2}}{R} + C\dfrac{dv_o}{dt} = 0 & \langle 2\rangle \end{cases}$$

$\langle 1\rangle - \langle 2\rangle$ $\dfrac{v_{g_1} - v_{g_2}}{R} + C\dfrac{dv}{dt} = 0$

The describing equation: $\dfrac{dv}{dt} + \dfrac{v_{g_1} - v_{g_2}}{RC} = 0$

This is a first-order equation.

$\because \dfrac{dv}{dt} = -\dfrac{v_{g_1} - v_{g_2}}{RC}$

$$v = -\frac{1}{RC}\int_{t_0}^{t_1}(v_{g_1} - v_{g_2})d\tau + v(t_0)$$

The circuit integrates the difference between two input signals.

7.7 Let $R_1 = 1K\Omega$, $C_1 = 1\mu F$, $C_2 = 2\mu F$, $C_3 = 3\mu F$ and $v_g = 5\cos 10^4 t$

$\dfrac{v_1 - v_g}{R_1} + C_1\dfrac{d(v_1 - v)}{dt} + C_2\dfrac{dv_1}{dt} = 0$, or

$\dfrac{v_1}{R_1} + (C_1 + C_2)\dfrac{dv_1}{dt} - C_1\dfrac{dv}{dt} = \dfrac{v_g}{R_1}$ $\langle 1\rangle$

and $C_2\dfrac{d(0 - v_1)}{dt} = -C_3\dfrac{dv}{dt}$ $\langle 2\rangle$

Assume initial conditions are all zero

$C_2 v_1 = -C_3 v$, $v_1 = -\dfrac{C_3}{C_2}v$

substitute v_1 into $\langle 1\rangle$

$\dfrac{C_1C_2 + C_1C_3 + C_2C_3}{C_2}\dfrac{dv}{dt} - \dfrac{C_3}{R_1C_2}v = \dfrac{v_g}{R_1}$

$\Rightarrow \dfrac{2\times10^{-12} + 3\times10^{-12} + 6\times10^{-12}}{2\times10^{-6}}\dfrac{dv}{dt} - \dfrac{3\times10^{-6}}{10^3\times2\times10^{-6}}v = \dfrac{5\cos 10^4 t}{10^3}$

$\Rightarrow \dfrac{11}{2}\times10^{-6}\dfrac{dv}{dt} - \dfrac{3}{2}\times10^{-3}v = 5\times10^{-3}\times\cos 10^4 t$

7.8 Let $R_1 = 10K\Omega$, $C_1 = 1\mu F$, $R_2 = 20K\Omega$, $R_3 = 40K\Omega$ and $v_g = 10\cos 10^3 t\ V$

By KCL

$$\begin{cases} \dfrac{v_1 - v_g}{R_1} + \dfrac{v_1}{R_1} + C_1\dfrac{d(v_1 - v_2)}{dt} = 0 & \langle 1\rangle \\ \dfrac{0 - v_2}{R_2} + \dfrac{0 - v}{R_3} = 0 \quad \text{or } v_2 = -\dfrac{R_2}{R_3}v \\ v_1 = v \end{cases}$$

plug v_1 and v_2 into $\langle 1\rangle$

$v\left(\dfrac{2}{R_1}\right) + C_1\dfrac{dv}{dt} + \dfrac{C_1R_2}{R_3}\dfrac{dv}{dt} = \dfrac{v_g}{R_1}$

$\Rightarrow C_1\left(1 + \dfrac{R_2}{R_3}\right)\dfrac{dv}{dt} + \dfrac{2}{R_1}v = \dfrac{v_g}{R_1}$

$\Rightarrow \dfrac{3}{2}\times10^{-6}\dfrac{dv}{dt} + \dfrac{1}{5}\times10^{-3}v = 10^{-2}\cos 10^3 t$

7.9 By definition

$\cosh 2x = \frac{1}{2}(e^{2x} + e^{-2x})$

$\sinh 2x = \frac{1}{2}(e^{2x} - e^{-2x})$

$x = A\cosh 2x + B\sinh 2x = \dfrac{(A+B)}{2}e^{2x} - \dfrac{(A-B)}{2}e^{-2x}$

$\dfrac{dx}{dt} = (A+B)e^{2x} - (A-B)e^{-2x}$

$\dfrac{d^2x}{dt^2} = 2(A+B)e^{2x} - 2(A-B)e^{-2x}$

Substitute x and its derivatives into the given equation

$$\frac{d^2x}{dt^2} - 4x = 2(A+B)e^{2x} - 2(A-B)e^{-2x} - 4\left[\frac{(A+B)}{2}e^{2x} - \frac{(A-B)}{2}e^{-2x}\right] = 0$$

Hence x is a solution of $\dfrac{d^2x}{dt^2} - 4x = 0$

7.10 $x_1 = e^{1000 - (\pi-1)t} = e^{1000} \times e^{-(\pi-1)t}$

$x_2 = e^{-1000 - (\pi+1)t} = e^{-1000} \times e^{-(\pi+1)t}$

The characteristic equation of the desired differential equation has to be the form of

$s^2 + [(\pi-1) + (\pi+1)]s + (\pi+1)(\pi-1) = 0$,

$s^2 + 2\pi s + \pi^2 - 1 = 0$

Hence the equation is:

$$\frac{d^2x}{dt^2} + 2\pi\frac{dx}{dt} + (\pi^2 - 1)x = 0$$

7.11 The characteristic equation

$s^2 + 18s + 81 = 0$, $(s+9)^2 = 0$, or $s = -9$

Let the general solution be:

$x = A_1e^{-9t} + A_2te^{-9t}$

$x(0) = A_1 = 3$

$\left.\dfrac{dx}{dt}\right|_{t=0} = -9A_1e^{-t} - 9A_2te^{-2t} = -27$,

$A_2 = 0$

Hence the natural response is

$x = 3e^{-9t}$

7.12 Try the forced solution of the form:

$x_f = A_1e^{-2t} + A_2t + A_3$

$\dfrac{dx_f}{dt} = -2A_1e^{-2t} + A_2$, $\dfrac{d^2x_f}{dt^2} = 4A_1e^{-2t}$

Substitute x_f, $\dfrac{dx_f}{dt}$ and $\dfrac{d^2x_f}{dt^2}$ into the forced equation:

$4A_1e^{-2t} + 6(-2A_1e^{-2t} + A_2) + 9(A_1e^{-2t} + A_2t + \ )$

$= 4e^{-2t} + t \Rightarrow A_1 = 4, A_2 = \frac{1}{9}, A_3 = -\frac{6}{81}$

Hence $x_f = 4e^{-2t} + \frac{1}{9}t - \frac{6}{81}$

7.13 Since $s_{1,2}=-1,-3$ and $f(t)=2e^{-3t}-e^{-t}$, try $x_f = Ate^{-3t}+Bte^{-t}$

$\frac{dx_f}{dt} = -3Ate^{-3t}+Ae^{-3t}-Bte^{-t}+Be^{-t}$

$\frac{d^2x_f}{dt^2} = 9Ate^{-3t}-6Ae^{-3t}+Bte^{-t}-2Be^{-t}$

$\therefore (9Ate^{-3t}-6Ae^{-3t}+Bte^{-t}-2Be^{-t}) + 4(-3Ate^{-3t}+Ae^{-3t}-Bte^{-t}+Be^{-t}) + 3(Ate^{-3t}+Bte^{-t}) = -2Ae^{-3t}+2Be^{-t} = 2e^{-3t}-e^{-t}$

Equating coefficients

e^{-3t}: $-2A=2$, $A=-1$

e^{-2t}: $2B=-1$, $B=-\frac{1}{2}$

$\therefore x_f = -t(te^{-3t}+\frac{1}{2}e^{-t})$

7.14 Try $x_f = A_1\sin 3t + A_2\cos 3t$

$\frac{dx_f}{dt} = 3A_1\cos 3t - 3A_2\sin 3t$

$\frac{d^2x_f}{dt^2} = -9A_1\sin 3t - 9A_2\cos 3t$

$\therefore (-6A_1-12A_2)\sin 3t + (12A_1-6A_2)\cos 3t = -\sin 3t$

Equating coefficients

$-6A_1-12A_2=-1$
$12A_1-6A_2=0$ or $A_1=\frac{1}{30}$, $A_2=\frac{1}{15}$

Charateristic equation $s^2+4s+3=0$
$s=-1,-3$

Let $x = k_1e^{-3t}+k_2e^{-t}+\frac{1}{30}\sin 3t+\frac{1}{15}\cos 3t$

$x(0)=k_1+k_2+\frac{1}{15}=0$ or $k_1=\frac{1}{12}$

$\frac{dx(0)}{dt} = -3k_1-k_2+\frac{1}{10}=0$ $\quad k_2=\frac{-9}{60}$

Hence the complete solution is

$x=\frac{1}{12}e^{-3t}-\frac{9}{60}e^{-t}+\frac{1}{30}\sin 3t+\frac{1}{15}\cos 3t$

7.15 Since $s_{1,2}=\pm j3$ and $f(t)=\sin 3t$

try $x_f = t(A\cos 3t+B\sin 3t)$

$\frac{dx_f}{dt} = t(-3A\sin 3t+3B\cos 3t)+A\cos 3t+B\sin 3t$

$\frac{d^2x_f}{dt^2} = t(-9A\cos 3t-9B\sin 3t)-6A\sin 3t+6B\cos 3t$,

$\frac{d^2x_f}{dt^2}+9x_f = -6A\sin 3t+6B\cos 3t = \sin 3t$

Equating coefficients

$A=-\frac{1}{6}$, $B=0$; $x_f=-\frac{1}{6}t\cos 3t$

$\because x = x_n + x_f$

7.15 Cont.

$x=A_1\cos 3t+A_2\sin 3t-\frac{1}{6}t\cos 3t$, $x(0)=0=A_1$

$\frac{dx}{dt} = -3A_1\sin 3t+3A_2\cos 3t-\frac{1}{6}\cos 3t - \frac{t}{2}\sin 3t$

$\frac{dx(0)}{dt} = 3A_2-\frac{1}{6}=0$, $A_2=\frac{1}{18}$

$\therefore x=\frac{1}{18}\sin 3t-\frac{1}{6}t\cos 3t$

7.16 $s^2+2\sqrt{2}s+2=0$, $s=-\sqrt{2}$

Try $x_f = At^2e^{-\sqrt{2}t}$

$\frac{dx_f}{dt} = 2Ate^{-\sqrt{2}t}-\sqrt{2}At^2e^{-\sqrt{2}t}$

$\frac{d^2x_f}{dt^2} = 2Ae^{-\sqrt{2}t}-4\sqrt{2}Ate^{-\sqrt{2}t}+2At^2e^{-\sqrt{2}t}$

$\therefore 2Ae^{-\sqrt{2}t}-4\sqrt{2}Ate^{-\sqrt{2}t}+2At^2e^{-\sqrt{2}t} + 4\sqrt{2}Ate^{-\sqrt{2}t}-4At^2e^{-\sqrt{2}t}+2At^2e^{-\sqrt{2}t} = 2Ae^{-\sqrt{2}t}$

Equating coefficients, $2A=1$, $A=\frac{1}{2}$

$x = x_n+x_f = k_1e^{-\sqrt{2}t}+k_2te^{-\sqrt{2}t}+\frac{1}{2}t^2e^{-\sqrt{2}t}$

$x(0)=k_1=1$

$\left.\frac{dx}{dt}\right|_{t=0} = -\sqrt{2}k_1+k_2 = k_2-\sqrt{2} = 1-\sqrt{2}$, $k_2=1$

$\therefore x = e^{-\sqrt{2}t}+(1-\sqrt{2})te^{-\sqrt{2}t}+\frac{1}{2}t^2e^{-\sqrt{2}t}$

7.17 Let i_1 = current in left 1-H inductor to right. At $t=0^-$ the inductors are short circuits,

$\therefore i_1(0^-) = \frac{40}{2+\frac{2(3)}{5}} = \frac{25}{2}A$;

$i(0^-) = \frac{2}{2+3}i_1(0^-) = 5A$

For $t>0$ the mesh equations are

(1) $\frac{di_1}{dt}+2(i_1-i)=0$; (2) $\frac{di}{dt}+3i+2(i-i_1)=0$

From (2), $i_1 = \frac{1}{2}\frac{di}{dt}+\frac{5}{2}i$. Substitution into (1): $\frac{d^2i}{dt^2}+7\frac{di}{dt}+6i=0 \Rightarrow$

$s^2+7s+6=0 \Rightarrow s_{1,2}=-1,-6$

$\therefore i = A_1e^{-t}+A_2e^{-6t}$

$i(0^+)=A_1+A_2$; from (2) $\frac{di(0^+)}{dt} = 2i_1(0^+)-5i(0^+)=0$

$\frac{di(0^+)}{dt} = -A_1-6A_2=0$, $\therefore A_1=6$, $A_2=-1$

$\therefore i = 6e^{-t}-e^{-6t}A$

7.18 At $t=0^-$, the capacitor is an open circuit; the 3Ω resistor and the inductor are shorted.

$\therefore i_L(0^-)=\frac{8}{1}=8A$, $v_C(0^-)=0$

For $t>0$, $v_{cf}=0$, KCL gives

(1) $\frac{v_C-8}{4}+\frac{1}{20}\frac{dv_C}{dt}+\frac{1}{5}\int_0^t v_C\,dt+i_L(0^-)=0$

or $\frac{d^2v_C}{dt^2}+5\frac{dv_C}{dt}+4=0$

From equation (1) $\frac{dv_C(0)}{dt}=-120$

the characteristic equation is

$s^2+5s+4=0$; $s_{1,2}=-1,-4$

$v_C=A_1e^{-t}+A_2e^{-4t}$

$v_C(0)=A_1+A_2=0$

$\frac{dv_C(0)}{dt}=-A_1-4A_2=-120$, Adding,

$-3A_2=-120 \Rightarrow A_2=40, A_1=-40$

$\therefore v_C=40(e^{-4t}-e^{-t})V$

$i=\frac{8-v_C}{4}=2+10(e^{-t}-e^{-4t})A$

7.19 At $t=0^-$, the capacitor is open-circuited and the inductor is a short circuit.

By current division and Ohm's law,

$i(0^-)=\frac{18}{2+\frac{2(8)}{10}}\frac{8}{2+8}=4A$, $v_C(0^-)=4(2)=8V$

Let i_L be the inductor current to the right and i_C be the capacitor current downward. For $t>0$, $v_C=2i$,

$i_C=\frac{1}{4}\frac{dv_C}{dt}=\frac{1}{2}\frac{di}{dt}$ and $i_L=i+i_C$

$i_L=i+\frac{1}{2}\frac{di}{dt}$. KVL around the left mesh yields: $8i_L+4\frac{di_L}{dt}+2i=0$

or $8(i+\frac{1}{2}\frac{di}{dt})+4\frac{d}{dt}(i+\frac{1}{2}\frac{di}{dt})+2i=0$

or $\frac{d^2i}{dt^2}+4\frac{di}{dt}+5i=0 \Rightarrow s^2+4s+5=0 \Rightarrow$

$s_{1,2}=-2\pm j$, $i=e^{-2t}(A_1\cos t+A_2\sin t)$

$i(0^+)=\frac{1}{2}v_C(0^+)=\frac{1}{2}v_C(0^-)=\frac{8}{2}=4=A_1$

$\frac{di(0^+)}{dt}=2[i_L(0^+)-i(0^+)]=2i(0^-)-v_C(0^+)$

$=2i(0^-)-v_C(0^+)=(2)(4)-8=0$

$=A_2-2A_1 \Rightarrow A_2=8$

$\therefore i=e^{-2t}(4\cos t-8\sin t)A$

7.20 v_L = inductor voltage, positive top

KCL gives (1) $\frac{1}{5}\frac{dv}{dt}+v-v_L=i_g$

(2) $\frac{-v+v_L}{4}+\int_0^t v_L\,dt=0$

Add (1) and (2), differentiate and substitute for v_L from (1):

$\frac{d^2v}{dt^2}+4\frac{dv}{dt}+5v=5\frac{di_g}{dt}+20i_g$

At $t=0^-$, $v(0^-)=i_L(0^-)=0$.

Therefore $v(0^+)=0$ and $v_L(0^+)=0$ by KVL. From (1) $\frac{dv(0^+)}{dt}=5i_g(0^+)$

(a) $i_g=2u(t)A$. For $t>0$, $\frac{di_g}{dt}=0$, $i_g=2A$:

$\frac{d^2v}{dt^2}+4\frac{dv}{dt}+5v=40 \Rightarrow s_{1,2}=-2\pm j$

$v=e^{-2t}(A_1\cos t+A_2\sin t)+8$

$v(0^+)=0=A_1+8$; $\frac{dv(0^+)}{dt}=-2A_1+A_2=10$

$A_1=-8$; $A_2=-6 \Rightarrow$

$v=8-e^{-2t}(8\cos t+6\sin t)V$

(b) $i_g=2e^{-t}u(t)$: For $t>0$, $i_g=2e^{-t}$ and $\frac{di_g}{dt}=-2e^{-t}$. Therefore

$\frac{d^2v}{dt^2}+4\frac{dv}{dt}+5v=30e^{-t}$

Try $v_f=Ae^{-t}$, then

$A(1-4+5)e^{-t}=30e^{-t} \Rightarrow A=15$.

$v=e^{-2t}(A_1\cos t+A_2\sin t)+15e^{-t}$

$v(0)=0=A_1+15 \Rightarrow A_1=-15$

$\frac{dv(0)}{dt}=5(2)=-2A_1+A_2-15 \Rightarrow A_2=-5$

$\therefore v=15e^{-t}-e^{-2t}(15\cos t+5\sin t)V$ $(t>0)$

7.21 At $t=0^-$, the capacitor is an open circuit; the 3Ω resistor and the inductor are shorted. $\therefore i_L(0^-)=\frac{8}{1}=8A$

$v_C(0^-)=0$. For $t>0$, $v_{cf}=0$, KCL gives: $\frac{v_C-8}{4}+\frac{1}{20}\frac{dv_C}{dt}+\frac{1}{5}\int_0^t v_C\,dt+i_L(0^-)=0$

or $\frac{d^2v_C}{dt^2}+5\frac{dv_C}{dt}+4=0$, $\frac{dv_C(0)}{dt}=-120$

the characteristic equation is

$s^2+5s+4=0$; $s_{1,2}=-1,-4$

$v_C=A_1e^{-t}+A_2e^{-4t}$

7.21 Cont.

$v_c(0) = A_1 + A_2 = 0$

$\frac{dv_c(0)}{dt} = -A_1 - 4A_2 = -120$, Adding,

$-3A_2 = -120 \Rightarrow A_2 = 40, A_1 = -40$

$\therefore v_c = 40(e^{-4t} - e^{-t})V$

$i = \frac{8 - v_c}{4} = 2 + 10(e^{-t} - e^{-4t})A$

7.22 At $t > 0$, KCL gives

(1) $\frac{v_c - 36e^{-4t}}{4} + \frac{1}{20}\frac{dv_c}{dt} + \frac{1}{5}\int_0^t v_c dt = 0$

$i_L(0^+) = 0$, then

(2) $\frac{d^2v_c}{dt^2} + 5\frac{dv_c}{dt} + 4v_c = -720e^{-4t}$

Try $v_{cf} = Ate^{-4t}$

$\frac{dv_{cf}}{dt} = -4Ate^{-4t} + Ae^{-4t}$

$\frac{d^2v_{cf}}{dt^2} = 16Ate^{-4t} - 8Ae^{-4t}$

substituting into (2) gives

$(16Ate^{-4t} - 8Ae^{-4t}) + 5(-4Ate^{-4t} + Ae^{-4t}) + 4(Ate^{-4t}) = -3Ae^{-4t} = -720e^{-4t}$

therefore $A = \frac{720}{3} = 240$ and

$v_{cf} = A_1e^{-t} + A_2e^{-4t} + 240te^{-4t}$

$v_c(0) = 0$; $\frac{dv_c(0)}{dt} = -5v_c(0) + 180 = 180$

$v_c(0) = A_1 + A_2 = 0$

$\frac{dv_c(0)}{dt} = -A_1 - 4A_2 + 240 = 180$ or $-A_1 - 4A_2 = -60$

Adding, $-3A_2 = -60 \Rightarrow A_2 = 20, A_1 = -20$

$v_c = 20e^{-4t} - 20e^{-t} + 240te^{-4t}$

$i = \frac{36e^{-4t} - v_c}{4} = 5e^{-t} + 4e^{-4t} - 60te^{-4t}A$

7.23 $\frac{dv}{dt} + \frac{dv}{dt} + \frac{1}{2}\int_{t_0}^t v d\tau + i(t_0) = 0$

$2\frac{d^2v}{dt^2} + \frac{1}{2}v = 0$ or $\frac{d^2v}{dt^2} + \frac{1}{4}v = 0$

$s^2 + \frac{1}{4} = 0$, $s_{1,2} = \pm j\frac{1}{2}$

$v = A_1 \sin\frac{t}{2} - A_2\cos\frac{t}{2}$

$v(0) = -A_2 = 1$, $A_2 = -1$

Since $2\frac{dv}{dt} + i = 0$

$i = -\frac{2dv}{dt} = -2\left[\frac{A_1}{2}\cos\frac{t}{2} - \frac{1}{2}\sin\frac{t}{2}\right] = -A_1\cos\frac{t}{2} + \sin\frac{t}{2}$

$i(0) = -A_1 = -2, A_1 = 2$,

Hence $v = 2\sin\frac{t}{2} + \cos\frac{t}{2}$

7.24 $\frac{d(v - v_g)}{dt} - \frac{dv}{dt} + \frac{1}{2}\int_{t_0}^t (v - v_g)d\tau + i(t_0) = 0$

$\frac{2d^2v}{dt^2} - \frac{d^2v_g}{dt} + \frac{1}{2}(v - v_g) = 0$ or

$\frac{d^2v}{dt^2} + \frac{1}{4}v = \frac{1}{4}v_g + \frac{1}{2}\frac{d^2v_g}{dt}$

$v_g = \sin$, $\frac{d^2v_g}{dt} = -\frac{1}{4}\sin\frac{t}{2}$

$\frac{d^2v}{dt^2} + \frac{1}{4}v = \frac{1}{4}\sin\frac{t}{2} - \frac{1}{8}\sin\frac{t}{2} = \frac{1}{8}\sin\frac{t}{2}$

Try $v_f = A_1 t\sin\frac{t}{2} + A_2 t\cos\frac{t}{2}$

$\frac{dv_f}{dt} = \frac{A_1}{2}t\cos\frac{t}{2} + A_1\sin\frac{t}{2} - \frac{A_2}{2}t\sin\frac{t}{2} + A_2\cos\frac{t}{2}$

$\frac{d^2v_f}{dt^2} = -\frac{A_1}{4}t\sin\frac{t}{2} + \frac{A_1}{2}\cos\frac{t}{2} + \frac{A_1}{2}\cos\frac{t}{2} - \frac{A_2}{4}t\cos\frac{t}{2} - \frac{A_2}{2}\sin\frac{t}{2} - \frac{A_2}{2}\sin\frac{t}{2}$

$\frac{d^2v_f}{dt^2} + \frac{1}{4}v_f = A_1\cos\frac{t}{2} - A_2\sin\frac{t}{2} = \frac{1}{8}\sin\frac{t}{2}$

Equating coefficients

$A_1 = 0$, $A_2 = -\frac{1}{8}$

$\therefore v_f = -\frac{1}{8}t\cos\frac{t}{2}V$

7.25 KVL gives : $i - i_g = \frac{1}{2}\frac{dv}{dt}$

$i + 2\frac{di}{dt} + v - 3(i_g - i) = 0$;

$4(i_g + \frac{1}{2}\frac{dv}{dt}) + 2\frac{d}{dt}(i_g + \frac{1}{2}\frac{dv}{dt}) + v = 3i_g$

or $\frac{d^2v}{dt^2} + 2\frac{dv}{dt} + v = -i_g - 2\frac{di_g}{dt} = 2e^{-t}$

$s^2 + 2s + 1 = 0 \Rightarrow s_{1,2} = -1, -1$; try

$v_f = At^2e^{-t}$ $\therefore 2Ae^{-t} = 2e^{-t} \Rightarrow A = 1$

$v = (A_1 + A_2t)e^{-t} + t^2e^{-t}$

$= (A_1 + A_2t + t^2)e^{-t}$

$v(0) = 4 = A_1$, $\frac{dv(0)}{dt} = 2[i(0) - i_g(0)]$

$= 2(2-2) = 0$

$\frac{dv(0)}{dt} = 0 = -A_1 + A_2 \Rightarrow A_2 = 4$

$v = (4 + 4t + t^2)e^{-t}V$

7.26

$2u(t)$ A, 2Ω, $\frac{8}{3}H$, v, i, $\frac{1}{8}F$

Using the Norton circuit of the first 3 elements, we obtain the parallel RLC circuit. The differential equation is $\frac{d^2i}{dt^2} + \frac{1}{CR}\frac{di}{dt} + \frac{1}{LC}i = \frac{i_g}{LC}$

For $t = 0^+$, $i(0^+) = i(0^-) = 0$,

$\frac{di(0^+)}{dt} = \frac{v(0^+)}{L} = \frac{v(0^-)}{L} = 0$,

7.26 Cont.

$$S_{1,2} = -\frac{1}{2RC} \pm \sqrt{\left(\frac{1}{2RC}\right)^2 - \frac{1}{LC}}, \quad \frac{1}{2RC} = 2$$

$$= -2 \pm \sqrt{4-3} = -1, -3$$

$i_f = i_g = 2A$

$i = A_1 e^{-t} + A_2 e^{-3t} + 2, \quad i(0^+) = 0 = A_1 + A_2 + 2$

$\frac{di(0^+)}{dt} = 0 = -A_1 - 3A_2$, Adding,

$-2A_2 = -2, \quad A_2 = 1, \quad A_1 = -3$

$\therefore i = e^{-3t} - 3e^{-t} + 2A \quad (t > 0)$

7.27 To simplifies calculation, rearrange the circuit as:

By Mesh analysis,

(1) $i = \cos 2t$, or $\frac{di_3}{dt} = -2\sin 2t$,

(2) $1(i_2 - i_1) + \frac{1}{2}\frac{d(i_2 - i_3)}{dt} = 0$, or

$$\frac{di_1}{dt} = \frac{di_2}{dt} + \frac{1}{2}\frac{d^2 i_2}{dt^2} + 2\cos 2t$$

(3) $\int i_1 dt + 1\cdot i_1 + (i_1 - i_2)\cdot 1 = \frac{1}{2}(i_2 - i_1)$

or $\int i_1 dt + \frac{5}{2} i_1 - \frac{3}{2} i_2 = 0$

Differentiate both sides of (3),

$i_1 + \frac{5}{2}\frac{di_1}{dt} - \frac{3}{2}\frac{di_2}{dt} = 0$, substitution into (2)

$$\left(i_2 + \frac{1}{2}\frac{di_2}{dt} + \sin 2t\right) + \frac{5}{2}\left(\frac{di_2}{dt} + \frac{1}{2}\frac{d^2 i_2}{dt^2} + 2\cos 2t\right) - \frac{3}{2}\frac{di_2}{dt} = 0, \text{ or}$$

$$\frac{5}{4}\frac{d^2 i_2}{dt^2} + \frac{3}{2}\frac{di_2}{dt} + i_2 = -\sin 2t - 5\cos 2t$$

Try $i_{2f} = A\sin 2t + B\cos 2t$

$$\frac{di_{2f}}{dt} = 2A\cos 2t - 2B\sin 2t$$

$$\frac{d^2 i_f}{dt^2} = -4A\sin 2t - 4B\cos 2t$$

$$\frac{5}{4}[-4A\sin 2t - 4B\cos 2t] + \frac{3}{2}[2A\cos 2t - 2B\sin 2t] + [A\sin 2t + B\cos 2t] = -\sin 2t - 5\cos 2t$$

Equating coefficients, $\begin{cases} -5A - 3B + A = -1 \\ -5B + 3A + B = -5 \end{cases}$

$A = -\frac{11}{25}, \quad B = \frac{23}{25}$

Hence $i_{2f} = -\frac{11}{25}\sin 2t + \frac{23}{25}\cos 2t$

$$\frac{di_{2f}}{dt} = -\frac{22}{25}\cos 2t - \frac{46}{25}\sin 2t$$

$$i_{1f} = i_{2f} + \frac{1}{2}\frac{di_{2f}}{dt} + \sin 2t = -\frac{19}{25}\sin 2t + \frac{12}{25}\cos 2t$$

$$i_f = i_{2f} - i_{1f} = \frac{8}{25}\sin 2t + \frac{11}{25}\cos 2t$$

7.28 Nodal analysis gives

(1) $\frac{1}{3}\frac{dv_1}{dt} + \frac{v_1 - v_2}{6} = 0$, or $2\frac{dv_1}{dt} + v_1 = v_2$, or

$$2\frac{d^2 v_1}{dt^2} + \frac{dv_1}{dt} = \frac{dv_2}{dt},$$

(2) $\frac{v_2 - v_1}{6} + \frac{v_2 - 10}{4} + \frac{1}{8}\frac{d(v_2 - \frac{11}{3}v_1)}{dt} = 0$, or

$$10v_2 - 4v_1 - 60 + 3\frac{dv_2}{dt} - 11\frac{dv_1}{dt} = 0$$

substitute v_2 and $\frac{dv_2}{dt}$ into (2)

$6\frac{d^2 v_1}{dt^2} + 12\frac{dv_1}{dt} + 6v_1 = 60$, $s^2 + 2s + 1 = 0$, $s = -1$

with forced response 10V.

$v_1 = k_1 e^{-t} + k_2 t e^{-t} + 10$

$\because$ zero initial stored energy,

$v_1(0) = v_2(0) = 0$, $v_1(0) = k_1 + 10 = 0$, $k_1 = -10$,

$v_2 = 2\frac{dv_1}{dt} + v_1 = (2k_2 - k_1)e^{-t} - k_2 t e^{-t} + 10$

$v_2(0) = 2k_2 - k_1 + 10 = 0, \quad k_2 = 10$

Hence $v_1(t) = 10 - 10e^{-t} + 10te^{-t}$ V

$i(t) = \frac{11}{3}\frac{v_1(t)}{10} = \frac{1}{3}(1 - e^{-t} + te^{-t})$A

7.29 KCL gives $1\cdot\frac{d(v_1 + \cos 3t)}{dt} + \frac{v_1}{1} = 0$,

or $\frac{dv_1}{dt} + v_1 = 3\sin 3t$ —(1)

$\frac{0 - v_1}{1} + \frac{0 - v}{\frac{1}{3}} + \frac{d(0 - v)}{dt} = 0$, or $v_1 = -3v - \frac{dv}{dt}$ —(2)

$\frac{dv_1}{dt} = -3\frac{dv}{dt} - \frac{d^2 v}{dt}$, substitute v_1 and

$\frac{dv_1}{dt}$ into (1), $-3\frac{dv}{dt} - \frac{d^2 v}{dt} - 3v - \frac{dv}{dt} = 3\sin 3t$,

or $\frac{d^2 v}{dt} + 4\frac{dv}{dt} + 3v = -3\sin 3t$.

Because the excitation is $-\cos 3t\, u(t)$,

$v(0) = \left.\frac{dv(0)}{dt}\right|_0 = 0$. The describing equation with the initial condition was solved in problem 7.14

From the solution of Problem 7.14

$v = \frac{1}{12}e^{-3t} - \frac{9}{60}e^{-t} + \frac{1}{30}\sin 3t + \frac{1}{15}\cos 3t$ V

7.30 Let v_1 be the nodal voltage at the top of 1/4 capacitor, then KCL gives

$v_1\left(\frac{1}{4} + \frac{1}{4} + \frac{1}{2}\right) + \frac{1}{4}\frac{dv_1}{dt} - \frac{v}{4} - \frac{6}{4} = 0$, or

$4v_1 + \frac{dv_1}{dt} - v = 6$

Node analysis at inverting op amp terminal gives: $\frac{v_1}{2} + \frac{1}{10}\frac{dv}{dt} = 0$,

or $v_1 = -\frac{1}{5}\frac{dv}{dt}$

7.30 Cont.

Substituting for v_1 in the first equation.

$$\frac{d^2v}{dt^2}+4\frac{dv}{dt}+5v=-30$$

Since $\frac{d^2v_f}{dt^2}=\frac{dv_f}{dt}=0$, $5v_f=-30$;

The characteristic equation is

$s^2+4s+5=0$; $s_1=-2\pm j$

$v=e^{-2t}(A_1\cos t+A_2\sin t)-6$

$$\frac{dv(0)}{dt}=-5v_1(0)=0$$

$v(0)=0=A_1-6 \Rightarrow A_1=6$

$$\frac{dv(0)}{dt}=-2A_1+A_2=0 \Rightarrow A_2=12$$

$\therefore v=\underline{e^{-2t}(6\cos t+12\sin t)-6\text{ V}}$

7.31 By KCL at the two terminals of 10kΩ resistor (1) $-10^{-6}\frac{dv_1}{dt}-\frac{v_1}{10^4}=0$, or $\frac{dv_1}{dt}+10^2v_1=0$,

$v_1=Ae^{-100t}$, (2) $\frac{v_1}{10^4}-10^{-5}\frac{dv}{dt}=0$, or $\frac{dv}{dt}=10v_1$

$$v=\int_0^t 10v_1\,d\tau+v(0)=10A\int_0^t e^{-100t}\,d\tau+v(0)$$

$$=-\frac{A}{10}e^{-100t}\Big|_0^t+v(0)$$

$$=-\frac{A}{10}e^{-100t}+\frac{A}{10}+v(0)$$

Since $v_1(t)=Ae^{-100t}$, $v_1(0)=A=5\text{ V}$

$v(t)=-\frac{5}{10}e^{-100t}+\frac{5}{10}+0=\underline{-0.5e^{-100t}+0.5\text{ V}}$

7.35 The characteristic equation

$s^3+7s^2+14s+8=0$

$(s+1)(s+2)(s+4)=0$, $s=-1,-2,-4$

Let the natural response be

$X=A_1e^{-t}+A_2e^{-2t}+A_3e^{-4t}$

The forced response is

$8X_f=24$ or $X_f=3$

The total response is

$X=\underline{A_1e^{-t}+A_2e^{-2t}+A_3e^{-4t}}$

7.32

```
*PROB.7_32
*CURRENT OUTPUT
R_R1      2 1 4
R_R2      1 3 6
R_R3      4 0 10
C_C1      1 4 .125
C_C2      3 0 .333
E_E1      4 0   3 0 3.33
V_V1      2 0
+PWL 0 0 0.001 10 10 10
*Control statement
.TRAN     1S 9.5S
.PRINT TRAN I(R_R3)
.END

*THE PLOT OF OUTPUT CURRENT
* I(R_R3) IS:
```

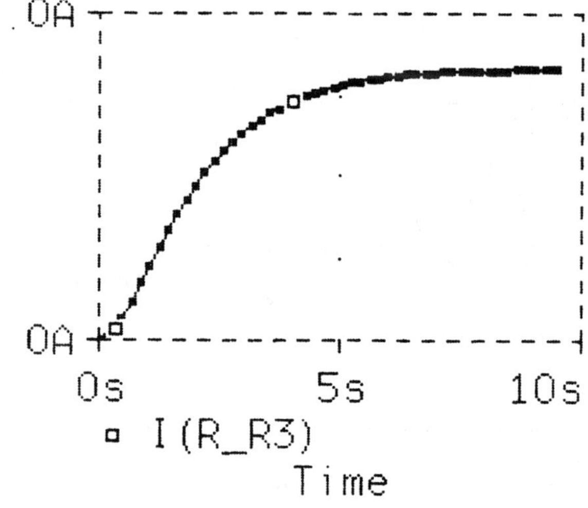

7.33

```
*PROB.7_33
* Schematics Netlist *
X_U1     0 1 2 OPAMP
X_U2     0 3 4 OPAMP
X_U3     6 7 5 OPAMP
R_R1     5 1 10k
R_R2     2 3 10k
R_R3     6 2 10k
R_R4     7 5 100k
R_R5     7 4 22k
R_R6     8 6 47k
C_C1     1 2 1u
C_C2     3 4 1u
V_V1     8 0 SIN (0 50 200 0 0 0
*Control statement
.INC C:\PS\OPAMP.LIB
.TRAN 10MS 100MS
*Output V(4)
.PRINT AC V(4)
.END
```

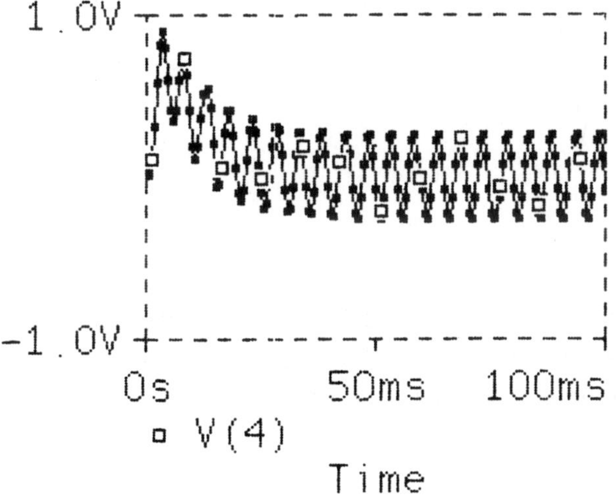

7.34

```
*PROB.7_34
*Schematics Netlist *
X_U1     0 1 2 OPAMP
X_U2     0 3 4 OPAMP
X_U3     0 5 6 OPAMP
R_R1     4 5 10k
R_R2     1 6 20k
R_R3     2 1 20k
R_R4     2 3 20k
C_C1     5 6 10u
C_C2     3 4 1u
.IC      V(3,4)=5V
*Control statement
.INC C:\PS\OPAMP.LIB
.OP
.TRAN 1S 10S
*Output V(6)
.PRINT AC V(6)
.END
```

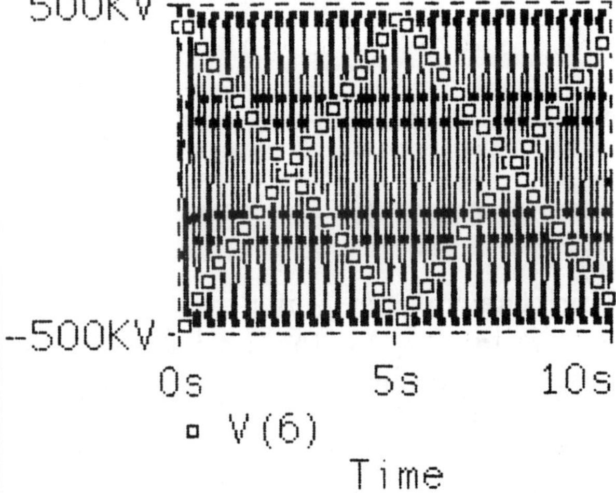

Chapter 8
Sinusoidal sources and phasors

8.1 Properties of sinusoids

8.1 **Consider two voltages $\nu_1 = 110\cos(120\pi t + 47^o)$ and $\nu_2 = -99\sin(120\pi t + 0.31)$. Find the amplitudes, frequencies, phase angles (in both degree and radian) and periods of the two signals. Determine which leads and by how many degrees ?**

8.2 **Find the quadrature representations of the two voltages of Prob.8.1.**

8.3 **Find i_4, using the properties of sinusoids, if $i_1 = 2\cos(3t)$A, $i_2 = 25\cos(3t + 36.9^o)$A, and $i_3 = 13\sin(3t + 157.4^o)$A.**

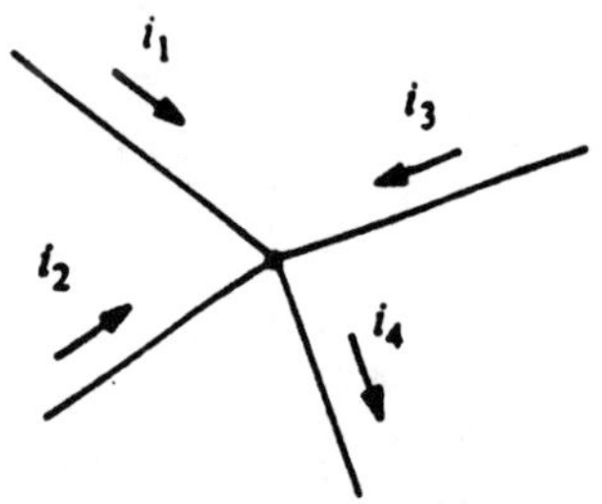

Problem 8.3

8.4 **Find the votage ν_1 if $\nu_2 = 5\cos(100\pi t\text{-}30^o)$V, ν_3 has the same amplitude as ν_2 but lags ν_2 by 15°, and ν_4 has no in-phase component but the quadrature component with amplitude 3. Assume all four signals have the same frequencies.**

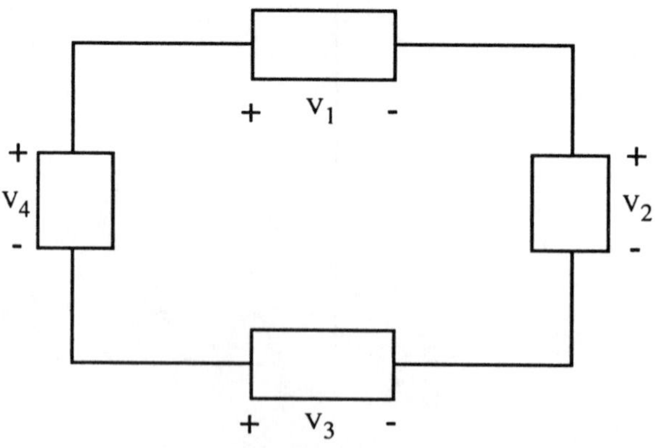

Problem 8.4

8.2 An RLC example

8.5 If the source is $10\cos(3000t)$mA and the output is $\nu = 8\cos(3000t - 36.9^o)$V, find R and C.

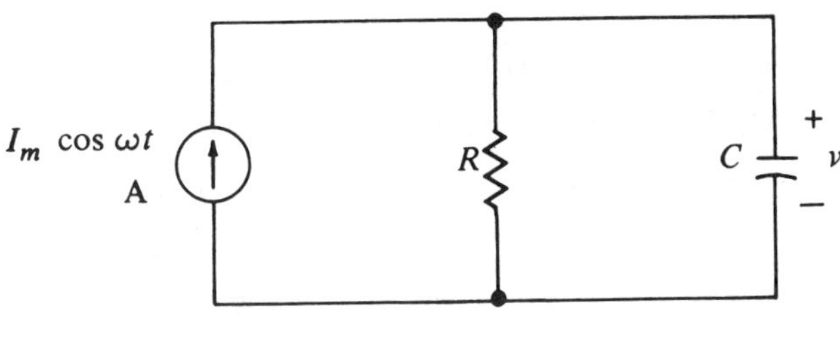

Problem 8.5

8.6 Write a differential equation for ν. Find the forced solution.

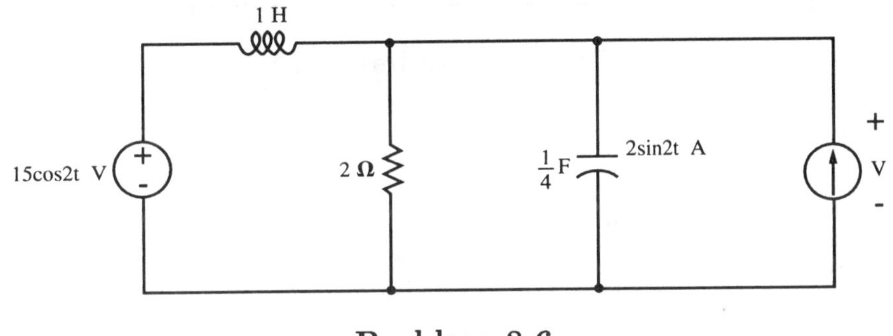

Problem 8.6

8.7 Solve Prob.8.5 if the independent current source is replaced by a new current source of $2\sin(3t)$A.

8.3 Complex sources

8.8 Express each complex number in rectangular form, polar form, and exponential polar form (a)$23\angle 19^o$ (b)$-1+j\sqrt{3}$ (c)$8e^{j45^o}$ (d)$36\angle 90^o$ (e)$4+j2$ (f)$0.25\angle -180^o$.

8.9 For each of the sinusoidal source functions shown, determine the complex exponential source which has the given source as it real part.
(a)$2\cos(3t)$ (b)$220\cos(100\pi t - 31^o)$ (c)$-\sin(7t + 44^o)$ (d)$\cos(4t + 20^o) + 2\sin(4t - 39^o)$
(e)$\cos(4t + 20^o) + 2\sin(6t - 39^o)$

8.10 For each of the sinusoidal source functions specified in Prob.8.9, determine the complex exponential source which has the given source as its imaginary part.

8.11 For the describing equation $\frac{dx}{dt} + 6x = 7\cos(2t)$, find the forced solution by replacing the sinusoidal forcing function by a complex exponential function. Check by using the trial forced solution $A\cos(7t) + B\sin(7t)$.

8.12 Solve the Prob.8.6 by replacing the sinusoidal sources $15\cos(2t)$**V** and $2\sin(2t)$**A** by complex sources.

8.13 Find the response ν_1 to the source $2e^{j8t}$**A** and use the result to find the response ν to (a)$2\cos(8t)$, and (b)$2\sin(8t)$**A**.

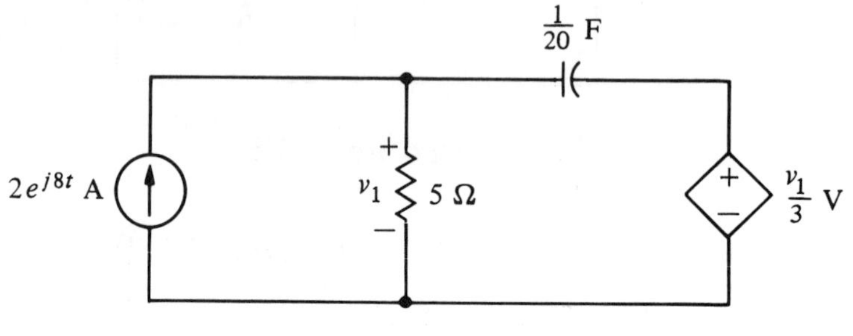

Problem 8.13

8.4 Phasors

8.14 Find the phasor representation of (a)$-4\sin(3t)$ (b)$7\cos(2t - 19^o)$ (c)$3\cos(5t - 26^o) + 4\sin(5t - 114^o)$.

8.5 Voltage-current relationship for phasors

8.15 If the phasor current is $2\angle 36^o$ mA, and $\omega = 10^4$rad/sec, find the phasor voltages V_R, V_C, R_L.

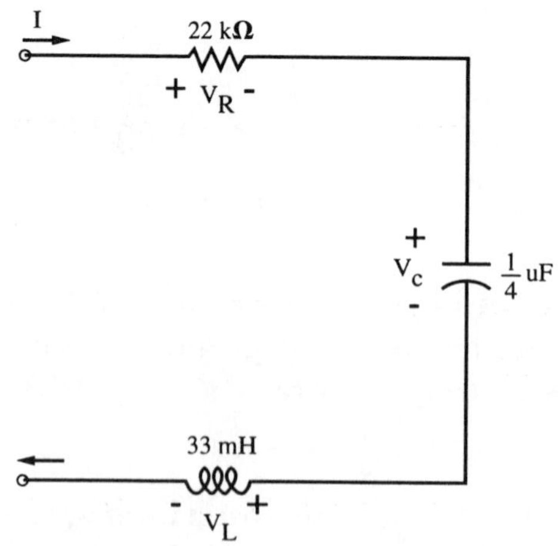

Problem 8.15

8.16 Find the conductance and susceptance if Z is (a)$12 + j5$ (b)$3 - j3$ (c)$5\angle 30^o$.

8.17 Find the resistance and reactance if Y is (a)$12 + j5$ (b)$3 - j3$ (c)$5\angle 30^o$.

8.6 Impedance and admittance

8.18 Find the impedance of the circuit shown if the time domain functions are (a)$\nu = -15\cos(2t) + 8\sin(2t)$V, $i = 1.7\cos(2t + 40^o)$A (b)$\nu = Re[je^{j2t}]$V, $i = Re[(1+j)e^{j(2t+30^o)}]$mA (c)$\nu = aV_m\cos(wt + \theta)$V, $i = V_m\cos(wt + \theta - \alpha)$A.

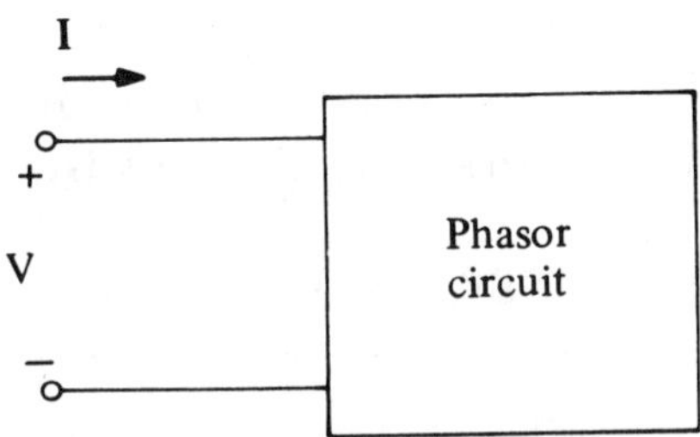

Problem 8.18

8.19 Solve Prob.8.4 by converting all voltages to phasor voltages and Kirchoff's voltage law.

8.20 For the circuit shown, find the impedance as a function of frequency ω.

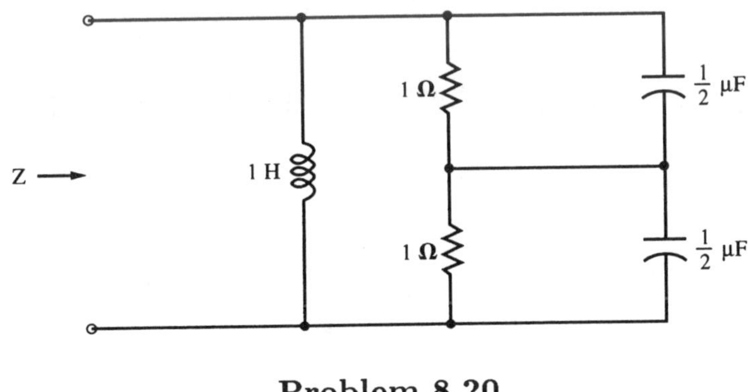

Problem 8.20

8.21 At what frequency is the conductance G of the two-terminal circuit of Prob.8.20 equal to $\frac{1}{2}$s.

8.7 Kirchoff's laws and impedance combinations

8.22 Find the forced current i using phasors and Kirchoff's Laws if $L = 1\text{H}$.

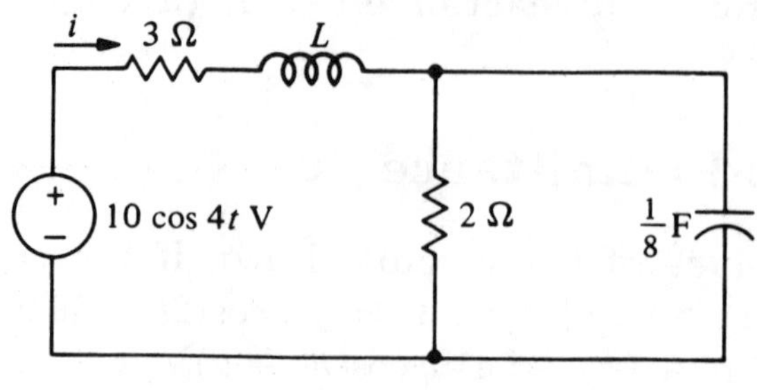

Problem 8.22

8.23 Determine L in Prob.8.22 so that the impedance seen by the source is real, and for this case find the power delivered by the source at $t = \frac{\pi}{4}$ sec.

8.24 Find the steady-state voltage v using phasors and voltage division.

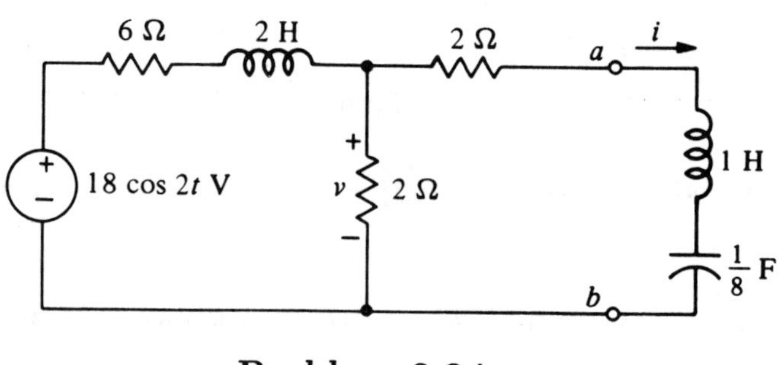

Problem 8.24

8.25 Find the forced response i of Prob.8.24 using phasors and current division.

8.8 Phasor circuit

8.26 Find the steady-state values of i and v.

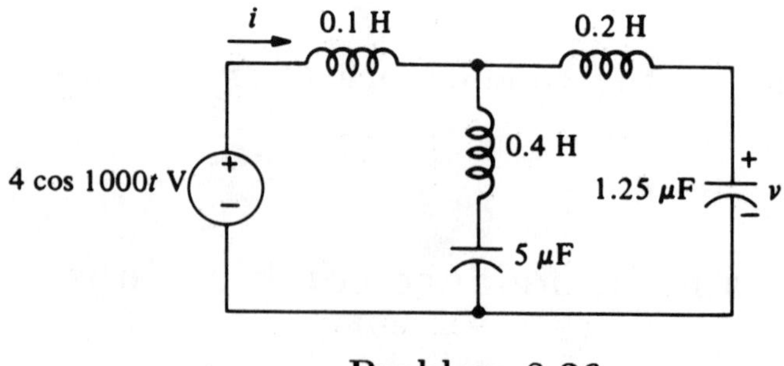

Problem 8.26

8.27 Find the steady-state value of v_x.

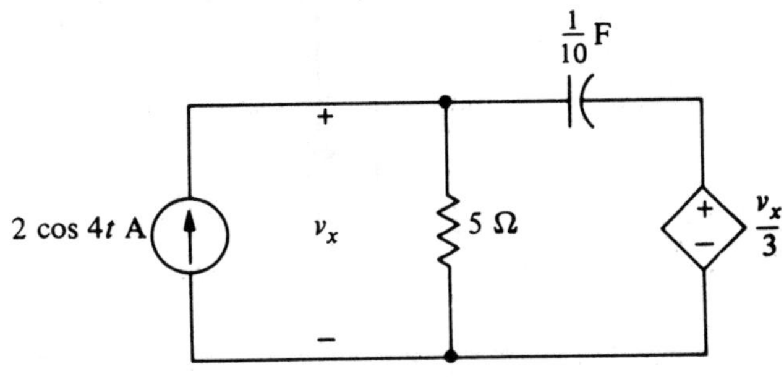

Problem 8.27

8.28 Find the steady-state value of v.

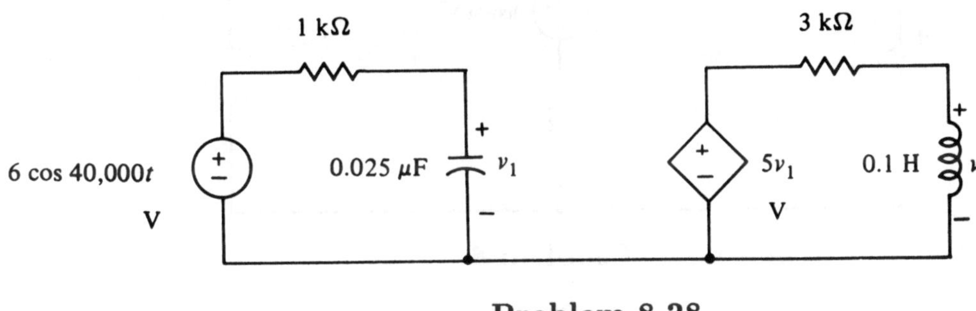

Problem 8.28

8.29 Find the steady-state voltage v if $v_g = 10\cos(10000t)$ V.

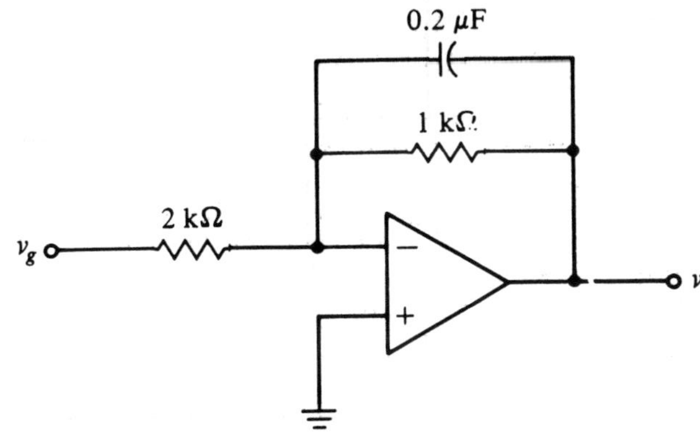

Problem 8.29

8.30 Find the ac steady-state current i_1 and i_2.

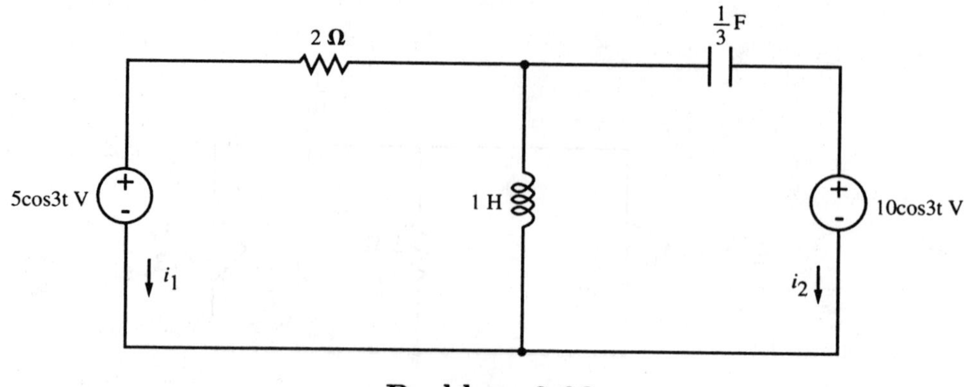

Problem 8.30

8.31 Find the ac steady-state voltage ν.

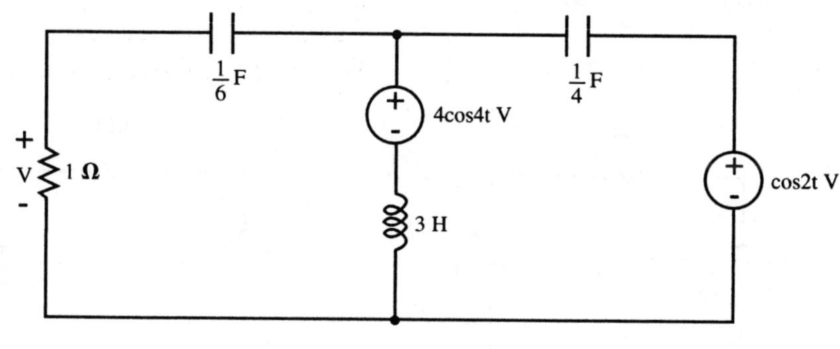

Problem 8.31

8.32 Find the input impedance seen by the voltage source, given $C = 1\mu$F, $R_1 = R_2 = 1$KΩ. Express the input impedance in terms of R, L and C elements.

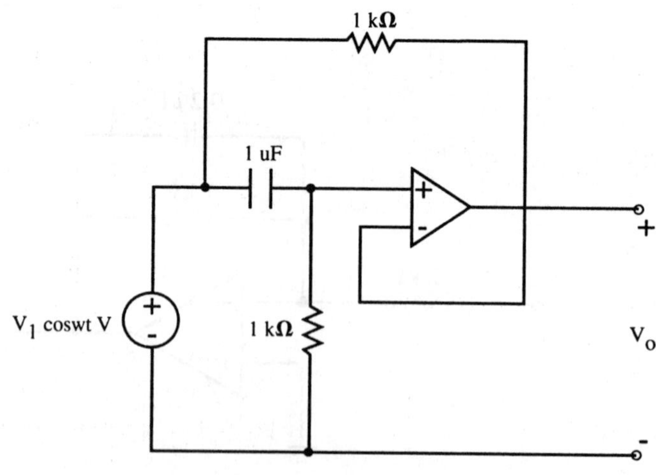

Problem 8.32

8.1 Amplitude: $v_1 = 110,\ v_2 = -99$

frequency: $v_1 = 60Hz,\ v_2 = 60Hz$

Phase angle: In degree $v_1 = 47°,\ v_2 = 17.7°$

In radians $v_1 = 0.82,\ v_2 = 0.31$

Period: $v_1 = v_2 = \frac{1}{60} = 0.01675$

$47 - 17.7 + 90 = 119.3°$

v_1 leads v_2 by $29.3°$

8.2 $v_1 = 110\cos(120\pi t + 47°)$

$= 110\cos(12\pi t + 0.82)$

$= 110\cos(0.82) * \cos(120\pi t) - 110\sin(0.82) \times \sin(120\pi t)$

$= 109.99\cos(120\pi t) - 1.57\sin(120\pi t)$

$v_2 = -99\sin(120\pi t + 0.31)$

$= -99\cos(120\pi t + 0.31 - \frac{\pi}{2})$

$= -99\cos(-1.26)\cos(120\pi t) - 99\sin(-1.26)\sin(120\pi t)$

$= -98.98\cos(120\pi t) + 2.18\sin(120\pi t)$

8.3 By KCL, $i_4 = i_1 + i_2 + i_3$

$i_4 = 25[\cos 3t \cos 36.9° - \sin 3t \sin 36.9°]$
$+ 13[\sin 3t \cos 157.4° + \cos 3t \sin 157.4°]$
$+ 2\cos 3t$

$= 27\cos 3t - 27\sin 3t$

$= 27\sqrt{2}\cos(3t + 45)A$

8.4 By KVL

$v_1 + v_2 = v_3 + v_4$

$v_1 + 5\cos(100\pi t - 30°) = 5\cos(100\pi t - 45°) + 3\sin(100\pi t)$

$v_1 + 5[\frac{\sqrt{3}}{2}\cos(100\pi t) + \frac{1}{2}\sin(\pi t)]$

$= 5[\frac{1}{\sqrt{2}}\cos(100\pi t) + \frac{1}{\sqrt{2}}\sin(100\pi t)]$
$+ 3\sin(100\pi t)$

$v_1 = -0.79\cos(100\pi t) + 4.04\sin(100\pi t)$

$= 4.11\cos(100\pi t + 79°)$

8.5 $C\frac{dv}{dt} + \frac{v}{R} = I_m \cos\omega t$

$\frac{dv}{dt} = 8(-3000\sin(3000t - 36.9°))$

8.5 Cont.

$-24000C\sin(\omega t - 36.9°) + \frac{8}{R}\cos(\omega t - 36.9°)$

$= \sqrt{\frac{64}{R^2} + C^2 576\times10^6} - \cos[\omega t - 36.9 - \tan^{-1}(-RC\times 3\times10^3)]$

$\therefore I_m = 10\times10^{-3} = \frac{\sqrt{64 + R^2C^2 \times 576\times10^6}}{R}$ or

$R = \frac{\sqrt{64 + (RC)^2 \times 576\times10^6}}{10\times10^{-3}}$

$\phi = 0 = 36.9° + \tan^{-1}[-(RC)(3\times10^3)]$ or

$RC = \frac{\tan(-36.9°)}{-3\times10^3} = 250\times10^{-6}$

$\therefore R = 1k\Omega,\ C = \frac{250\times10^{-6}}{R} = 250nF$

8.6 $\int_{t_0}^{t}(v - 15\cos 2t)d\tau + i(t_0) + \frac{v}{2} + \frac{1}{4}\frac{dv}{dt} = 2\sin 2t$

$\frac{1}{4}\frac{d^2v}{dt^2} + \frac{1}{2}\frac{dv}{dt} + v = 19\cos 2t$ —(1)

Try forced solution $v_f = A\cos 2t + B\sin 2t$

$\frac{dv}{dt} = -2A\sin 2 + 2B\cos 2t$

$\frac{d^2v}{dt^2} = -4A\cos 2t - 4B\sin 2t$

substitute v, $\frac{dv}{dt}$ and $\frac{d^2v}{dt^2}$ into (1)

$-A\sin 2t + B\cos 2t = 19\cos 2t$,

Equating coefficients: $A = 0,\ B = 19$.

Hence the forced solution is

$v = 19\sin 2t\ V$

8.7 The differential equation with the new current source is:

$\frac{1}{4}\frac{d^2v}{dt^2} + \frac{1}{2}\frac{dv}{dt} + v = 6\cos 3t - 15\sin 2t$

Try the forced solution:

$v = A_1\cos 2t + A_2\sin 2t + B_1\cos 3t + B_2\sin 3t$

$\frac{dv}{dt} = -2A_1\sin 2t + 2A_2\cos 2t - 3B_1\sin 3t + 3B_2\cos 3t$

$\frac{d^2v}{dt^2} = -4A_1\cos 2t - 4A_2\sin 2t - 9B_1\cos 3t - 9B_2\sin 3t$

substitute v, $\frac{dv}{dt}$ and $\frac{d^2v}{dt^2}$ into the describing equation.

$-A_1\sin 2t + A_2\cos 2t + (-\frac{5}{4}B_1 + \frac{3}{2}B_2)\cos 3t$
$+ (-\frac{5}{4}B_2 - \frac{3}{2}B_1)\sin 3t = 6\cos 3t - 15\sin 2t$

Equating coefficients:

$A_1 = 15,\ A_2 = 0,\ B_1 = -\frac{12}{61},\ B_2 = \frac{144}{61}$

$v = 15\cos 2t - \frac{120}{61}\cos 3t + \frac{144}{61}\sin 3t\ V$

8.8 (a) polar form: $23\angle 19°$
exponential form: $23e^{j19°}$
rectangular form: $21.7+j7.49$

(b) polar form: $2\angle 120°$
exponential form: $2e^{j120°}$
rectangular form: $-1+j\sqrt{3}$

(c) Polar form: $8\angle 45°$
exponential form: $8e^{j45}$
rectangular form: $4\sqrt{2}+j4\sqrt{2}$

(d) polar form: $36\angle 90°$
exponential form: $36e^{j90°}$.
rectangular form: $j36$

(e) polar form: $\sqrt{20}\angle 26.6°$
exponential: $\sqrt{20}e^{j26.6°}$
rectangular form: $4+j2$

(f) Polar form: $0.25\angle -180°$
exponential form: $0.25e^{j(-180°)}$
rectangular form: -0.25

8.9 (a) $2e^{j3t}$

(b) $220e^{j(100\pi t-31°)}$

(c) $-e^{j(7t-46°)}$

(d) $e^{j(4t+20°)}+2e^{j(4t-129°)}=1.25e^{j(4t+75°)}$

(e) $e^{j(4t+20°)}+2e^{j(6t-129°)}$

8.10 (a) $2e^{j(3t+90°)}$

(b) $220e^{j(100\pi t+59°)}$

(c) $-e^{j(7t+44°)}$

(d) $\cos(4t+20°)+2\sin(4t-39°)$ is the real part of $1.25e^{j(4t+75°)}$
$\therefore$ $\cos(4t+20°)+2\sin(4t-39°)$ is the imaginary part of $1.25e^{j(4t+165°)}$

(e) $e^{j(4t+110°)}+2e^{j(6t-39°)}$

8.11 Try the complex exponential function: $x=Ae^{j2t}$, $\frac{dx}{dt}=j2Ae^{j2t}$

$j2Ae^{j2t}+6Ae^{j2t}=7e^{j2t}$

$A(6+j2)=7$, or $A=\frac{7}{6+j2}=\frac{21-j7}{20}$

$x=\frac{21-j7}{20}Ae^{j2t}=1.11e^{j(2t-18.4°)}$

Hence $x=1.11\cos(2t-18.4°)$

8.12 Replace $15\cos 2t$ by $15e^{j2t}$ V
$2\sin 2t$ A by $2e^{j(2t-90°)}$

$\int_{t_0}^{t}(v-15e^{j2t})dz+i(t_0)+\frac{v}{2}+\frac{1}{4}\frac{dv}{dt}=2e^{j(2t-90°)}$

$\frac{1}{4}\frac{d^2v}{dt^2}+\frac{1}{2}\frac{dv}{dt}+v=19e^{j2t}$

substitute the trial form $v=Ae^{j2t}$

$\frac{dv}{dt}=j2Ae^{j2t}$, $\frac{d^2v}{dt^2}=-4Ae^{j2t}$

$-Ae^{j2t}+jAe^{j2t}+Ae^{j2t}=19e^{j2t}$

$jA=19$, $A=-j19=19\angle -90°$.

Hence the forced response

$v=(19\angle -90°)e^{j2t}=19e^{j(2t-90°)}$

$Re[19e^{j2t-90°}]=19\cos(2t-90°)$
$=19\sin 2t$ V

8.13 KCL: $\frac{v_1}{5}+\frac{1}{20}\frac{d}{dt}(v_1-\frac{v_1}{3})=2e^{j8t}$, or

$\frac{dv_1}{dt}+6v_1=60e^{j8t}$. Try $v_1=Ae^{j8t}$

then $Ae^{j8t}(j8t\quad)=60e^{j8t}$

$A=\frac{60}{60+j8}=6e^{-j53.1°}$

$\therefore$ $v_1=6e^{j(8t-53.1°)}$ V

(a) $v=Re\,v_1=6\cos(8t-53.1°)$ V

b) $v=Im\,v_1=6\sin(8t-53.1°)$ V

8.14 (a) $-4\sin 3t=-4\cos(3t+90°)$
phasor $=-4\angle 90°$.

(b) $7\angle 19°$

(c) $3\cos(5t-26°)+4\sin(5t-114°)$
$=-0.94\cos(5t)-0.35\sin(5t)$
$=-1.003\cos(5t-20.4°)$

phasor $=-1.003\angle -20.4°$

8.15 $V_R = IZ_R = (2\times10^{-3}\angle 36°)(22\times10^{4}) = 44\angle 36°$ V

$V_C = IZ_C = (2\times10^{-3}\angle 36°)\left(\frac{1}{j10^4\times\frac{1}{4}\times10^{-6}}\right)$

$= (2\times10^{-3}\angle 36°)(4\times10^{2}\angle -90°) = 0.8\angle -54°$ V

$V_L = IZ_L = (2\times10^{-3}\angle 36°)(j10^4\times35\times10^{-3})$

$= (2\times10^{-3}\angle 36°)(0.35\angle 90°) = 0.7\angle 126°$

8.16 (a) $Y = G + jB = \frac{1}{12+j5} = 0.071 - j0.030$

(b) $Y = G + jB = \frac{1}{3-j3} = 0.167 + j0.167$

(c) $Y = G + jB = \frac{1}{5\angle 30°} = 0.173 - j0.100$

8.17 (a) $Z = R + jX = \frac{1}{12+j5} = 0.071 - j0.030$

(b) $Z = R + jX = \frac{1}{3-j3} = 0.167 + j0.167$

(c) $Z = R + jX = \frac{1}{5\angle 30°} = 0.173 - j0.100$

8.18 (a) $Z = V/I = \frac{-15-j8}{1.7\angle 40°} = 10\angle 111.9\ \Omega$

(b) $v = \text{Re}[j(\cos 2t + j\sin 2t)] = -\sin 2t$

$= \cos(2t + 90°)$ V

$i = \text{Re}\{(1+j)[\cos(2t+30°) + j\sin(2t+30°)]\}$

$= \cos(2t+30°) - \sin(2t+30°)$

$= \sqrt{2}\cos(2t+75°)$ mA

$Z = \frac{V}{I} = \frac{1\angle 90°}{\sqrt{2}\angle 75°} = \frac{1}{\sqrt{2}}\angle 15°\ k\Omega$

(c) $Z = \frac{aV_m\angle\theta}{V_m\angle\theta - \alpha} = a\angle\alpha\ \Omega$

8.19 $V_1 + 5\angle -30° = 5\angle -45° + 3\angle 90°$

$V_1 = -5\angle -30° + 5\angle -45° + 3\angle 90°$

$= -5\left(\frac{\sqrt{3}}{2} - j\frac{1}{2}\right) + 5\left(\frac{\sqrt{2}}{2} - j\frac{\sqrt{2}}{2}\right) + j(-3)$

$= -0.795 + j(-4.035)$

$= 4.05\angle 79°$

8.20 $Z_{R-C} = 2\left(1 // \frac{1}{j\omega\frac{1}{2}}\right) = 2\left(1 // \frac{2}{j\omega}\right) = 2\left(\frac{2}{2+j\omega}\right)$

$Z = (Z_L // Z_{R-C}) = \frac{j4\omega}{4 + j2\omega + \omega^2} = \frac{8\omega^2 + j(4\omega^3 + 16\omega)}{\omega^4 + 12\omega^2 + 16}$

8.21 $Z = \frac{j4\omega}{4 + j2\omega + \omega^2}$,

$Y = \frac{1}{Z} = \frac{4 + \omega^2 + j2\omega}{j4\omega} = \frac{2\omega - j(4+\omega^2)}{4\omega^2}$

$G = \frac{2\omega}{4\omega^2} = \frac{1}{2} \Rightarrow 2\omega = \omega^2 \Rightarrow$

$\omega = 1$ rad/sec

8.22 The impedance seen by the source is:

$Z = 3 + j(4)(1) + \frac{2}{1 + j(4)(2)(\frac{1}{8})} = 4 + j3$

$= 5\angle 36.9°$

$I = \frac{V}{Z} = \frac{10}{5\angle 36.9°} = 2\angle -36.9°$ A

$i = 2\cos(4t - 36.9°)$ A

8.23 $Z = 3 + j4L + 1 - j = 4 + j(4L - 1)$

$\therefore 4L - 1 = 0 \Rightarrow L = 1/4$ H

$I = \frac{1}{4} = 2.5\angle 0°$ A

$\therefore i = 2.5\cos 4t$ A, $v = 10\cos 4t$

$p = vi = 2.5\cos[4(\frac{\pi}{2})] \times 10\cos[4(\frac{\pi}{4})]$

$= 25$ W at $t = \pi/4$ s

8.24 By voltage division

$V = \frac{\frac{(2)(2+j2-j4)}{2+2-j2}}{6 + j4 + 1.2 - j0.4}(18) = 2\sqrt{2}\angle -45°$

$v = 2\sqrt{2}\cos(2t - 45°)$ V

8.25 Phasor circuit diagram

18; 2; j4; 2Ω; 2Ω; j2; -j4; i

$Z = 2 + j4 + [2 // (2 + j2 - j4)]$

$= 2 + j4 + [2 // (2 - j2)] = 2 + j4 + \left[\frac{2(2-j2)}{2+2-j2}\right]$

$= 2 + j4 + \frac{2-j2}{2-j}$

$I = \left(\frac{18}{2 + j4 + \frac{2-j2}{2-j}}\right)\left(\frac{2}{4-j2}\right) = \frac{18}{(2+j4)(2-j) + (2-j2)}$

$= \frac{9(6-j)}{37} = 1.47\angle -9.5°$

8.26 The impedance seen by the source is

$$Z = j100 + \frac{(j400-j200)(j200-j800)}{j400-j200+j200-j800} = j400\,\Omega$$

$$I = \frac{4\angle 0^\circ}{Z} = \frac{4}{j400}\,A = 10\angle -90^\circ\,mA$$

$i = 10\cos(1000t - 90^\circ) = \underline{10\sin 1000t\ mA}$

The current through the 0.2 H inductor is

$$I_1 = \frac{j400-j200}{j400-j200+j200-j800} = j5mA$$

$V = -j800\,I = 4\angle 0^\circ$

$v = \underline{4\cos 1000t\ V}$

8.27 By KCL

$$V_x\left(\frac{1}{5}+j\frac{4}{10}\right) - \frac{V}{3}\left(j\frac{4}{10}\right) = 2 \text{ or}$$

$$V_x = \frac{60}{6+j8} = 6\angle -53.13^\circ$$

$\underline{v_x = 6\cos(4t - 53.13^\circ)\ V}$

8.28 By voltage division

$$V_1 = \frac{6}{1+j(40000)(10^3)(0.025\times10^{-6})} = 3\sqrt{2}\angle -45^\circ,$$

and $$V = \frac{j(40000)(0.1)(5V)}{3\times10^3 + j4000} = 12\sqrt{2}\angle -8.13^\circ$$

$v = \underline{12\sqrt{2}\cos(40{,}000t - 8.13^\circ)\,V}$

8.29 KCL at the inverting input terminal of the op amp yields

$$(j2+1)V + \frac{V_g}{2} = 0$$

$$V = \frac{1}{2(1+j2)}V_g = \frac{5}{\sqrt{5}}\angle -63.4^\circ + 180^\circ$$

$$= \sqrt{5}\angle 116.6^\circ$$

$\underline{v = \sqrt{5}\cos(1000t + 116.6^\circ)\ V}$

8.30 Phasor diagram

$$\begin{cases} 2I_1 + j3(I_1 - I_2) = 5 \\ -jI_2 + 10 + (I_2 - I_1)j3 = 0 \end{cases}$$

or

$$\begin{cases} (2+j3)I_1 - (j3)I_2 = 5 \\ (-j3)I_1 + (-j+j3)I_2 = -10 \end{cases}$$

$$I_1 = \frac{-16-j12}{5} = 4\angle 36.9^\circ \qquad I_2 = \frac{j7-24}{5} = 5\angle -16.3^\circ$$

$\underline{i_1 = -4\cos(3t + 36.9^\circ)A}$, $\underline{i_2 = 5\cos(3t - 16.3^\circ)A}$

8.31 By superposition

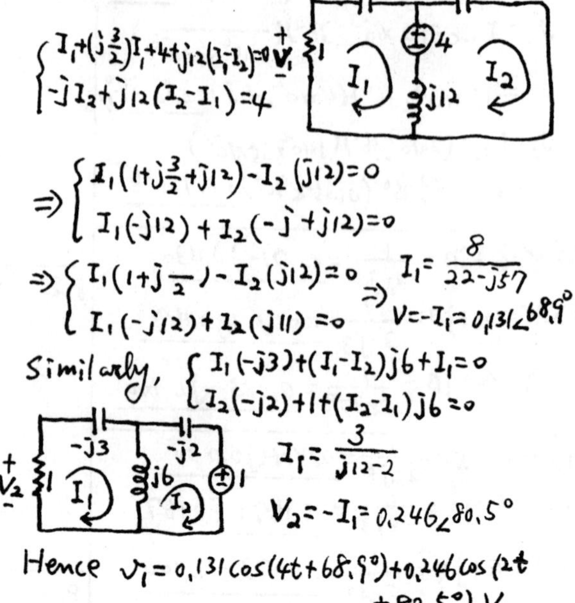

$$\begin{cases} I_1 + (j\tfrac{3}{2})I_1 + 4 + j12(I_1 - I_2) = 0 \\ -jI_2 + j12(I_2 - I_1) = 4 \end{cases}$$

$$\Rightarrow \begin{cases} I_1(1 + j\tfrac{3}{2} + j12) - I_2(j12) = 0 \\ I_1(-j12) + I_2(-j + j12) = 0 \end{cases}$$

$$\Rightarrow \begin{cases} I_1(1 + j\tfrac{27}{2}) - I_2(j12) = 0 \\ I_1(-j12) + I_2(j11) = 0 \end{cases} \Rightarrow I_1 = \frac{8}{22 - j57}$$

$V = -I_1 = 0.131\angle 68.9^\circ$

Similarly, $$\begin{cases} I_1(-j3) + (I_1 - I_2)j6 + I_1 = 0 \\ I_2(-j2) + 1 + (I_2 - I_1)j6 = 0 \end{cases}$$

$$I_1 = \frac{3}{j12 - 2}$$

$V_2 = -I_1 = 0.246\angle 80.5^\circ$

Hence $\underline{v_1 = 0.131\cos(4t + 68.9^\circ) + 0.246\cos(2t + 80.5^\circ)\ V}$

8.32 Phasor diagram

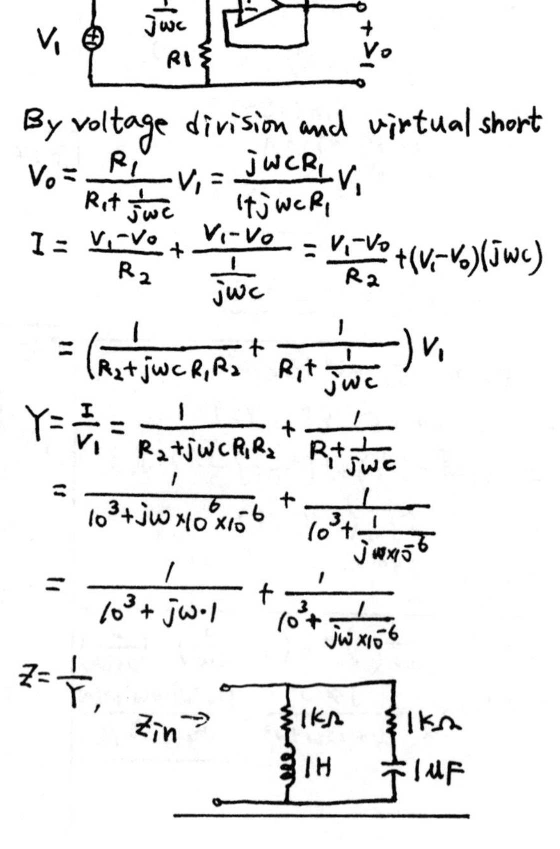

By voltage division and virtual short

$$V_o = \frac{R_1}{R_1 + \frac{1}{j\omega C}}V_1 = \frac{j\omega CR_1}{1 + j\omega CR_1}V_1$$

$$I = \frac{V_1 - V_o}{R_2} + \frac{V_1 - V_o}{\frac{1}{j\omega C}} = \frac{V_1 - V_o}{R_2} + (V_1 - V_o)(j\omega C)$$

$$= \left(\frac{1}{R_2 + j\omega CR_1R_2} + \frac{1}{R_1 + \frac{1}{j\omega C}}\right)V_1$$

$$Y = \frac{I}{V_1} = \frac{1}{R_2 + j\omega CR_1R_2} + \frac{1}{R_1 + \frac{1}{j\omega C}}$$

$$= \frac{1}{10^3 + j\omega\times10^6\times10^{-6}} + \frac{1}{10^3 + \frac{1}{j\omega\times10^{-6}}}$$

$$= \frac{1}{10^3 + j\omega\cdot 1} + \frac{1}{10^3 + \frac{1}{j\omega\times10^{-6}}}$$

$Z = \frac{1}{Y}$, $Z_{in} \rightarrow$ 1kΩ, 1H, 1kΩ, 1μF

Chapter 9
AC Steady-State Analysis

9.1 Circuit Simplification

9.1 **For the circuit shown, specify Z_1, Z_2, and Z_3 such that $V_1 = 2\sqrt{2}\angle 70^o$V, $V_2 = 2\sqrt{2}\angle 25^o$V, $V_3 = 2\sqrt{2}\angle -20^o$V.**

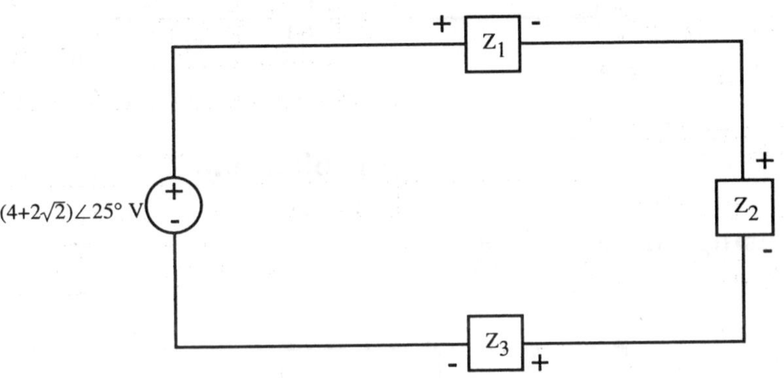

Problem 9.1

9.2 **For the circuit shown, specify y_1, y_2, and y_3 such that $I_1 = 6\angle 30^o$A, $I_2 = 6\angle -30^o$A, $I_3 = 6\angle 0^o$A.**

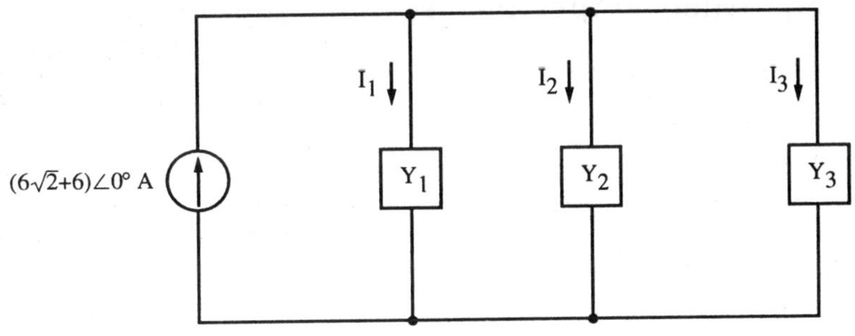

Problem 9.2

9.3 Find v_1 using voltage division.

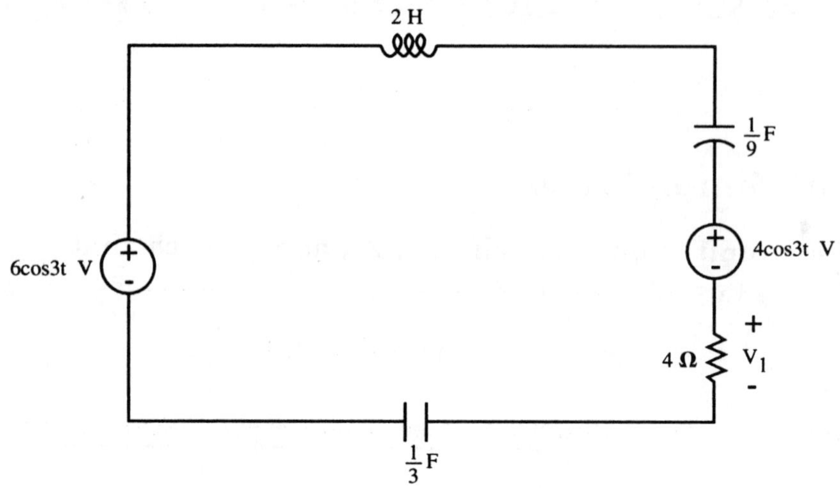

Problem 9.3

9.4 Find i_1 using current division.

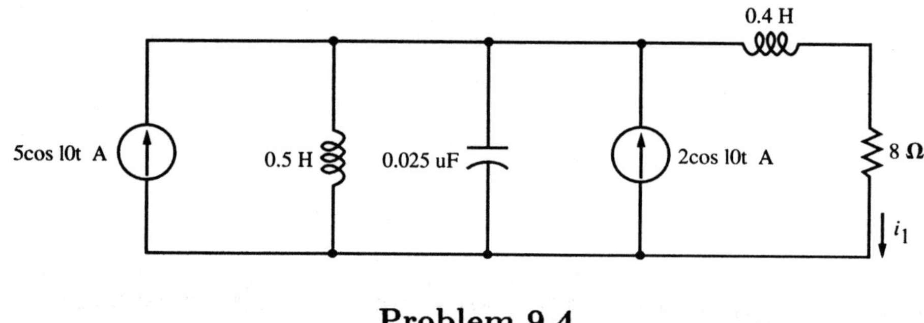

Problem 9.4

9.5 Find the equivalent impedance seen by the independent current source.

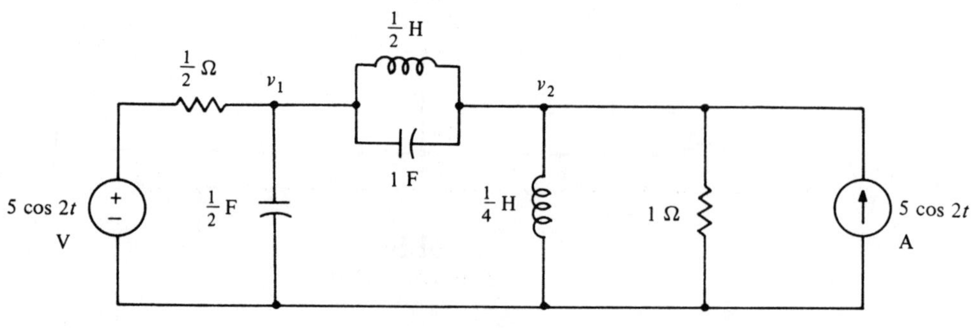

Problem 9.5

vsapce0.2in

9.6 Using proportionality to find the steady-state current i.

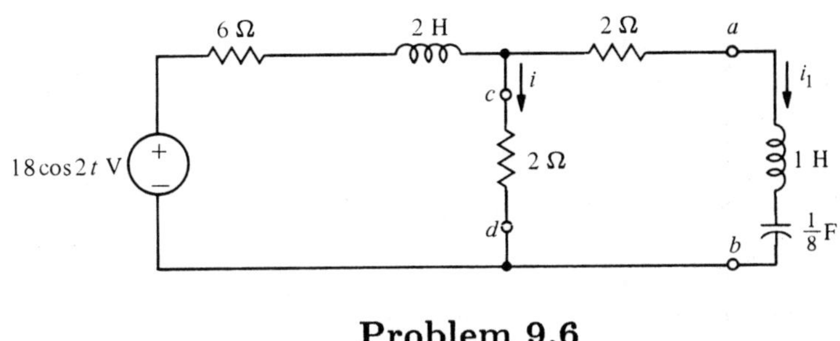

Problem 9.6

9.7 Using superposition principle to find the steady-state current i.

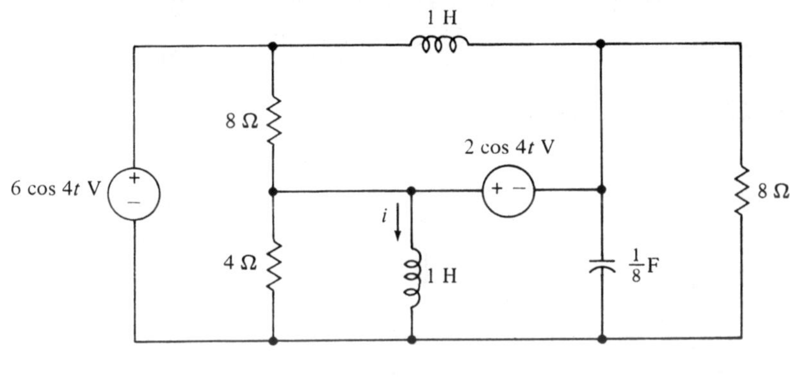

Problem 9.7

9.8 Find the Thevenin and Norton equivalents of the subcircuit shown.

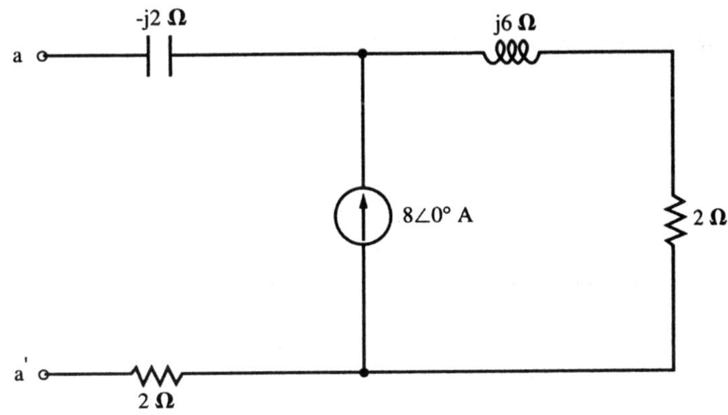

Problem 9.8

9.9 **Find the Thevenin and Norton equivalents of the subcircuit shown.**

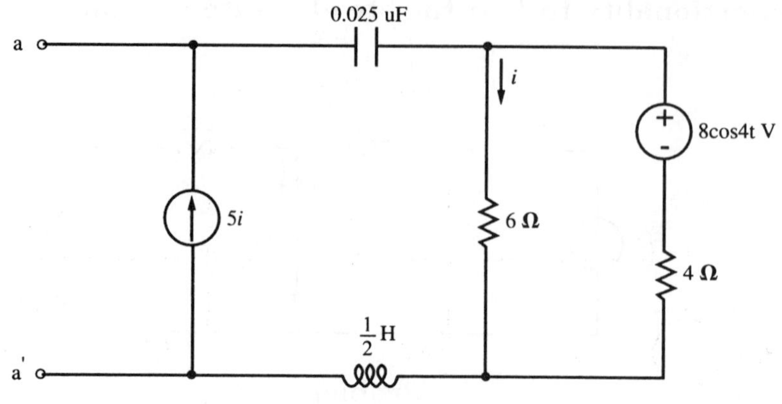

Problem 9.9

9.10 **Find the ac steady-state current i. (Hint: use the Thevenin and Norton equivalents to simplify the given network from right to left)**

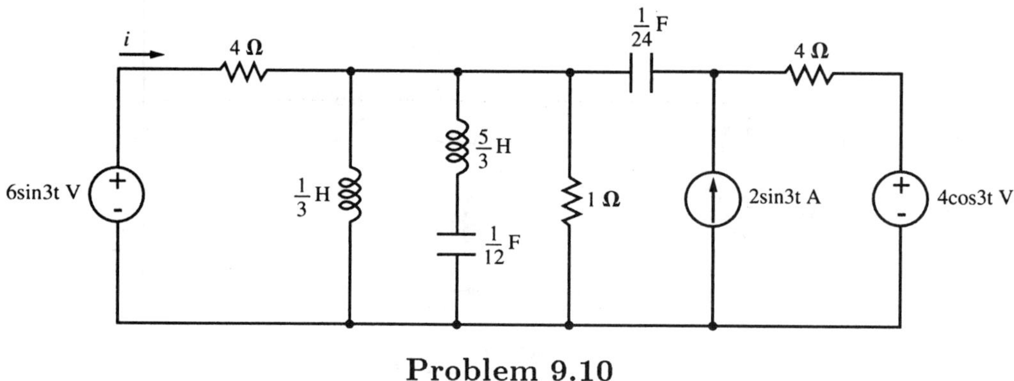

Problem 9.10

9.2 Nodal Analysis

9.11 **Find the forced response i using nodal analysis.**

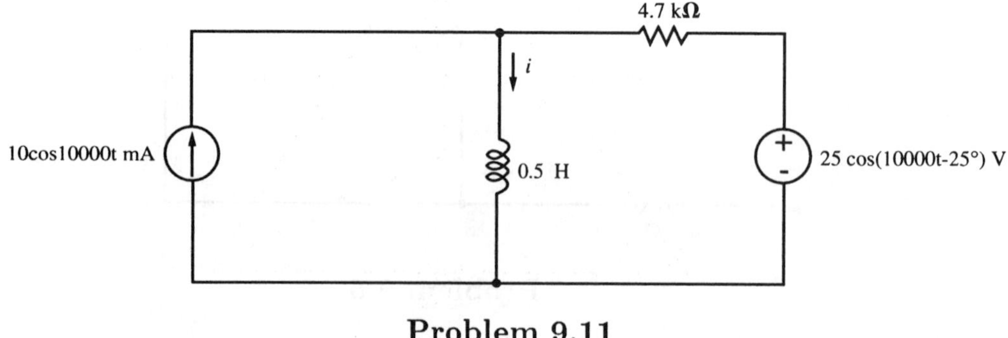

Problem 9.11

9.12 Find the forced response ν using nodal analysis.

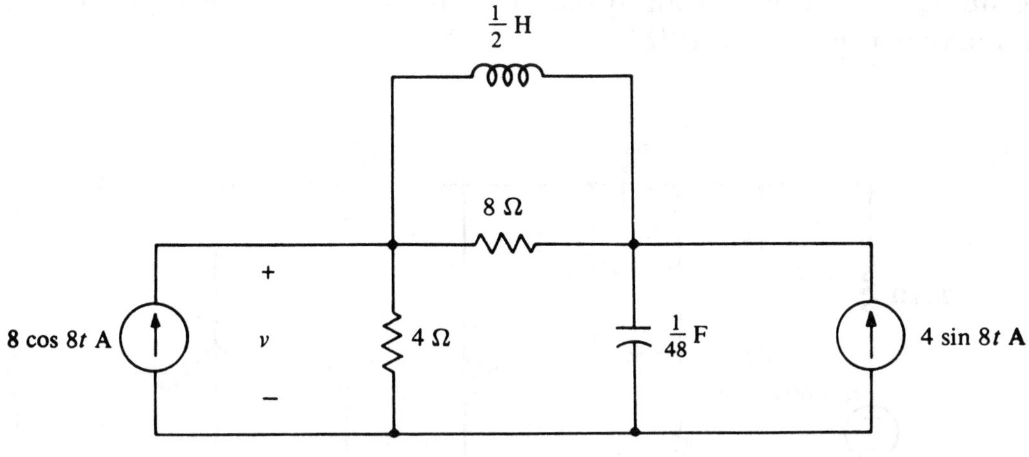

Problem 9.12

9.13 Find the steady-state voltage ν.

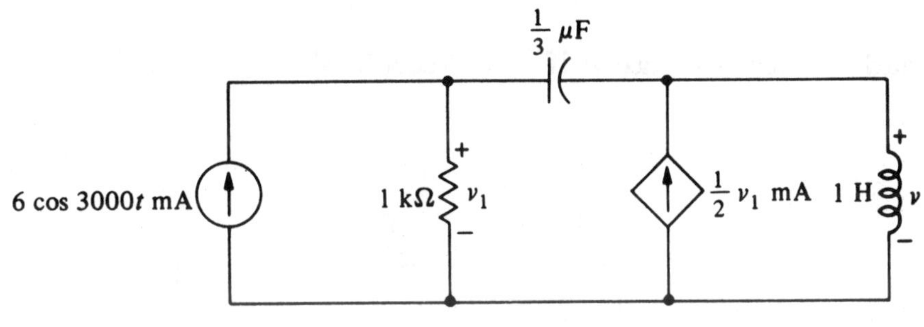

Problem 9.13

9.14 Find the current phasor I. The controlled source has transconductance g = 3.

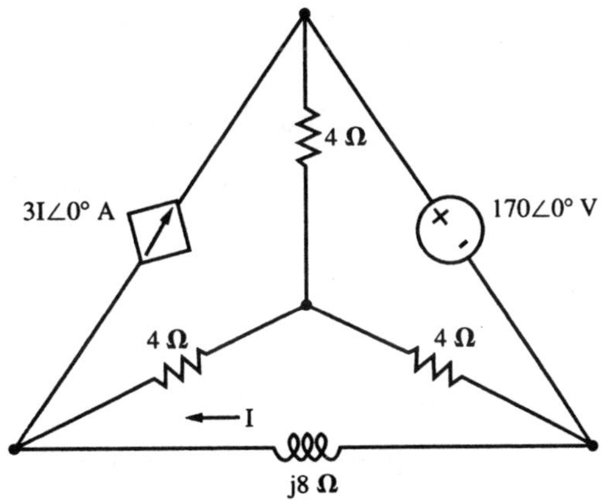

Problem 9.14

9.15 Solve Prob.9.10 using nodal analysis.

9.16 Find V_{out} using nodal analysis. Two dependent current sources both have transresistances $r = 10\Omega$.

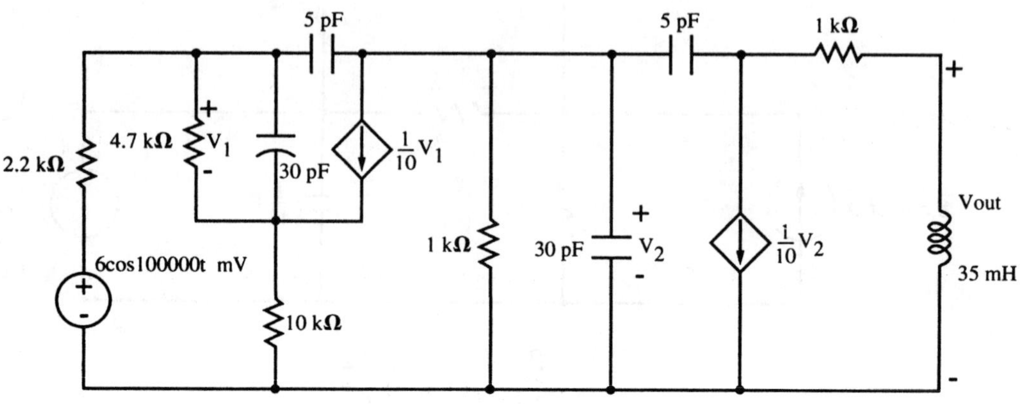

Problem 9.16

9.17 Find the steady-state voltage ν if $\nu_g = 6\cos(5t)$**V.**

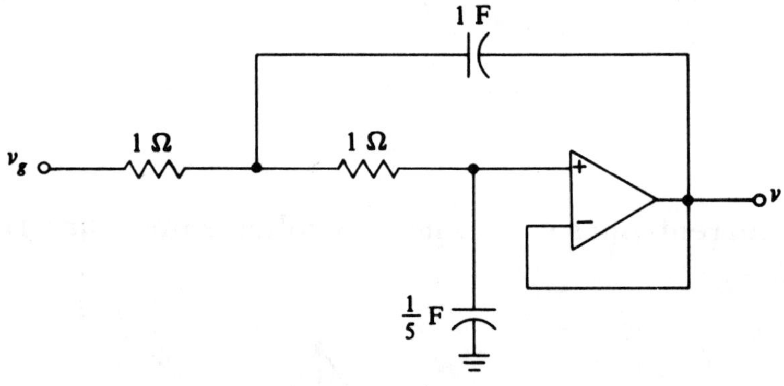

Problem 9.17

9.18 Find the steady-state current i.

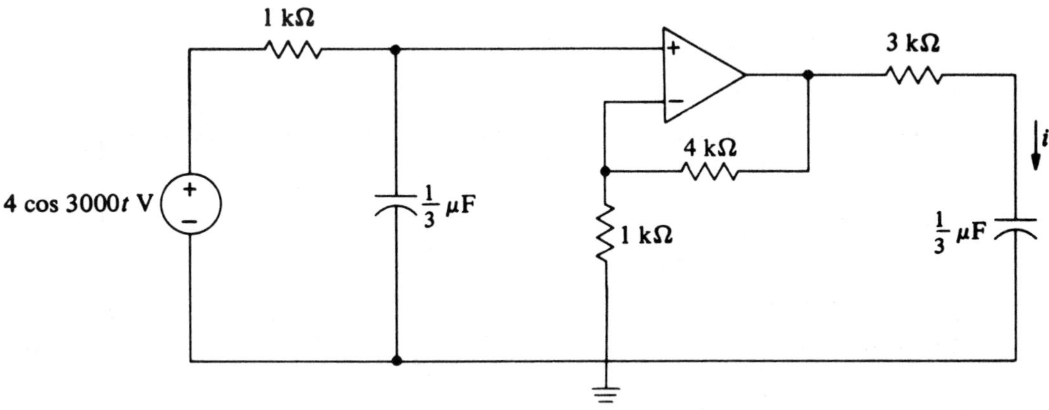

Problem 9.18

9.19 Find the steady-state voltage ν if $\nu_g = 5\cos(2t)$V.

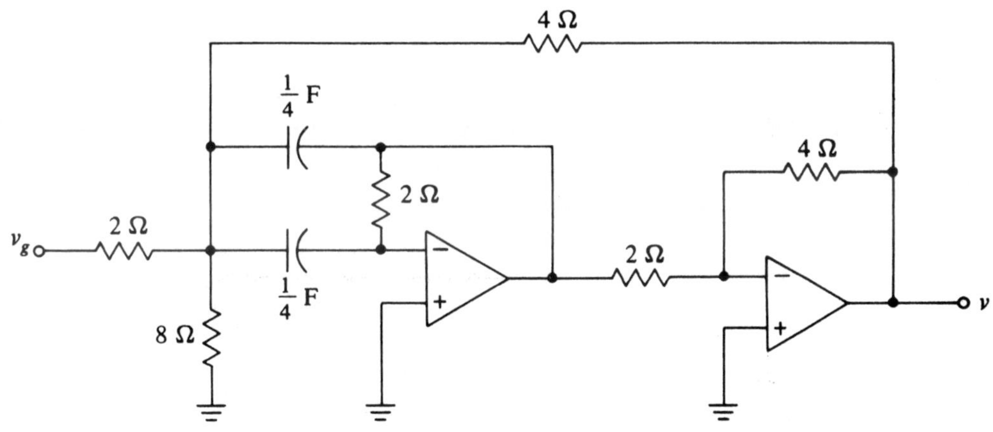

Problem 9.19

9.20 **Find the steady-state voltage ν if $\nu_g = 6\cos(t)$V.**

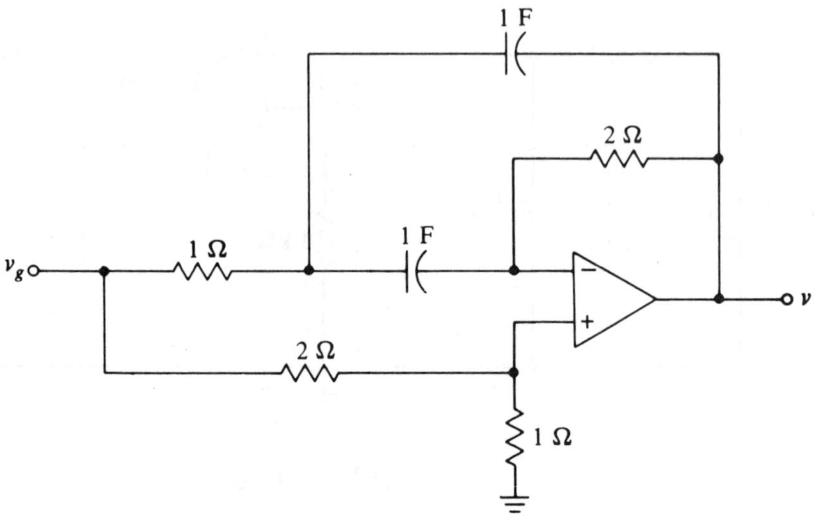

Problem 9.20

9.3 Mesh Analysis

9.21 **Find the impedance looking into terminals $a - a'$ using mesh analysis.**

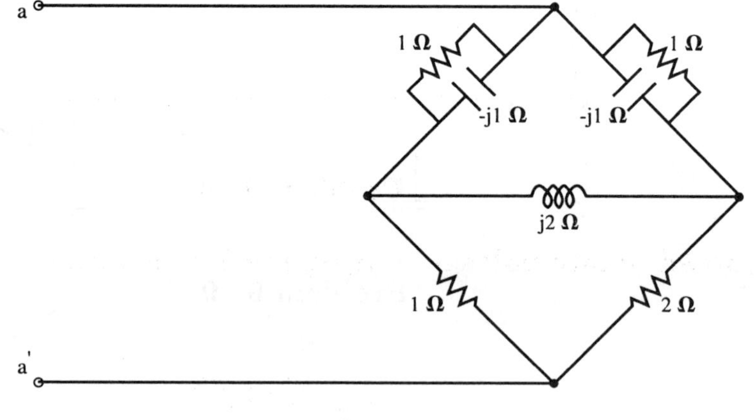

Problem 9.21

9.22 **What phasor current source connected to $a - a'$ in Prob.9.21 would result in a current of $4\angle 37^{\circ}$ leftward through the inductor.**

9.23 Find the steady-state voltage V_2 using mesh analysis, given $V_1 = 6\sin(2t+10^o)$ V.

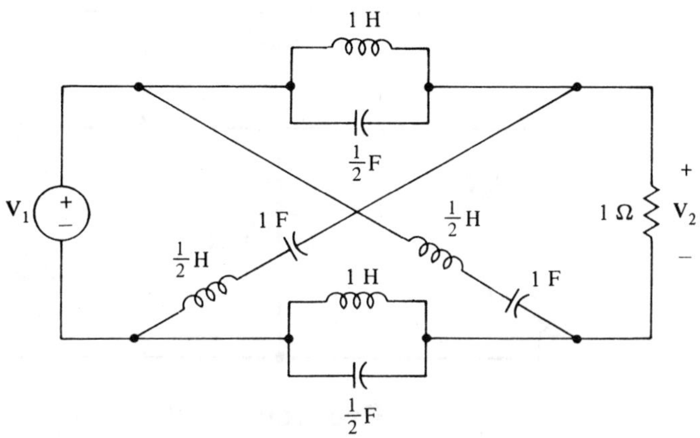

Problem 9.23

9.24 Find the steady-state response v using mesh analysis.

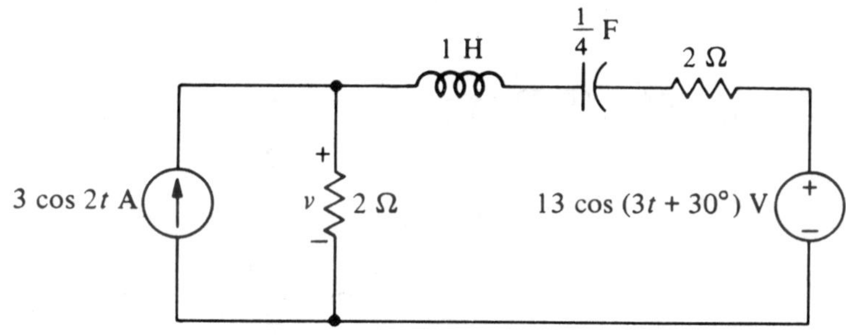

Problem 9.24

9.25 Find the steady-state voltage v using mesh analysis.

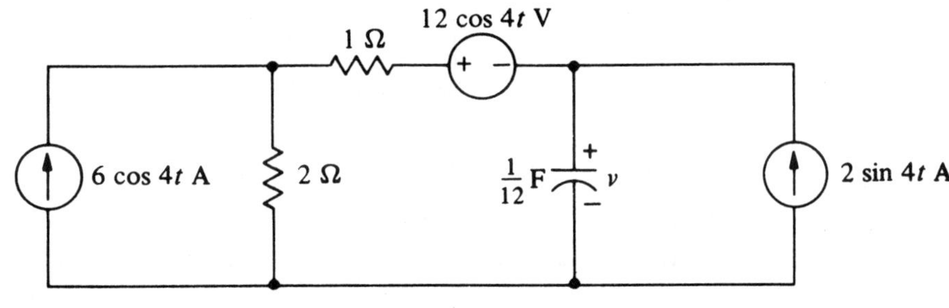

Problem 9.25

9.26 Find the steady-state voltage v using mesh analysis if $i_{g1} = 6\cos(4t)$A and $i_{g2} = 2\cos(4t)$A.

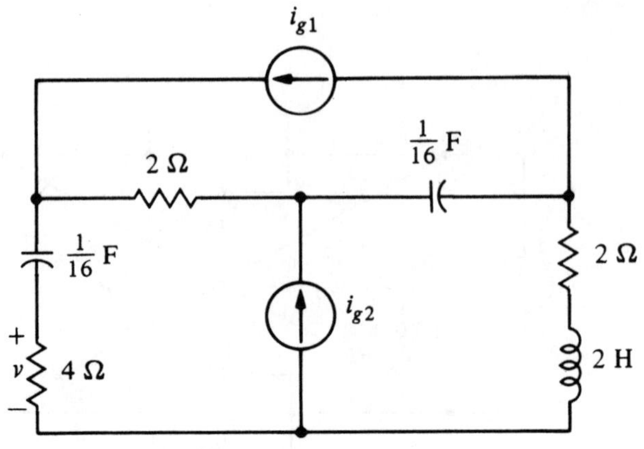

Problem 9.26

9.4 Sources with different frequencies

9.27 Find the steady-state voltage v using mesh analysis .

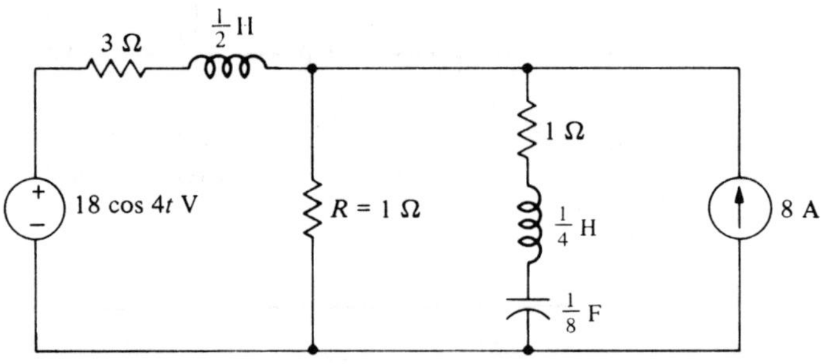

Problem 9.27

9.28 Find the steady-state current i if $i_g = 20\cos(t) - 39\cos(2t) + 18\cos(3t)$A.

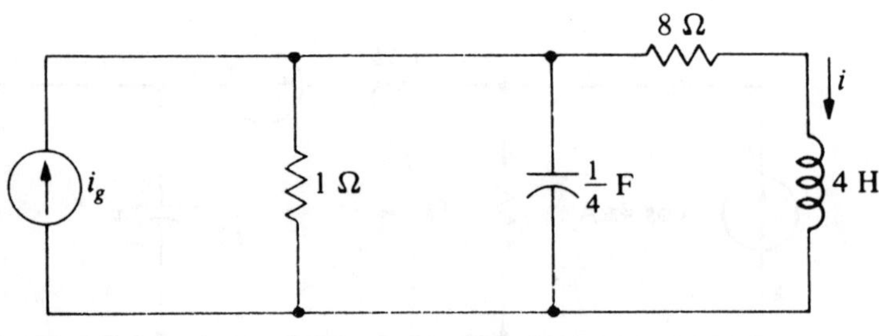

Problem 9.28

9.29 Find the steady-state current i using mesh analysis.

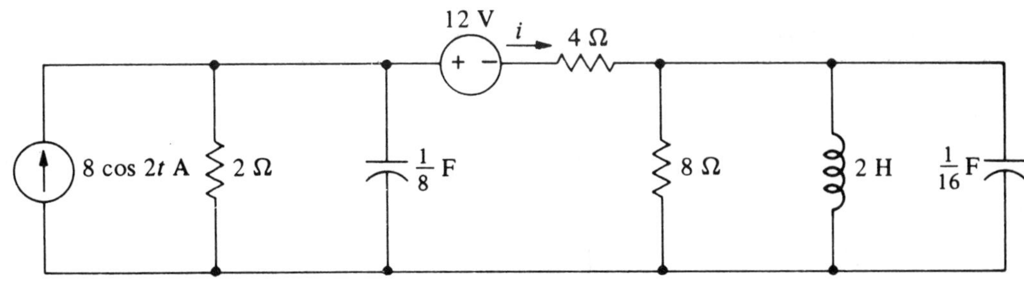

Problem 9.29

9.30 Replace the phasor circuit except for the 1Ω resisitor between terminals a-b by its Thevenin equivalent and find the steady-state value of i. The source is $v_g = 6\cos(2t)$V.

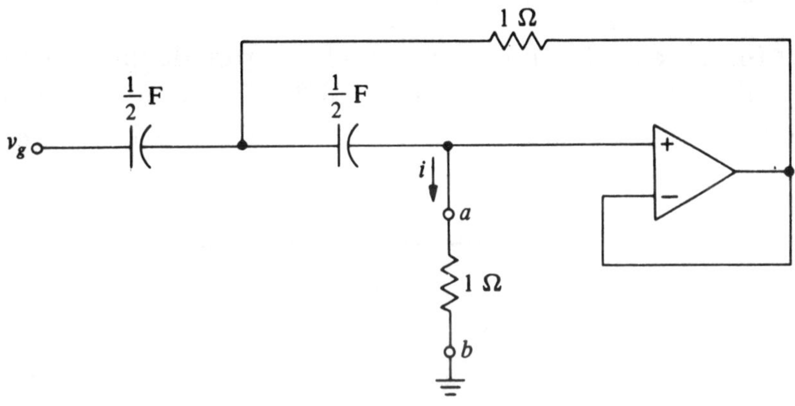

Problem 9.30

9.5 Phasor diagram

9.31 Sketch the admittance of the three elements in a phasor diagram . What must C be if the admittance seen by terminals $a - a'$ is $0.36\angle 56.3^o$ Simens. Assume $\omega =$ 10rad/sec.

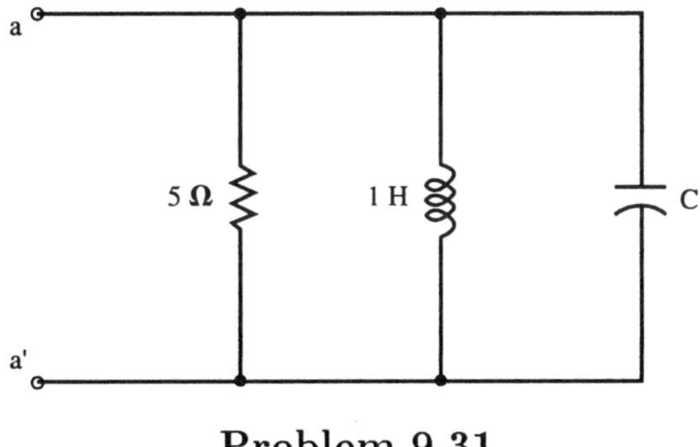

Problem 9.31

9.32 Sketch the locus of the admittance shown in Prob.9.31 in the complex plane as C varies from 0F to 0.05F.

9.33 Solve for V_m using a phasor diagram if $R = 2\Omega$, $L = 3H$ and $\omega = 1rad/sec$, $I = 1\angle 0^{o}A$.

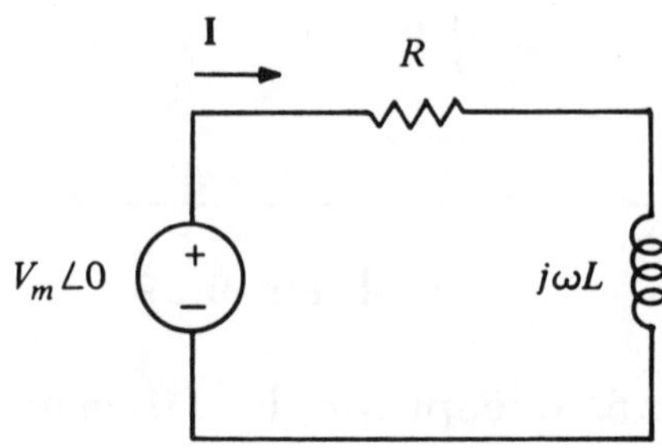

Problem 9.33

9.34 Use SPICE to solve Prob.9.16.

9.35 Find $v(t)$ for 10 evenly spaced frequencies per decade, from $\omega = 100rad/sec$ to $\omega = 10^4 rad/sec$. Use only one SPICE run and use the ideal voltage controlled voltage source op amp model of Fig.3.7 with $A = 100000$.

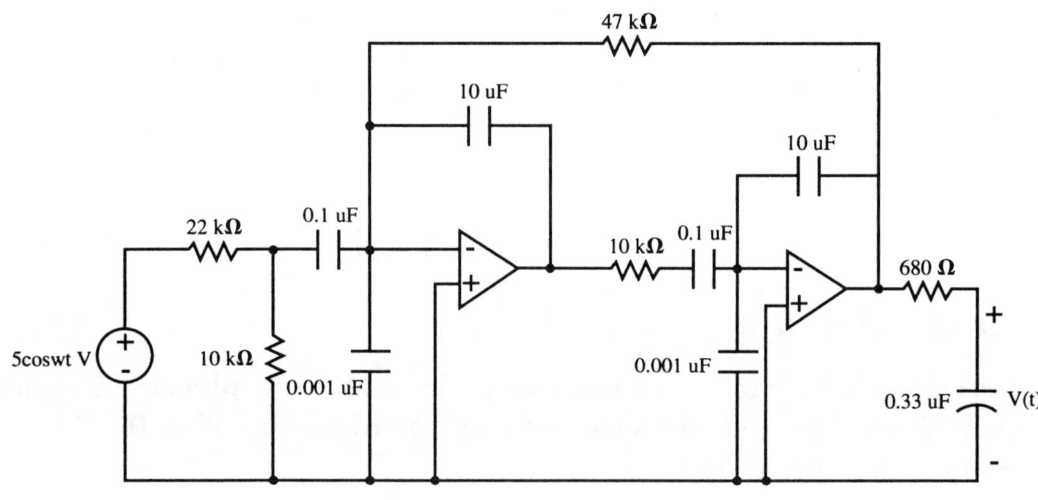

Problem 9.35

9.1 Try loop current $= 1\angle 0°$ A

$Z_1 = \frac{V_1}{1} = 2\sqrt{2}\angle 70° = \underline{0.97 + j2.66\,\Omega}$

$Z_2 = \frac{V_2}{1} = 2\sqrt{2}\angle 25° = \underline{2.56 + j1.20\,\Omega}$

$Z_3 = \frac{V_3}{1} = 2\sqrt{2}\angle -20° = \underline{2.66 - j0.97\,\Omega}$

9.2 Let the voltage at the top of the current source be 1 V.

$Y_1 = \frac{I_1}{1} = 6\angle 30° = \underline{5.20 + j3\,\Omega}$

$Y_2 = \frac{I_2}{1} = 6\angle -30° = \underline{5.20 - j3\,\Omega}$

$Y_3 = \frac{I_3}{1} = 6\angle 0° = \underline{6\,\Omega}$

9.3 $V_1 = \frac{4}{j6 + (-j3) + 4 + (-j)}(6-4) = \frac{8}{j2+4}$

$= \underline{1.79\angle -26.6°}$

Hence $v_1 = \underline{1.79\cos(3t - 26.6°)\,V}$

9.4 By superposition, two current sources can be considered as 1 source with amplitude 7.

$i_1 = \frac{\frac{1}{j4+8}}{\frac{1}{j5} + \frac{1}{-j4} + \frac{1}{j4+8}} \times 7 = \frac{35}{j+5}$

$= \underline{6.86\angle -11.3°\,A}$

9.5 Phasor diagram

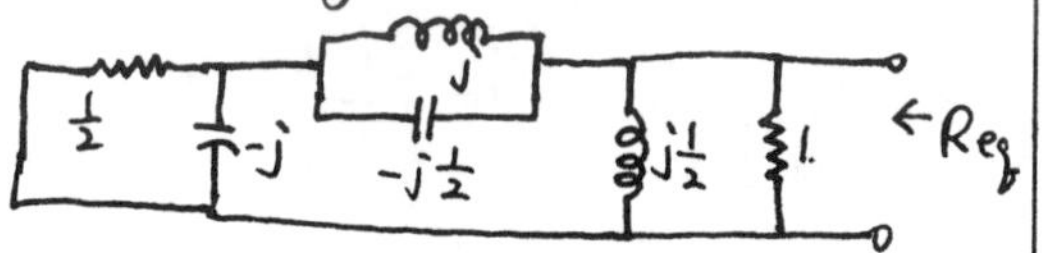

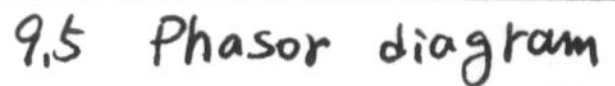

$R_{eq} = \{[\frac{1}{2} // (-j)] + [j // (-j\frac{1}{2})]\} // j\frac{1}{2} // 1$

$= \left(\frac{-j\frac{1}{2}}{\frac{1}{2} - j} + \frac{\frac{1}{2}}{j\frac{1}{2}}\right) // j\frac{1}{2} // 1$

$= \frac{-j6-2}{5} // j\frac{1}{2} // 1$

$= \frac{1}{\frac{5}{-2-j6} + \frac{1}{j\frac{1}{2}} + 1} = \frac{6-j2}{2-j9} = \underline{0.68\angle 59.1°\,\Omega}$

9.6 phasor diagram

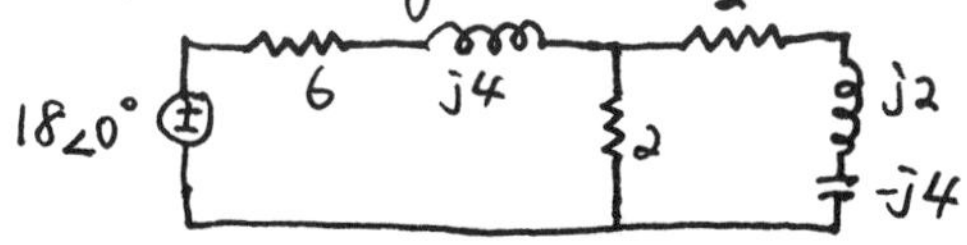

9.6 cont. Try current $i = 1\angle 0°$ A

$V_{cd} = 2\angle 0°$

$R = 2 // (2 + j2 - j4) = 2 // (2 - j2) = \frac{2-j2}{2-j}$

By proportionality, the potential of the voltage source

$V = \frac{6 + j4 + \frac{2-j2}{2-j}}{\frac{2-j2}{2-j}} \times 2\angle 0° = \frac{18+j}{1-j} = 12.8\angle 48.2°$

ratio: $\frac{18\angle 0°}{12.8\angle 48.2°} = 1.4\angle -48.2°$

$I = (1.4\angle -48.2°)(\quad) = 1.4\angle -48.2°$ A

$i = \underline{1.4\cos(2t + 48.2)\,A}$

9.7 To use superposition, one needs to find the two phasor circuit diagrams.

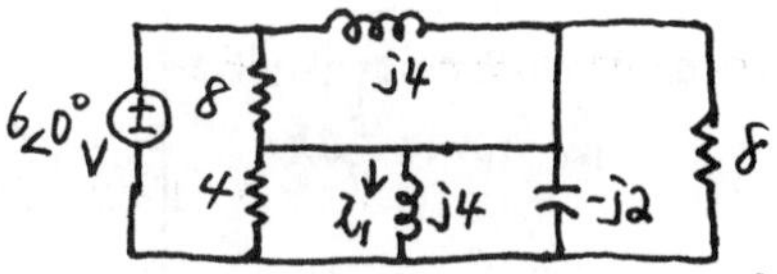

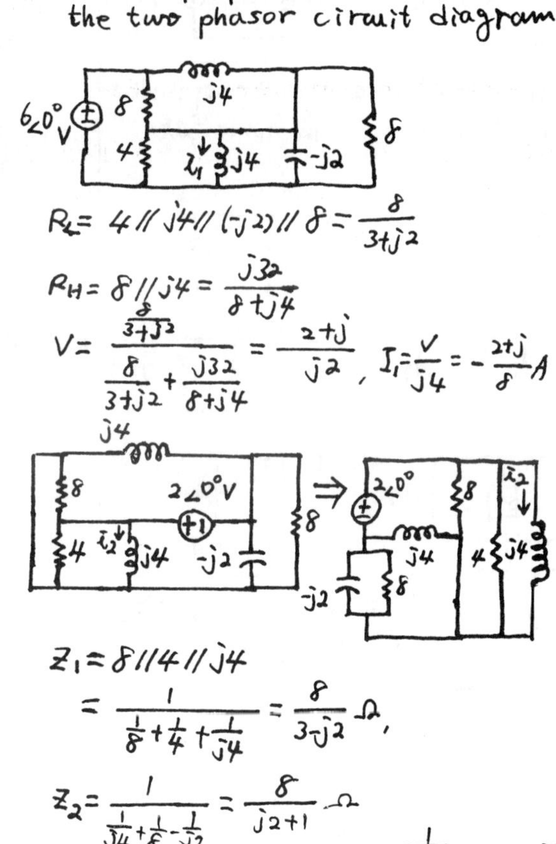

$R_L = 4 // j4 // (-j2) // 8 = \frac{8}{3+j2}$

$R_H = 8 // j4 = \frac{j32}{8+j4}$

$V = \frac{\frac{8}{3+j2}}{\frac{8}{3+j2} + \frac{j32}{8+j4}} = \frac{2+j}{j2}, \quad I_1 = \frac{V}{j4} = -\frac{2+j}{8}$ A

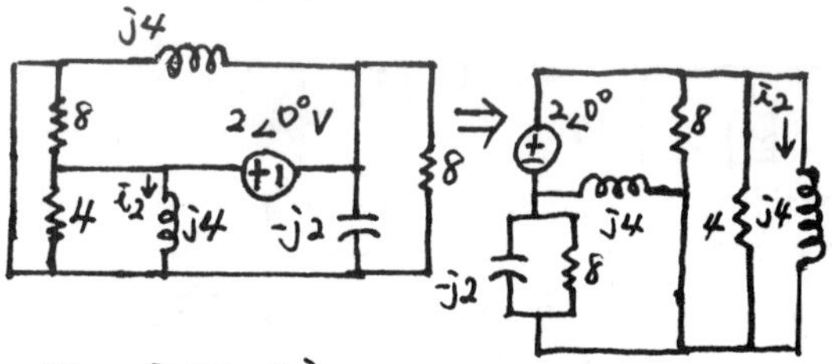

$Z_1 = 8 // 4 // j4$

$= \frac{1}{\frac{1}{8} + \frac{1}{4} + \frac{1}{j4}} = \frac{8}{3-j2}\,\Omega,$

$Z_2 = \frac{1}{\frac{1}{j4} + \frac{1}{8} - \frac{1}{j2}} = \frac{8}{j2+1}\,\Omega$

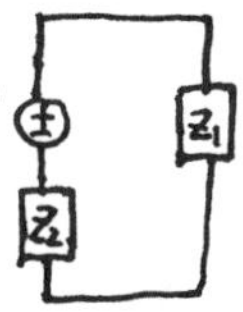

$I_2 = \left(\frac{2}{\frac{8}{3-j2} + \frac{8}{j2+1}}\right)\left(\frac{\frac{1}{j4}}{\frac{1}{8} + \frac{1}{4} + \frac{1}{j4}}\right) = \frac{4-j7}{24-j16}$

$I = I_1 + I_2 = -\frac{2+j}{8} + \frac{4-j7}{24-j16}$

$= \frac{-2-j3}{12-j8} = \frac{3.6\angle 56.4°}{14.4\angle -33.7°} = 0.25\angle 90.1°$

Hence $i = \underline{0.25\cos(4t - 90.1°)}$ A

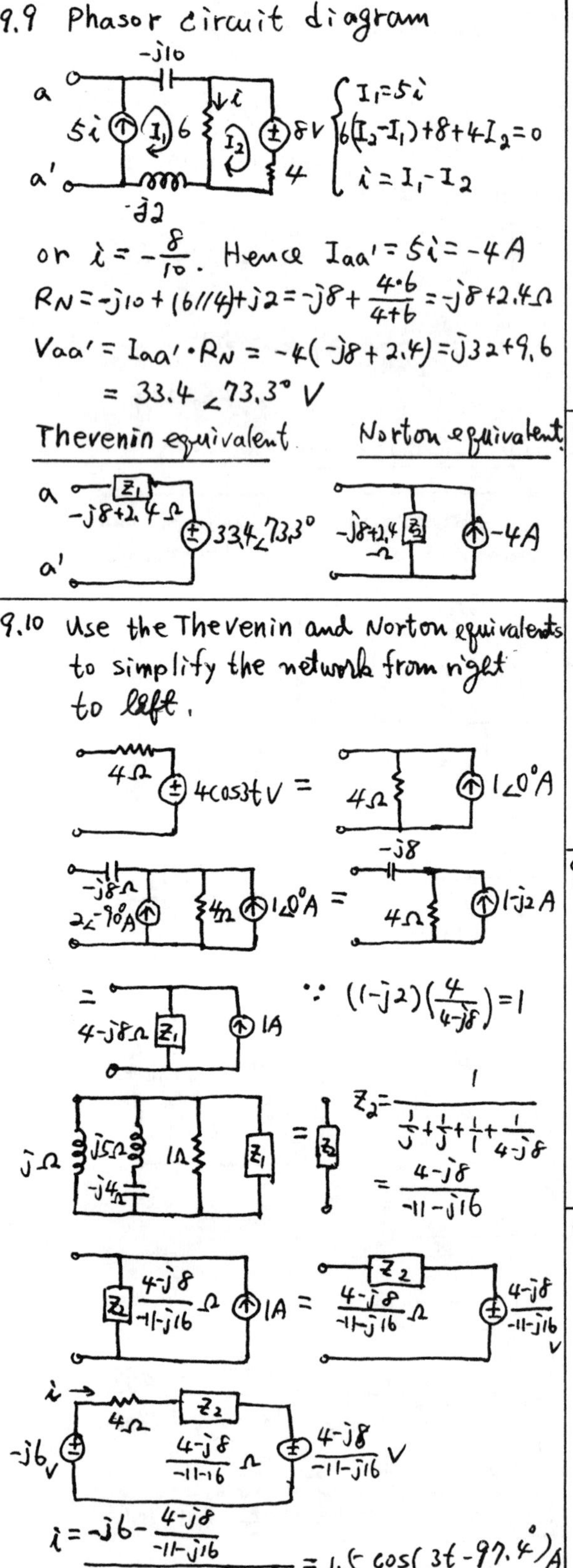

9.9 Phasor circuit diagram

$$\begin{cases} I_1 = 5i \\ 6(I_2 - I_1) + 8 + 4I_2 = 0 \\ i = I_1 - I_2 \end{cases}$$

or $i = -\frac{8}{10}$. Hence $I_{aa'} = 5i = -4A$

$R_N = -j10 + (6 \| 4) + j2 = -j8 + \frac{4 \cdot 6}{4+6} = -j8 + 2.4\,\Omega$

$V_{aa'} = I_{aa'} \cdot R_N = -4(-j8 + 2.4) = j32 + 9.6$

$= 33.4\angle 73.3^\circ$ V

Thevenin equivalent — Norton equivalent

9.10 Use the Thevenin and Norton equivalents to simplify the network from right to left.

$\because (1-j2)\left(\frac{4}{4-j8}\right) = 1$

$Z_2 = \frac{1}{\frac{1}{j} + \frac{1}{j} + \frac{1}{1} + \frac{1}{4-j8}} = \frac{4-j8}{-11-j16}$

$$i = \frac{-j6 - \frac{4-j8}{-11-j16}}{(4-j8)/(-11-j16) + 4} = 1.5\cos(3t - 97.4^\circ)\text{A}$$

9.11 $\frac{V}{j0.5 \times 10^4} + \frac{V - 25\angle -25^\circ}{4.7 \times 10^3} = 10 \times 10^{-3}\angle 0^\circ$

$V(-j2 \times 10^{-4} + 2.12 \times 10^{-4}) = (5.32\angle -25^\circ + 10)/10^3$

$V = \frac{4.82 + j(2.25) + 10}{0.212 - j0.2} = \frac{14.82 - j2.25}{0.212 - j0.2}$

$= \frac{14.99\angle -8.6^\circ}{0.29\angle -43.3^\circ} = 51.69\angle 34.7^\circ$ V

$I = \frac{V}{Z} = \frac{51.69\angle 34.7^\circ}{j0.5 \times 10^4} = 10.34\angle -55.3^\circ$ mA

$i = 10.34\cos(10000t - 55.3^\circ)$ mA

9.12 By KCL:

$V(\frac{1}{4} + \frac{1}{8} - j\frac{1}{4}) - V_1(\frac{1}{8} - j\frac{1}{4}) = 8$ or

$V(3 - j2) + V_1(-1 + j2) = 64$

$-V(\frac{1}{8} - j\frac{1}{4}) + V_1(\frac{1}{8} + j\frac{1}{6} - j\frac{1}{4}) = -j4$ or

$V(-3 + j6) + V_1(3 - j2) = -96$

$V = \begin{vmatrix} 8 & -1+j2 \\ -j96 & 3-j2 \end{vmatrix} \Big/ \begin{vmatrix} 3-j2 & -1+j2 \\ -3+j6 & 3-j2 \end{vmatrix} = \frac{-j224}{14} = -j16$

$\therefore v = 16\sin 8t$ V

9.13 By KCL using mA

$V_1(1+j) - V(j) = 6$

$-V_1(j) + V(j - j\frac{1}{3}) = V_1(\frac{1}{2})$ or

$V_1(-\frac{1}{2} - j) + V(j\frac{2}{3}) = 0$

$V = \begin{vmatrix} 1+j & 6 \\ -\frac{1}{2}-j & 0 \end{vmatrix} \Big/ \begin{vmatrix} 1+j & -j \\ -\frac{1}{2}-j & j\frac{2}{3} \end{vmatrix} = \frac{3+j6}{\frac{1}{3}+j\frac{1}{6}}$

$= 18\angle 36.9^\circ$

$v = 18\cos(3000t + 36.87^\circ)$ V

9.14 $\frac{V_1 - 170}{4} + \frac{V_1 - V_2}{4} + \frac{V_1}{4} = 0$ or $V_2 = 3V_1 - 170$

$\frac{V_2 - V_1}{4} + \frac{V_2}{j8} = 3I = 3\left(-\frac{V_2}{j8}\right)$

$(V_2 - V_1)j2 + V_2 = -3V_2$

$V_2(j2 + 4) = V_1(j2)$,

$(3V_1 - 170)(j2 + 4) = V_1(j2)$,

$V_1 = \frac{680 + j340}{12 + j4} = 60.3\angle 8.2^\circ$

$V_2 = 3V_1 - 170 = 180.9\angle 8.2^\circ - 170$

$I = -\frac{V_2}{j8} = (180.9\angle 8.2^\circ - 170)/8\angle 90^\circ = 3.4\angle -18.4^\circ$ A

9.15

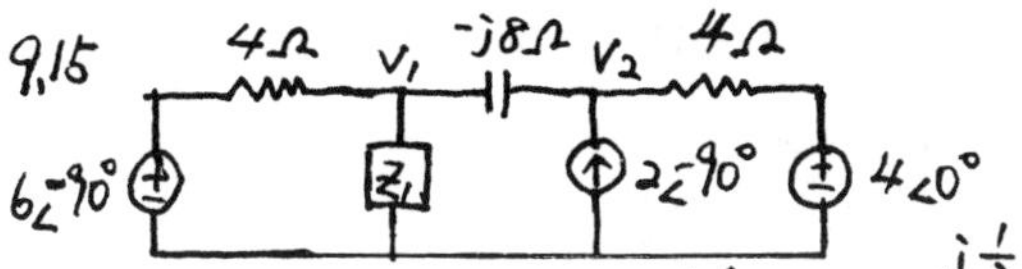

$Z_1 = j1 \| (j5 - j4) \| 1 = j\frac{1}{2} \| 1 = \frac{j\frac{1}{2}}{1+j\frac{1}{2}}$

$\frac{V_1 - j6}{4} + \frac{V_1(1+j\frac{1}{2})}{j(\frac{1}{2})} + \frac{V_1 - V_2}{-j8} = 0 \Rightarrow V_1(15+j10) = -V_2 + 12$

$\frac{V_2 - V_1}{-j8} + \frac{V_2 - 4}{4} = -j2 \Rightarrow V_2 - V_1 + (-j2)(V_2 - 4) = (-j2)(-j8)$

or $V_2(1-j2) - V_1 = -16 - j8$

$\Rightarrow [-V_1(15+j10) - 12](1-j2) - V_1 = -16 - j8$

$\Rightarrow V_1 = -V_1(35 - j20) + (28 - j16)$

$\Rightarrow V_1(9 - j5) = 7 - j4$

$\Rightarrow V_1 = \frac{8.06\angle -29.7°}{10.30\angle -29.0°} = 0.78\angle -0.7° \approx 0.78$

$I = \frac{-j6 - 0.78}{4} = -0.195 - j1.5 = 1.51\angle 97.4°$

Hence $i = \underline{1.51\cos(3t - 97.4°)\text{ A}}$

9.16

$Z_{30pF} = \frac{1}{j\omega(30pF)} = \frac{1}{j(10^5)(30\times10^{-12})} = -j3.3\times10^5\,\Omega$

$Z_{5pF} = \frac{1}{j\omega(5pF)} = \frac{1}{j\cdot(10^5)(5\cdot10^{-12})} = -j19.8\times10^5\,\Omega$

Thus for practical purpose, the impedance across the four small capacitors can be treated as open circuit.

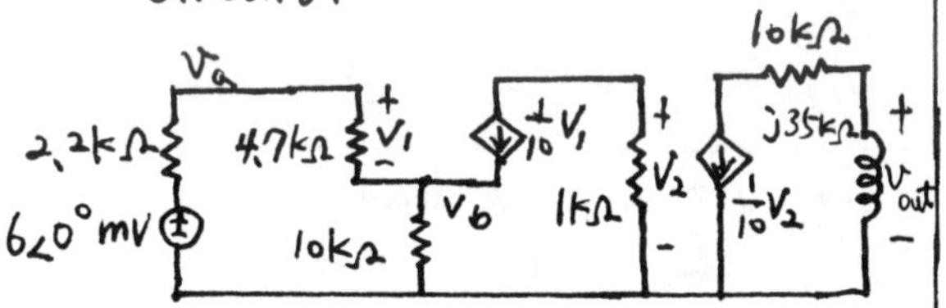

$\frac{V_a - 6}{2.2\times10^3} + \frac{V_a - V_b}{4.7\times10^3} = 0 \Rightarrow V_a = 0.32V_b + 4.087$

$\frac{V_b - V_a}{4.7\times10^3} + \frac{V_b}{10\times10^3} = \frac{V_a - V_b}{10} \Rightarrow V_b = 0.9979V_a$

$V_a = \frac{4.087}{(1 - 0.32\times0.9979)} = 6.004\text{ mV}$

$V_b = 0.9979\times6.004 = 5.9918\text{ mV}$

Hence $V_1 = V_a - V_b = 6.004 - 5.9918 = 0.0122\text{ mV}$

$V_2 = \frac{0.0122}{10}\times10^3 = 0.00122\text{ V}.$

$V_{out} = \frac{0.00122}{10}\times(j35\times10^3) = \underline{4.4\text{ V}}$

9.17 Let V_1 be the phasor node voltage at the junction of the 1Ω resistors and V at the input terminals of opamp. KCL gives:

$V_1(1+1+j5) - V(1+j5) = V_g$ or

$V_1(2+j5) - V(1+j5) = 6$

KCL at the noninverting terminal of op amp gives

$-V_1 + V(1+j1) = 0$ or $V_1 = V(1+j1)$

$\therefore V = \frac{6}{(1+j1)(2+j5) - (1+j5)} = 1.34\angle -153°$

$v = \underline{1.342\cos(5t - 153.4°)\text{ V}}$

9.18 Let v = op amp output voltage and v_1 be the voltage at the input terminals.

$V_1 = \frac{1(4)}{1+j}$ (voltage division)

Since $V = (1 + \frac{4}{1})V_1 = 5V_1$

then $V = \frac{5(4)}{1+j} = 10 - j10$

$I = \frac{V}{3000 - j1000} = 4.47\angle -26.6°\text{ mA}$

$i = \underline{4.47\cos(3000t - 26.6°)\text{ mA}}$

9.19 Let V_1 = node voltage at junction of 2Ω and 8Ω resistor, V_2 = output of 1st op amp. Then KCL at V_1 gives.

$V_1\left(\frac{1}{2} + \frac{1}{8} + \frac{1}{4} + j\frac{1}{2} + j\frac{1}{2}\right) - V_2\left(j\frac{1}{2}\right) - V\frac{1}{4} = \frac{5}{2}$

or $V_1(7+j8) - V_2(j4) - V_2 = 20$

KCL at inverting terminal of first op amp gives $V_1 = jV_2$ and KCL at inverting terminal of 2nd op amp gives $V_2 = -V/2$ therefore

$V_1 = -j\,V/2$ and

$V[(7+j8)(-j/2) + j4/2 - 2] = 20$ or

$V = \frac{20}{2 - j3/2} = 8\angle 36.87°\text{ V}$

$v = \underline{8\cos(2t + 36.87°)\text{ V}}$

9.20 Let V_i be the voltage at the input terminals of the op amp, then at the noninverting terminal KCL gives $V_1(\frac{1}{2}+1)=V_g\frac{1}{2} \Rightarrow V_1=2V$

KCL at inverting terminal gives $V_1(\frac{1}{2}+j)-V_2(j)-V(\frac{1}{2})=0$ or $V_2(j)+V\frac{1}{2}=1+j2$ where V_2 is the node voltage to the right of the 1Ω resistor. KCL at V_2 yields

$V_2(1+j+j)-V_1(j)-V(j)=V_g$ or

$V_2(1+j2)-Vj=6+j2$

$$V=\begin{vmatrix} j & 1+j2 \\ 1+j2 & 6+j2 \end{vmatrix} \Big/ \begin{vmatrix} j & \frac{1}{2} \\ 1+j2 & -j \end{vmatrix} = \frac{1+j2}{\frac{1}{2}-j}=2\angle 126.9^\circ \text{ V}$$

$v=2\cos(t+126.9^\circ)$ V

9.21 Connect a $1\angle 0^\circ$A current source to terminals a-a'. KVL gives:

(1) $\{2[\frac{-j}{1-j}]+j2\}I_1-(j2)(I_2)+\frac{j}{1-j}=0$

or $\{-j2+(j2)(1-j)\}I_1-(j)(j2)I_2-j=0$

or $2I_1-(2+j2)I_2+j=0$

(2) $-j2I_1+(3+j2)I_2-1=0$, or $I_1=(1-j\frac{3}{2})I_2+j\frac{1}{2}$

Substitute (2) into (1)

$2[(1-j\frac{3}{2})I_2+j\frac{1}{2}]-(2+j2)I_2+j=0$

$\Rightarrow (2-j3)I_2-(2+j2)I_2+j2=0$

$\Rightarrow I_2(-j5)+j2=0 \Rightarrow I_2=\frac{2}{5}$

and $I_1=(1-j\frac{3}{2})\frac{2}{5}+j\frac{1}{2}=\frac{2}{5}-j\frac{1}{10}$

$V=(\frac{-j}{1-j})(\frac{2}{5}-j\frac{1}{10})+(2)\frac{2}{5}$

$=(-\frac{1}{10})(\frac{j}{1-j})(4-j)+\frac{4}{5}=\frac{4}{5}-(\frac{1}{10})(\frac{(1+j)(1+4j)}{2})$

$=\frac{4}{5}-\frac{1}{20}(1-4+j5)=\frac{19}{20}-j\frac{1}{4}=0.98\angle -14.7^\circ$

$Z_{aa'}=\frac{V}{I}=\frac{0.98\angle -14.7^\circ}{1}=0.98\angle -14.7^\circ\ \Omega$

9.22 From the solution of Prob. 9.21

$I_1-I_2=\frac{2}{5}-j\frac{1}{10}-\frac{2}{5}=-j0.1=0.1\angle -90^\circ$A

By proportionality

$1\angle 0^\circ \times \frac{4\angle 37^\circ}{0.1\angle -90^\circ}=40\angle 127^\circ$A

9.23 Redraw the phasor circuit diagram

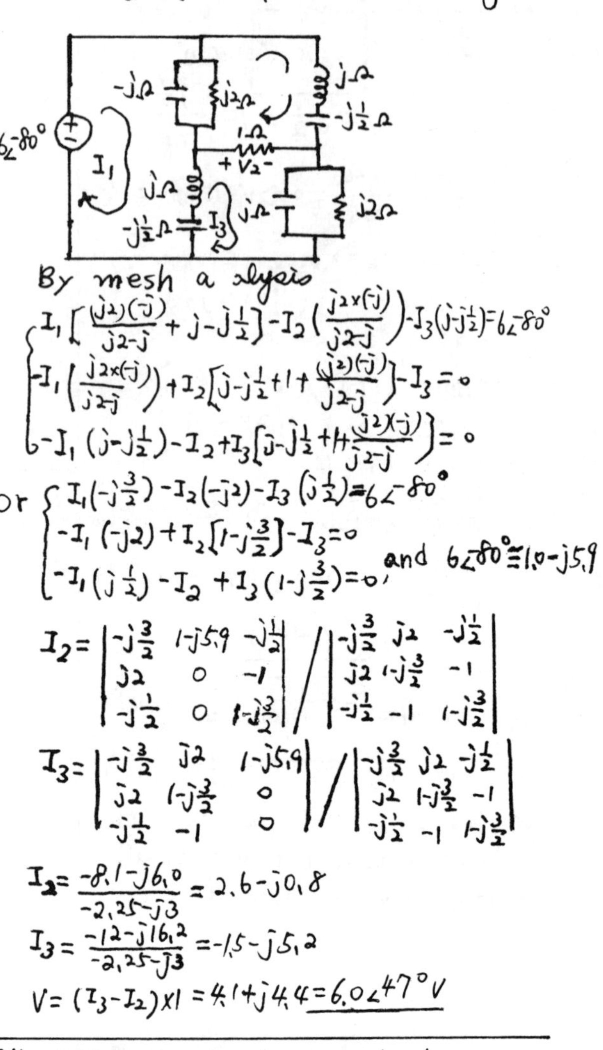

By mesh analysis

$I_1[\frac{(j2)(-j)}{j2-j}+j-j\frac{1}{2}]-I_2(\frac{j2\times(-j)}{j2-j})-I_3(j-j\frac{1}{2})=6\angle -80^\circ$

$-I_1(\frac{j2\times(-j)}{j2-j})+I_2[j-j\frac{1}{2}+1+\frac{(j2)(-j)}{j2-j}]-I_3=0$

$-I_1(j-j\frac{1}{2})-I_2+I_3[j-j\frac{1}{2}+1+\frac{j2\times(-j)}{j2-j}]=0$

or

$I_1(-j\frac{3}{2})-I_2(-j2)-I_3(j\frac{1}{2})=6\angle -80^\circ$

$-I_1(-j2)+I_2[1-j\frac{3}{2}]-I_3=0$

$-I_1(j\frac{1}{2})-I_2+I_3(1-j\frac{3}{2})=0$, and $6\angle -80^\circ \cong 1.0-j5.9$

$$I_2=\begin{vmatrix} -j\frac{3}{2} & 1-j5.9 & -j\frac{1}{2} \\ j2 & 0 & -1 \\ -j\frac{1}{2} & 0 & 1-j\frac{3}{2} \end{vmatrix} \Big/ \begin{vmatrix} -j\frac{3}{2} & j2 & -j\frac{1}{2} \\ j2 & 1-j\frac{3}{2} & -1 \\ -j\frac{1}{2} & -1 & 1-j\frac{3}{2} \end{vmatrix}$$

$$I_3=\begin{vmatrix} -j\frac{3}{2} & j2 & 1-j5.9 \\ j2 & 1-j\frac{3}{2} & 0 \\ -j\frac{1}{2} & -1 & 0 \end{vmatrix} \Big/ \begin{vmatrix} -j\frac{3}{2} & j2 & -j\frac{1}{2} \\ j2 & 1-j\frac{3}{2} & -1 \\ -j\frac{1}{2} & -1 & 1-j\frac{3}{2} \end{vmatrix}$$

$I_2=\frac{-8.1-j6.0}{-2.25-j3}=2.6-j0.8$

$I_3=\frac{-12-j16.2}{-2.25-j3}=-1.5-j5.2$

$V=(I_3-I_2)\times 1=4.1+j4.4=6.0\angle 47^\circ$V

9.24 With the current source dead voltage division yields

$V_1=\frac{2(13)\angle 30^\circ}{2+2+j3-j4/3}=\frac{26\angle 30^\circ}{4+j5/3}=6\angle 7.38^\circ$

$\therefore v_1=6\cos(3t-7.38^\circ)$V

With the voltage source dead KCL gives:

$V_2(\frac{1}{2}+\frac{1}{2+j2-j2})=3$

$V_2=3$

$\therefore v_2=3\cos(2t)$

$v=v_1+v_2=3\cos(2t)+6\cos(3t-7.38^\circ)$V

9.25 Let $\underline{I}$ be the phasor current in the center mesh. KVL yields

$$\underline{I}(1+2-j3)-6(2)+j3(-j2)+12=0$$

$$\underline{I}=\frac{6}{3-j3}=1+jA$$

$$\underline{V}=(\underline{I}-j2)(-j3)=3\sqrt{2}\angle-135^\circ\ V$$

$$v=3\sqrt{2}\cos(4t-135^\circ)\ V$$

9.26

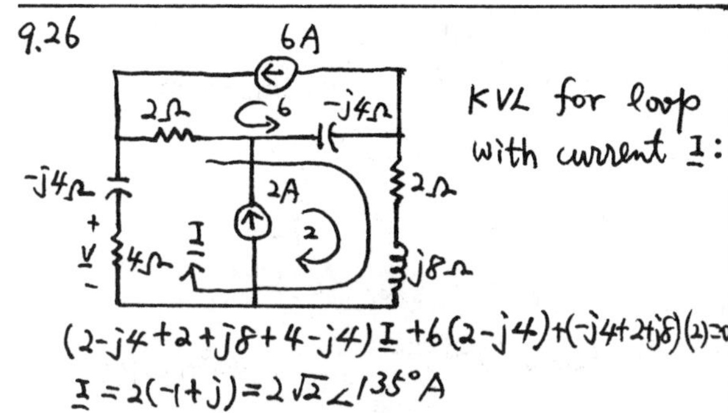

KVL for loop with current $\underline{I}$:

$$(2-j4+2+j8+4-j4)\underline{I}+6(2-j4)+(-j4+2+j8)(2)=0$$

$$\underline{I}=2(-1+j)=2\sqrt{2}\angle135^\circ A$$

$$\underline{V}=-4\underline{I}=8\sqrt{2}\angle-45^\circ V$$

$$v=8\sqrt{2}\cos(4t-45^\circ)V$$

9.27 By superposition, we consider the output of each source.

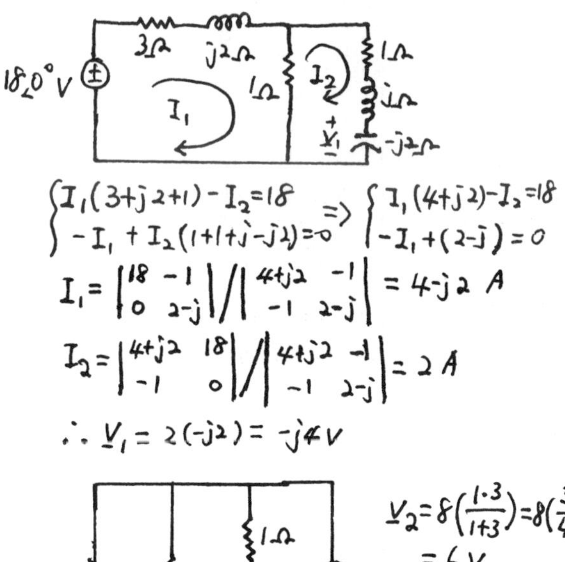

$$\begin{cases}\underline{I}_1(3+j2+1)-\underline{I}_2=18\\-\underline{I}_1+\underline{I}_2(1+1+j-j2)=0\end{cases}\Rightarrow\begin{cases}\underline{I}_1(4+j2)-\underline{I}_2=18\\-\underline{I}_1+(2-j)=0\end{cases}$$

$$\underline{I}_1=\begin{vmatrix}18&-1\\0&2-j\end{vmatrix}\Big/\begin{vmatrix}4+j2&-1\\-1&2-j\end{vmatrix}=4-j2\ A$$

$$\underline{I}_2=\begin{vmatrix}4+j2&18\\-1&0\end{vmatrix}\Big/\begin{vmatrix}4+j2&-1\\-1&2-j\end{vmatrix}=2\ A$$

$$\therefore\ \underline{V}_1=2(-j2)=-j4V$$

$$\underline{V}_2=8\left(\frac{1\cdot3}{1+3}\right)=8\left(\frac{3}{4}\right)=6V$$

$$\underline{V}=\underline{V}_1+\underline{V}_2=6-j4=7.2\angle-33.7^\circ$$

$$v=7.2\cos(4t-33.7^\circ)V$$

*

9.28 $i_{g0}=20\cos t$ produces i_1;
$i_{g1}=-39\cos(2t)$ produces i_2;
$i_{g2}=18\cos(3t)$ produces i_3;

By current division

$$\underline{I}_i=\frac{\frac{1}{8+j4\omega}\underline{I}g_i}{1+j\frac{\omega}{4}+\frac{1}{8+j4\omega}}=\frac{\underline{I}g_i}{9-\omega^2+j6\omega}$$

$\omega=1$ rad/s : $\underline{I}g_1=20A\Rightarrow\underline{I}_1=\frac{20}{8+j6}=2\angle-36.9^\circ A$

$\omega=2$ rad/s : $\underline{I}g_2=-36A\Rightarrow\underline{I}_2=-\frac{39}{5+j12}=3\angle112.6^\circ A$

$\omega=3$ rad/s : $\underline{I}g_3=18A\Rightarrow\underline{I}_3=\frac{18}{j18}=1\angle-90^\circ A$

$$i=i_1+i_2+i_3$$
$$=2\cos(t-36.9^\circ)+3\cos(t+112.6^\circ)+\sin3tA$$

9.29 With the current source dead

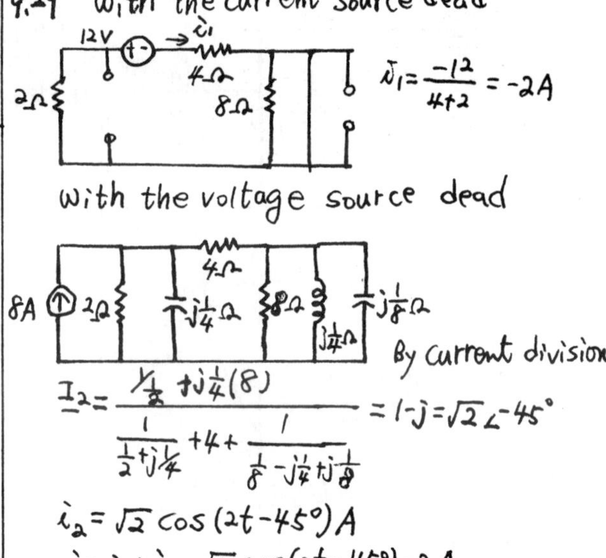

$$i_1=\frac{-12}{4+2}=-2A$$

With the voltage source dead

By current division

$$\underline{I}_2=\frac{\frac{1}{2}+j\frac{1}{4}(8)}{\frac{1}{\frac{1}{2}+j\frac{1}{4}}+4+\frac{1}{\frac{1}{8}-j\frac{1}{4}+j\frac{1}{8}}}=1-j=\sqrt{2}\angle-45^\circ$$

$$i_2=\sqrt{2}\cos(2t-45^\circ)A$$

$$i=i_1+i_2=\sqrt{2}\cos(2t-45^\circ)-2A$$

9.30

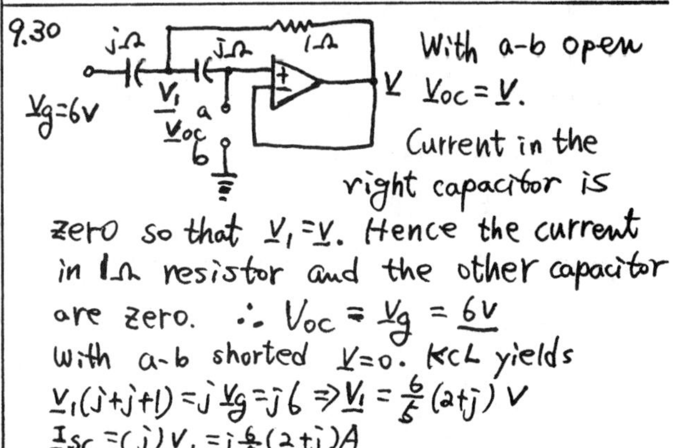

With a-b open $V_{oc}=\underline{V}$.

Current in the right capacitor is zero so that $\underline{V}_1=\underline{V}$. Hence the current in 1Ω resistor and the other capacitor are zero. $\therefore\ V_{oc}=V_g=6V$

With a-b shorted $\underline{V}=0$. KCL yields

$$\underline{V}_1(j+j+1)=jV_g=j6\Rightarrow\underline{V}_1=\frac{6}{5}(2+j)V$$

$$\underline{I}_{sc}=(j)\underline{V}_1=j\frac{6}{5}(2+j)A$$

9.30 Cont.

$$Z_{th} = \frac{V_{oc}}{I_{sc}} = \frac{6}{j\frac{6}{5}(2+j)} = \underline{-1-j2\,\Omega}$$

The current $\underline{I}$ from a to b is

$$\underline{I}_1 = \frac{V_{oc}}{Z_{th}+1} = \frac{6}{-1-j2+1} = j3A$$

$$i = \underline{3\cos(2t+90°)\,A}$$

9.31

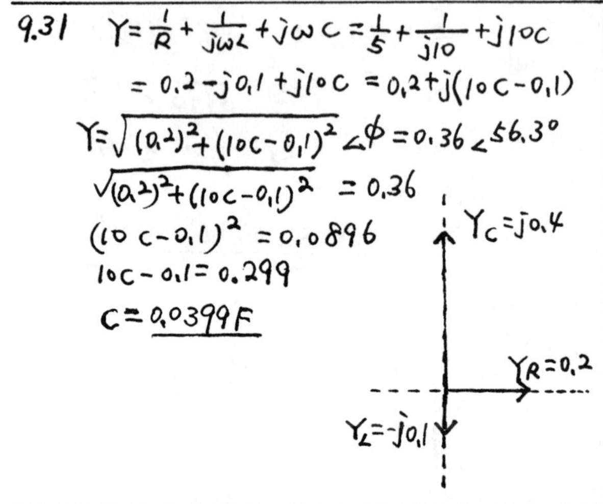

$$Y = \frac{1}{R} + \frac{1}{j\omega L} + j\omega C = \frac{1}{5} + \frac{1}{j10} + j10C$$

$$= 0.2 - j0.1 + j10C = 0.2 + j(10C - 0.1)$$

$$Y = \sqrt{(0.2)^2 + (10C-0.1)^2}\angle\phi = 0.36\angle 56.3°$$

$$\sqrt{(0.2)^2 + (10C-0.1)^2} = 0.36$$

$$(10C - 0.1)^2 = 0.0896$$

$$10C - 0.1 = 0.299$$

$$C = \underline{0.0399F}$$

9.32

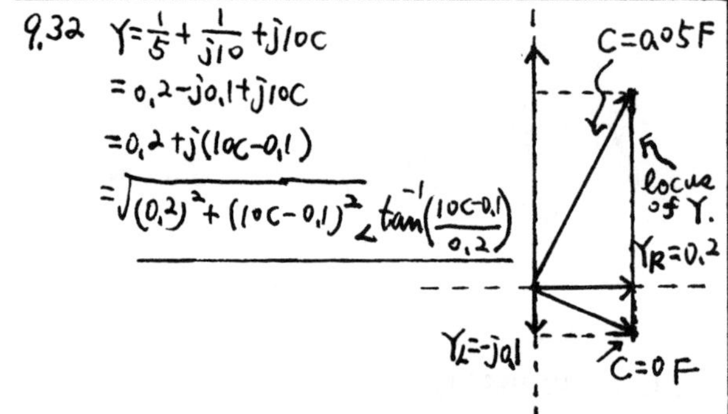

$$Y = \frac{1}{5} + \frac{1}{j10} + j10C$$

$$= 0.2 - j0.1 + j10C$$

$$= 0.2 + j(10C - 0.1)$$

$$= \underline{\sqrt{(0.2)^2 + (10C-0.1)^2}\angle \tan^{-1}\left(\frac{10C-0.1}{0.2}\right)}$$

9.33

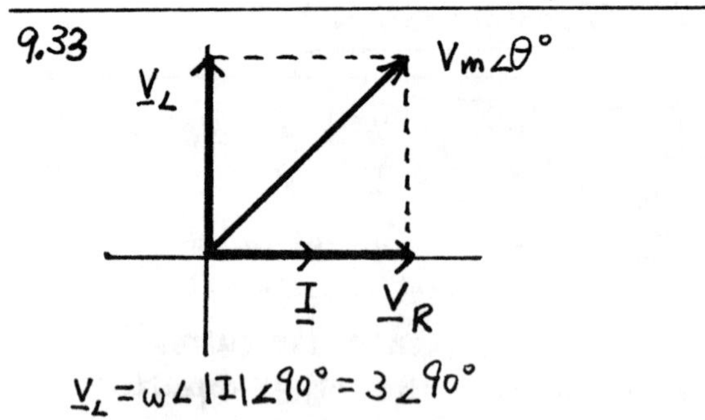

$$\underline{V}_L = \omega L\,|I|\angle 90° = 3\angle 90°$$

$$\underline{V}_R = R|I| = 2$$

$$\underline{V}_m = \underline{V}_R + \underline{V}_L = \sqrt{3^2+2^2}$$

$$= \underline{\sqrt{13}\angle 56.31°}$$

9.34

```
 * Schematics Netlist *
 ****      CIRCUIT DESCRIPTION
******************************
R_R1       2 1 2.2k
R_R2       3 1 4.7k
R_R3       0 3 10k
R_R4       0 4 1k
R_R5       6 5 10k
L_L1       6 0 35mH
C_C1       1 0 30PF
C_C2       1 4  5PF
C_C3       4 5  5PF
G_G1       5 0 4 0 .1
G_G2       4 3 1 3 .1
V_V1  2 0  SIN(0 .006 100000 0 0 0
*Control  statement
.TRAN     10US 100US UIC
*Output   V(6)
.PRINT TRAN V(6)
.END
**** 11/18/94 18:58:24 *********
 * Schematics Netlist *
 ****      TRANSIENT ANALYSIS
******************************
   TIME          V(6)
   0.000E+00    2.574E-06
   1.000E-05   -2.222E-01
   2.000E-05   -1.623E-01
   3.000E-05   -1.557E-01
   4.000E-05   -1.590E-01
   5.000E-05   -1.589E-01
   6.000E-05   -1.587E-01
   7.000E-05   -1.588E-01
   8.000E-05   -1.588E-01
   9.000E-05   -1.588E-01
   1.000E-04   -1.702E-01
```

9.35

```
 * Schematics Netlist *
 ****      CIRCUIT DESCRIPTION
******************************
X_U1     0 1 2 OPAMP
X_U2     0 3 4 OPAMP
C_C1     5 3 100n
C_C3     3 4 10u
C_C4     6 1 100n
C_C5     1 2 1n
C_C6     0 1 1n
C_C7     0 3 1n
C_C8     0 7 .33u
R_R1     4 7 680
R_R2     2 5 10k
R_R3     1 4 47k
R_R4     8 6 22k
R_R5     0 6 10k
V_V2     8 0 DC 0 AC 5
*Control statement
.INC C:\PS\OPAMP.LIB
**** INCLUDING C:\PS\OPAMP.LIB
*the op amp subcircuit:
.SUBCKT OPAMP 1 2 3
*NODE 1:+, 2:- 3:output
RIN    1     2      1MEG
E1     4   0   1   2   100K
RO     4     3      30
.ENDS
**** RESUMING PS4174.CIR ****
.AC DEC 10 15.9 1592.35
*Output V(7)
.PRINT AC V(7)
.END
```

FREQ	V(7)
1.590E+01	6.381E-01
2.002E+01	7.505E-01
2.520E+01	8.605E-01
3.172E+01	9.583E-01
3.994E+01	1.035E+00
5.028E+01	1.082E+00
6.330E+01	1.097E+00
7.969E+01	1.077E+00
1.003E+02	1.019E+00
1.263E+02	9.253E-01
1.590E+02	7.994E-01
2.002E+02	6.525E-01
2.520E+02	5.009E-01
3.172E+02	3.613E-01
3.994E+02	2.455E-01

Chapter 10
AC Steady-State Power

10.1 Average power

10.1 Find the average power delivered to a 4Ω resistor carrying a current of

$$\begin{aligned} (a)\ i(t) &= 8|\sin(4t)|A \\ (b)\ i(t) &= 8\sin(4t)A \qquad 0 \le t \le \frac{\pi}{4}s \\ &= 0A \qquad \frac{\pi}{4} \le t \le \frac{\pi}{2}s \end{aligned}$$

10.2 A current of $i = 163 + 163\cos(wt)$A flows into a 1Ω resistor. Determine the instantaneous power $P(t)$, its period T_p and the average power P.

10.3 The same current i of Prob.10.2 flows into the positive terminal of a two-terminal device with impedance $Z = 1 + j\Omega$ at frequency ω rad/s. Determine the instantaneous power $P(t)$, its period T_p and the average power P absorbed by the device.

10.4 Find the average power absorbed by each resistor, the capacitor, and the source.

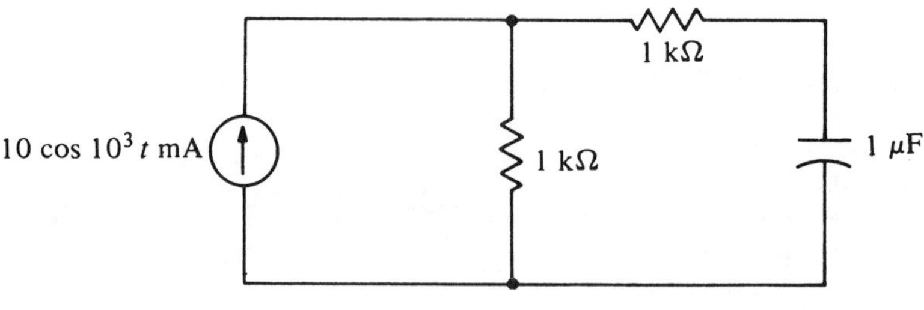

Problem 10.4

10.5 Find the average power delivered by the voltage source of the circuit shown.

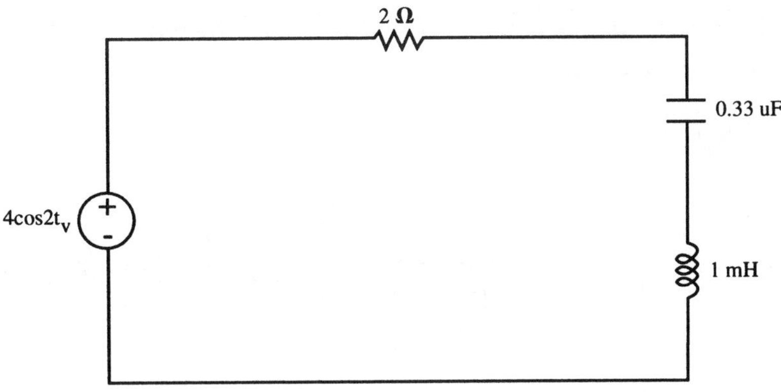

Problem 10.5

10.6 Determine the power delivered by the current source of the circuit shown.

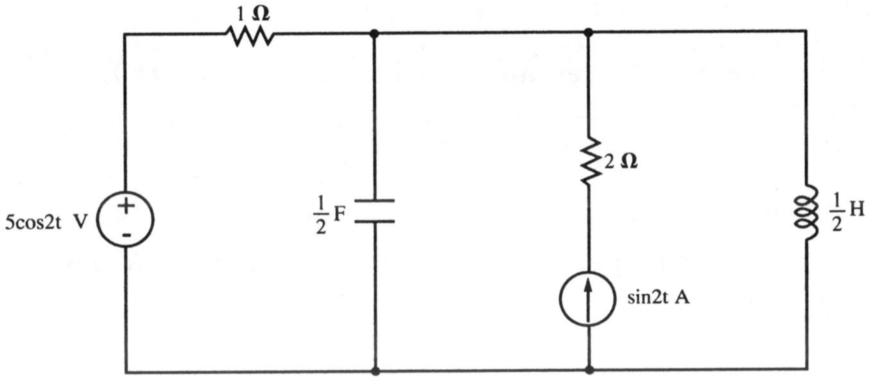

Problem 10.6

10.7 Find the average power delivered to the $\frac{1}{2}\Omega$ resistor.

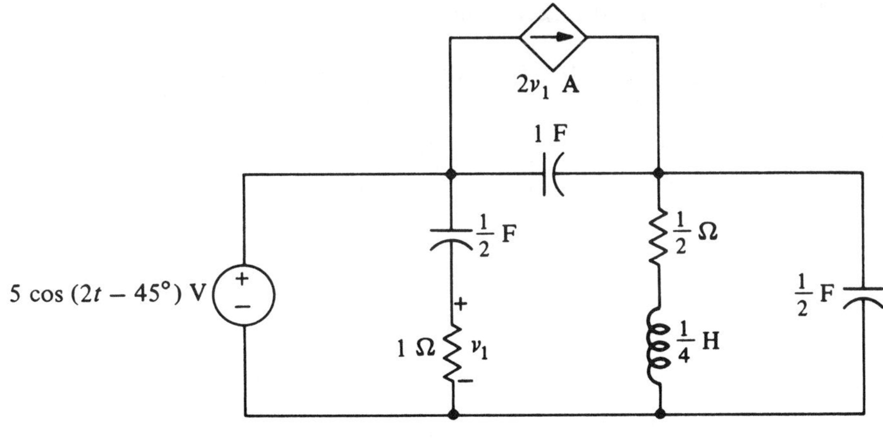

Problem 10.7

10.8 One cycle of periodic current is given by $i = 0.1(1 - e^{-t})$A, $0 \le t \le 2$ if the current flows in a 10Ω resistor, find the average power.

10.9 A periodic current i with period T = 2sec flows into a $1K\Omega$ resistor, find the the average power dissipated by the resistor.

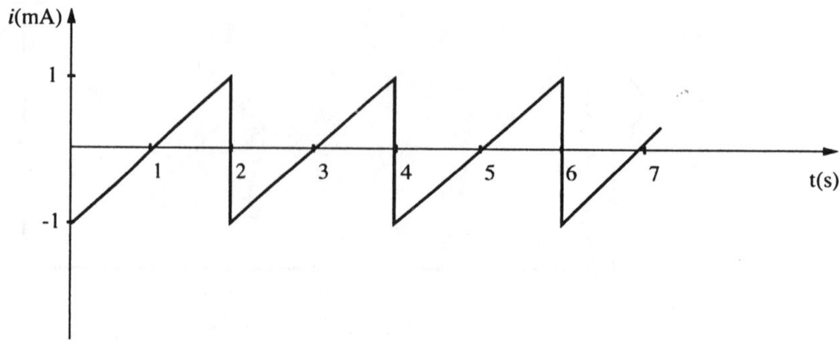

Problem 10.9

10.10 Find the average power absorbed by a resistor $R\Omega$, if it carries a current $i = I_m(1 + m\cos(\omega_m t))\sin(\omega_c t)$A. Assume $\omega_c > \omega_m$; m, ω_m and ω_c are constants; ω_m and ω_c are positive rational numbers. [Hint: given $f_1(t)$ and $f_2(t)$ are periodic functions with periods T_1 and T_2 respectively, then $f_1(t) + f_2(t)$ is also periodic of period T if positive integers K and L exist such that $T = KT_1 = LT_2$.]

10.2 RMS Values

10.11 Find the RMS value of the current i in Prob.10.9.

10.12 Find the RMS value of (a) $i = I(4 - 3\cos(377t))$A (b) $v = 12\sin(\omega t) + 5\cos(\omega t + 45°)$V (c) $v = 60\cos(\omega t) + 120\sin(2\omega t + 30°)$V

10.13 Find the rms value of a periodic for which one cycle is given by

$$\begin{aligned} i &= \sqrt{t}A, & 0 \le t \le 1s \\ &= 0A, & 1 \le t \le 3s. \end{aligned}$$

10.14 Find I_{rms}.

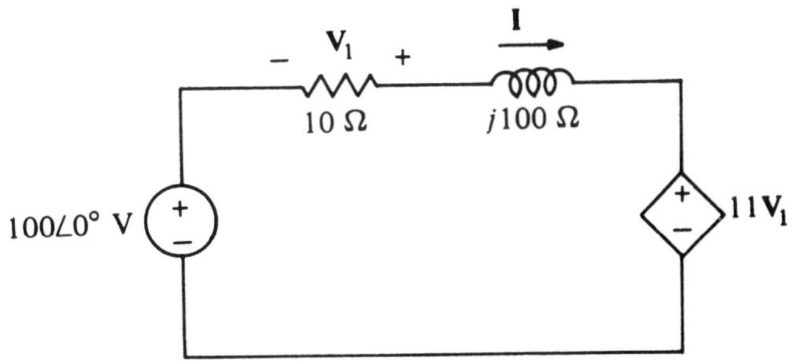

Problem 10.14

10.15 Find the rms voltage V_{rms} across the current source of the circuit shown.

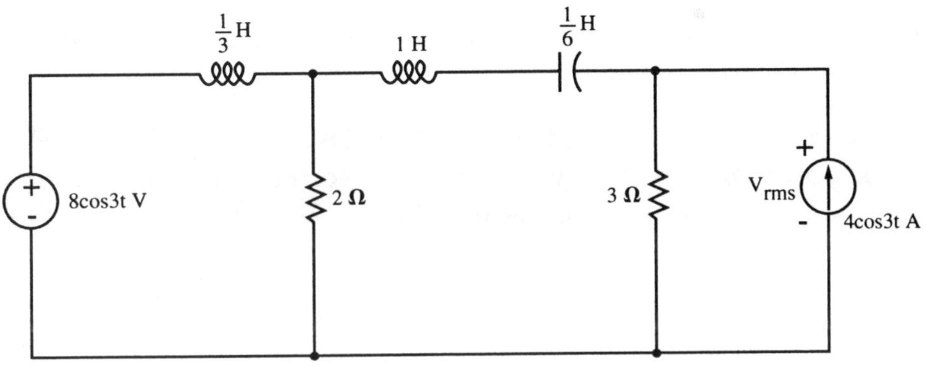

Problem 10.15

10.3 Complex Power

10.16 Find the complex power delivered to a load which has 0.707 lagging power factor and absorbs 40W.

10.17 Three passive loads, Z_1, Z_2 and Z_3 are receiving complex power values of $2 + j3$, $3 - j1$, and $1 + j6$VA, respectively. If these loads and a voltage source of $20\angle 0°$V rms are connected in series, find the rms value of the current that flows and the power factor seen by the source.

10.18 For the circuit shown, Z_1 draws $2 + j0$VA and Z_2 draws 10VA at a power factor of 0.6 lagging. If $I_g = 2\sqrt{2}\angle 15°$A rms find V_{rms}.

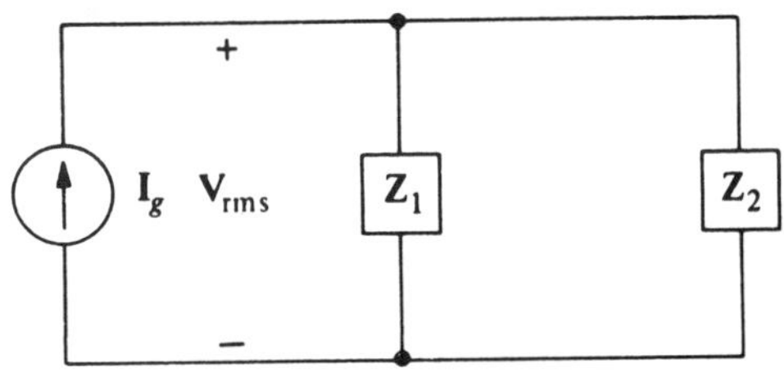

Problem 10.18

10.19 Find the complex power delivered by the source and the power factor seen by the source.

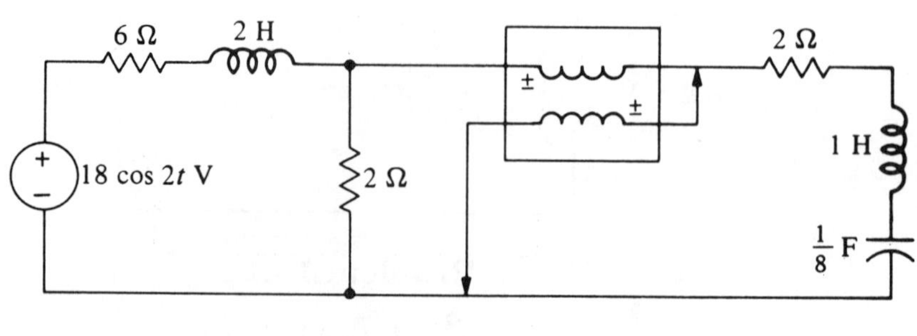

Problem 10.19

10.20 Find the complex power and the average power delivered by the voltage source of Prob.10.15.

10.21 For the circuit shown R_1=R_2=R_3=22KΩ, C_1=C_2=0.01uF, (a) find the complex power delivered by the voltage source $v_g = 7.07\cos(2\pi \times 10^4 t)$V; (b) does the answer to (a) depends on the source phase angle ϕ if $v_g = 7.07\cos(2\pi \times 10^4 t - \phi)$V.

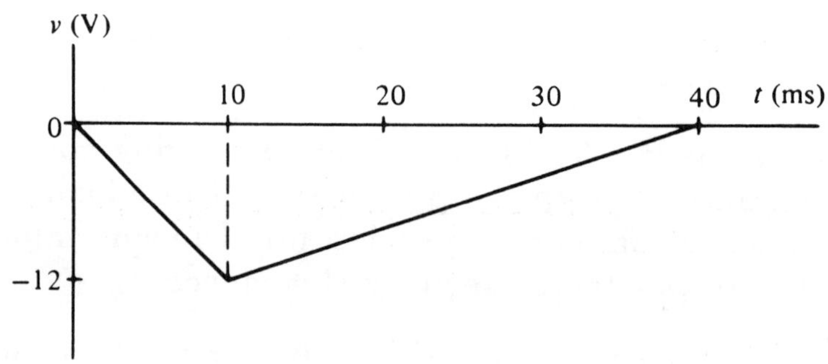

Problem 10.21

10.4 Superposition and power

10.22 Find the average power delivered to the resistor if R=1KΩ and $v_{g_1} = 5\cos(10t)$V and $v_{g_2} = 5\sin(10t)$V.

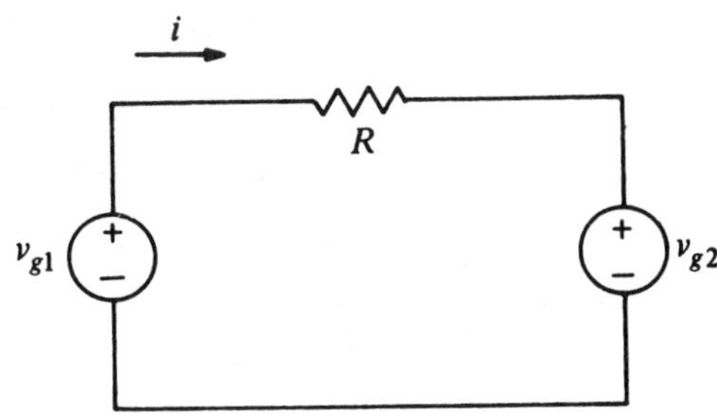

Problem 10.22

10.23 Find the average power delivered to R.

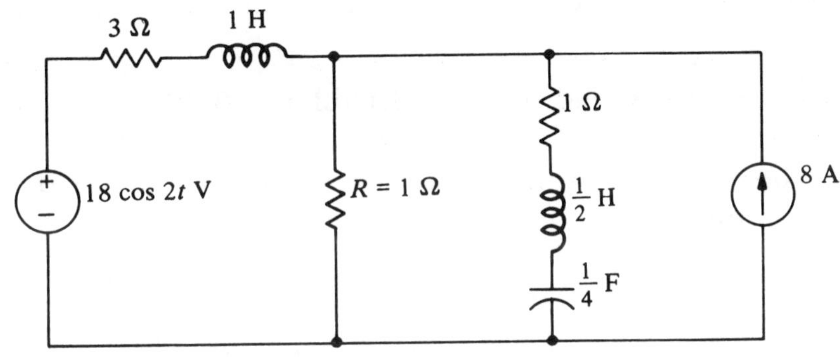

Problem 10.23

10.24 Find the average power delivered to the 1Ω resistor.

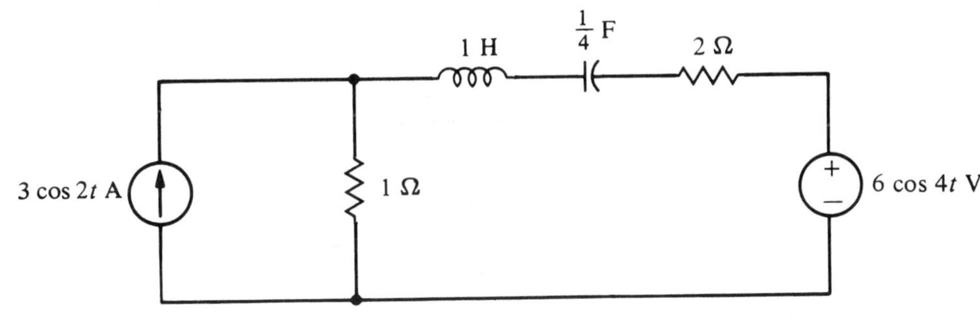

Problem 10.24

10.25 Show that superposition holds for (a) reactive power Q (b) complex power S when sources are of different frequencies.

10.5 Maximum power transfer

10.26 For the circuit shown, find (a) the load impedance Z_L which absorbe maximum power (b) the value of the maximum power delivered to the load.

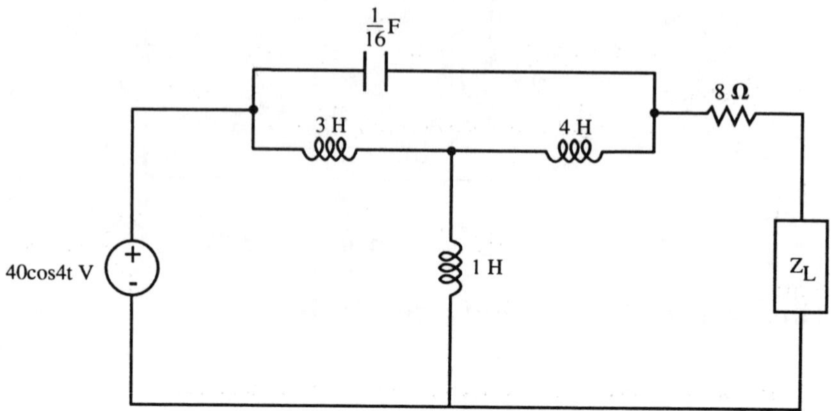

Problem 10.26

10.27 Find the value of transresistance γ for which the power delivered by the independent source is maximized.

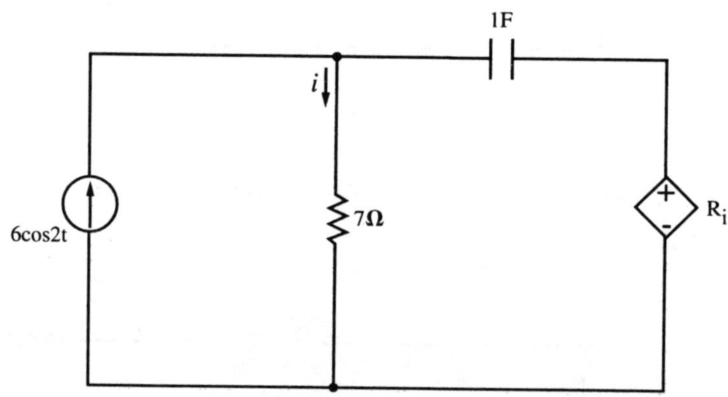

Problem 10.27

10.28 Construct the load impedance Z_L using **R, L, C** elements such that maximum power is transfered from the rest of circuit to the load.

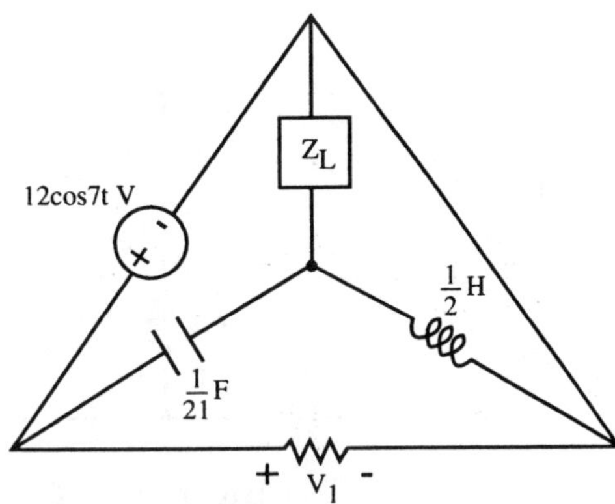

Problem 10.28

10.29 (a)Find the amplitude **A** of the voltage source which delivers **10W** to the purely resistive load $Z_L = 8\Omega$; (b)adjust Z_L to absorb **10W** using the lowest value of **A**.

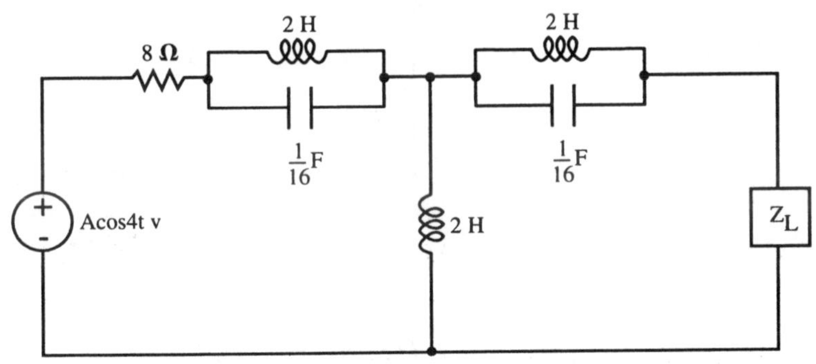

Problem 10.29

10.7 Reactive power and power factor

10.30 Determine (a) the apparent power delivered by the voltage source to the rest of the circuit (b) the power factor seen by the voltage source.

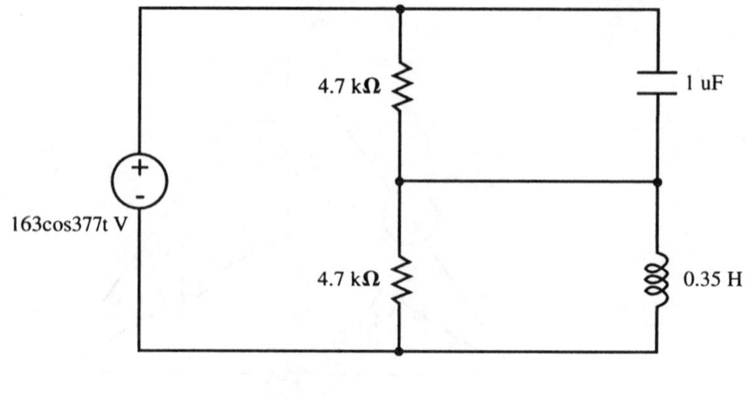

Problem 10.30

10.31 Find the reactive power absorbed by the inductor if (a) ω=3 rad/sec; (b) ω=0.

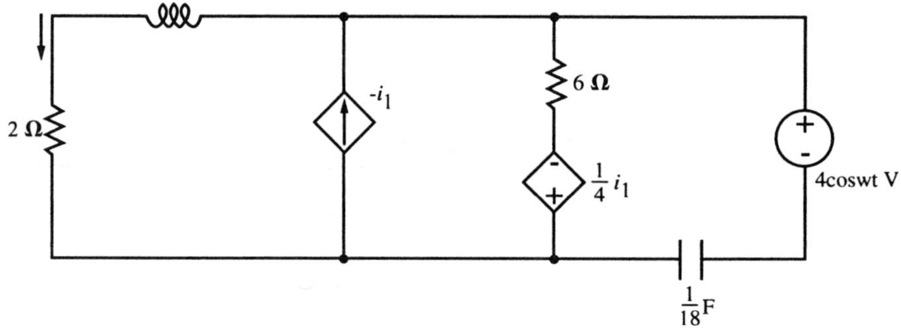

Problem 10.31

10.32 Find the total reactive power absorbed by the inductor and the capacitor. (Hint: find the complex power delivered by the source).

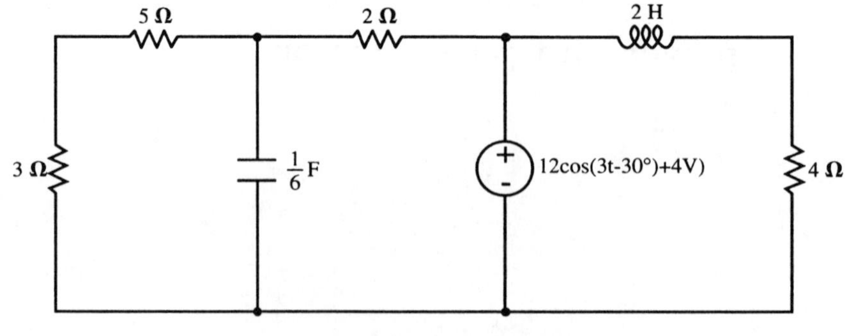

Problem 10.32

10.33 Correct the power factor seen by the voltage source of Prob.10.30 to (a) 0.95 leading (b) 0.95 lagging without reducing the voltage across each element.

10.34 A load consists of a 200Ω resistor in series with a 0.1H inductor. Find the parallel capacitance necessary to adjust to a power factor of 0.95 lagging if ω=1000 rad/s.

10.35 A load consists of a 200Ω resistor in series with a 0.1H inductor. Find the parallel capacitance necessary to adjust to a power factor of 0.95 leading if ω=1000 rad/s.

10.36 Find the average power delivered to the load Z_L. Assume C=1uF.

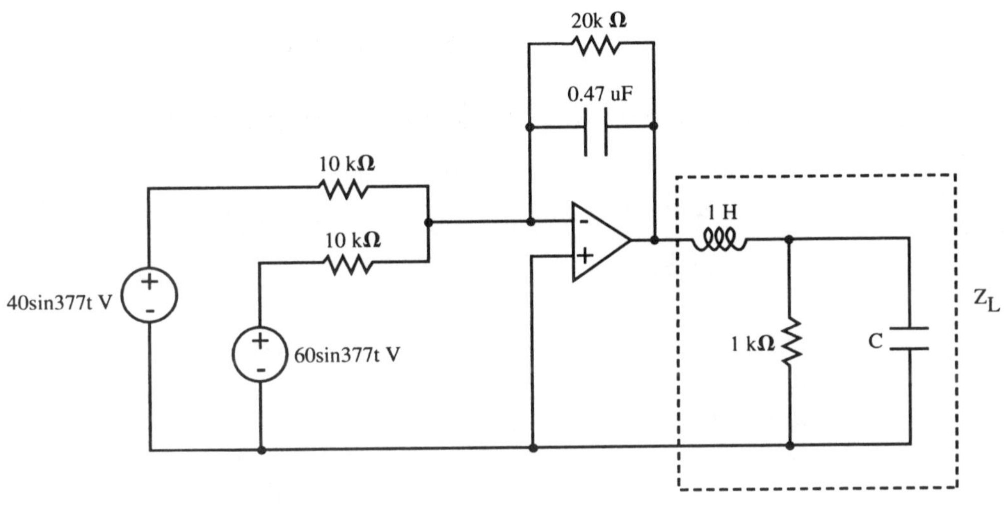

Problem 10.36

10.37 Determine the value of the tuning capacitor C in Prob.10.36. such that maximum power is delivered to the load Z_L.

10.38 Use SPICE to help determine the value of C for which the load Z_L absorbed maximum power in Prob.10.37.

10.1 (a) $P = \frac{1}{2\pi/4}\int_0^{2\pi/4} (8|\sin 4t|)^2 dt$

$= \frac{128}{\pi}\int_0^{\pi/2}\left(\frac{1}{2} - \frac{1}{2}\cos 8t\right)dt$

$= \frac{64}{\pi}\left(\frac{\pi}{2} - \frac{1}{8}\sin 8t\Big|_0^{\pi/2}\right)$

$= \underline{32\,W}$

(b) $P = \frac{1}{2\pi/4}\int_0^{\pi/4} (8\sin 4t)^2 dt$

$= \frac{128}{\pi}\int_0^{\pi/4}\left(\frac{1}{2} - \frac{1}{2}\cos 8t\right)dt$

$= \frac{64}{\pi}\left(\frac{\pi}{4} - \frac{1}{8}\sin 8t\right.$

$= \underline{16W}$

10.2 $P(t) = i^2R = \underline{1\cdot(163 + 163\cos\omega t)^2 W}$

$\underline{T_p = \frac{2\pi}{\omega} s}$

$P_{Avg} = \frac{1}{T_p}\int_0^{T_p}(163 + 163\cos\omega t)^2 dt$

$= \frac{1}{T_p}\int_0^{T_p} 163^2(1 + 2\cos\omega t + \cos^2\omega t)dt$

$= \frac{1}{T_p}\int_0^{T_p} 163^2\left(1 + 2\cos\omega t + \frac{1}{2} + \frac{1}{2}\cos 2\omega t\right)dt$

$= (163)^2\left(\frac{3}{2}\right) = \underline{39853.5W}$

10.3 Superposition does not hold for instantaneous power and average power.

$P(t) = i^2R = 163^2(1 + \cos\omega t)^2(1+j)$

$= 163^2\left(1 + 2\cos\omega t + \frac{1}{2} + \frac{1}{2}\cos 2\omega t\right)(1+j)$

$= 163^2\left(\frac{3}{2} + 2\cos\omega t + \frac{1}{2}\cos 2\omega t\right)(1+j)$

$= 163^2\left[\frac{3}{2} + 2\sqrt{2}\cos(\omega t + 45°) + \frac{1}{2}\cos(2\omega t + 45°)\right]$

$= \underline{39854 + 53138\cos(\omega t + 45°) + 18790\cos(2\omega t + 45°)\ W}$

$\underline{T_p = \frac{2\pi}{2\omega} = \frac{\pi}{\omega} s}$

$P_{Avg} = I_{dc}^2\, Re\{Z\} + \frac{1}{2} I_m^2\, Re\{Z\}$

$= 163^2\left[1 + \frac{1}{2}(1)\cos^2\omega t\right]$

$= 26569\left(1 + \frac{1}{2}\cos^2\omega t\right)$

$= 26569\left(1 + \frac{1}{4} + \frac{1}{4}\cos 2\omega t\right)$

$= \underline{33211.3 + 6642.3\cos 2\omega t\ W}$

10.4

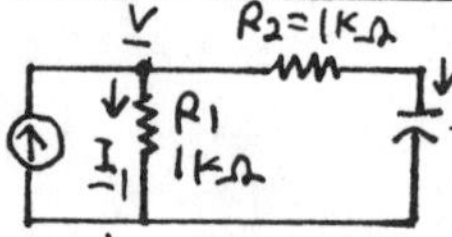

By current division

$I_1 = \frac{1-j}{1+1-j}(10mA) = 6 - j2 = \sqrt{40}\angle -18.43°mA$

10.4 Cont.

$I_2 = 10 - I_1 = 10 - 6 + j2 = \sqrt{20}\angle 26.57°mA$

$V = I_1R_1 = \sqrt{40}\angle -18.43°V$

$P_{R_1} = \frac{I_{m1}^2}{2}(1K\Omega) = \underline{20mW}$

$P_{R_2} = \frac{I_{m2}^2}{2}(1K\Omega) = \underline{10\,mW}$

$P_{capacitor} = \frac{I_{m2}}{2} Re\{Z\} = \underline{0W}$

$P_{source} = \frac{V_m}{2}(10mA)\cos(-18.43° - 0°)$

$= \underline{-30mW}$

10.5 Both capacitor and inductor are loseless elements

$P = \frac{1}{2}R I_m^2 = \frac{1}{2}(2)(4)^2 = \underline{16W}$

10.6

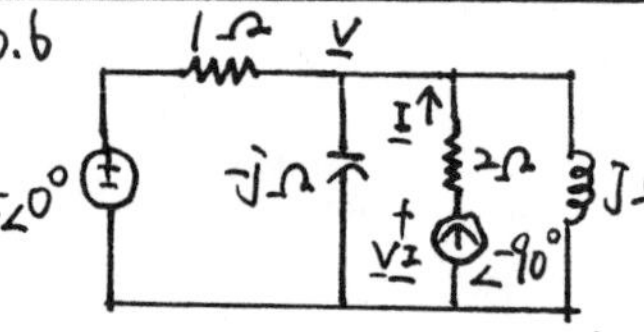

$\frac{V-5}{1} + \frac{V}{-j} + \frac{V}{j} = -j$

or $\underline{V} = 5 - j$

$V_I = V + 2I = 5 - j - 2j = 5 - 3j = 5.8\angle -31°$

$I = 1\angle -90°$

$P_s = \frac{1}{2}(5.8)(1)\cos(-31° + 90°) = \frac{1}{2}(5.8)(0.52)$

$= \underline{1.49W}$

10.7 v = capacitor voltage of right $\frac{1}{2}$F capacitor. By KCL

$j2(V - 5\angle -45°) + \frac{V}{\frac{1}{2} + j\frac{1}{2}} + jV = 2V_1,$

where $V_1 = \frac{(1)(5\angle -45°)}{1-j} = \frac{5}{\sqrt{2}}V$

Then $V = 5\sqrt{2}\angle -36.9°V$

$|I_{1/2\Omega}| = \frac{|V|}{\left|\frac{1}{2} + j\frac{1}{2}\right|} = 10A;$

$P = \frac{(I_{1/2\Omega})^2}{2}\left(\frac{1}{2}\right) = \underline{25W}$

10.8 $P = \frac{1}{2}\int_0^2 i^2 R\,dt = \frac{1}{2}\int_0^2 (10)(0.1)^2(1 - e^{-t})^2 dt$

$= (0.05)\int_0^2 (1 - 2e^{-t} + e^{-2t})dt$

$= (0.05)\left[2 + 2e^{-t} - 2 - \frac{e^{-4}}{2} + \frac{1}{2}\right]$

$= (0.05)\left[\frac{1}{2} + 2e^{-2} - \frac{e^{-4}}{2}\right] = \underline{38.08mW}$

10.9 $P_{Avg} = \frac{1}{2}\int_0^2 i^2 R\,dt = \frac{1}{2}\int_0^2 10^3[(t-1)10^{-3}]^2 dt$

$= \frac{10^{-3}}{2}\left(\frac{1}{3}\right)(t-1)^3\Big|_0^2 = \frac{10^{-3}}{6}\left[1^3-(-1)^3\right]$

$= \frac{1}{3}\times 10^{-3} = \underline{0.333\text{ mW}}.$

10.10 Instantaneous power $p(t)$ is

$p(t) = i^2R = [I_m(1+m\cos W_m t)\ \sin W_c t]^2 R$

$= \left\{(I_m \sin W_c t)^2 + \left[\frac{I_m m}{2}\cos(w_c+w_m)t\right]^2 + \left[\frac{I_m m}{2}\cos(w_c-w_m)t\right]^2\right\} R$

Let $T_1 = \frac{2\pi}{w_c}$, $T_2 = \frac{2\pi}{w_c+w_m}$ and $T_3 = \frac{2\pi}{w_c - w_m}$

$\because$ w_c, w_m are positive rational numbers, there exists T such that $T = k_1T_1 = k_2T_2 = k_3T_3$ where k_1, k_2 and k_3 are positive integers.

Average power P_{Avg}

$P_{Avg} = \frac{1}{T}\int_0^T R\left\{I_m^2\sin^2 W_c t + \frac{I_m^2 m^2}{4}\cos^2(w_c+w_m) + \frac{I_m^2 m^2}{4}\cos^2(w_c-w_m)t\right\}dt$

$= \frac{1}{T}\int_0^T RI_m^2\left\{\left(\frac{1}{2}-\frac{1}{2}\cos 2w_c t\right) + \frac{m^2}{4}\left[\frac{1}{2}-\frac{1}{2}\cos 2(w_c+w_m)t\right] + \frac{m^2}{4}\left[\frac{1}{2}-\frac{1}{2}\cos 2(w_c-w_m)t\right]\right\}dt$

$= \frac{1}{T}\int_0^T RI_m^2\left(\frac{1}{2}+\frac{m^2}{8}+\frac{m^2}{8}\right)dt$

$= \left(1+\frac{m^2}{4}+\frac{m^2}{4}\right)\frac{1}{2}RI_m^2 = \underline{\frac{1}{2}\left(1+\frac{m^2}{2}\right)RI_m^2\text{ W}}$

10.11 $I_{rms}^2 = \frac{1}{T}\int_0^T i^2 dt = \frac{1}{2}\int_0^2 (t-1)^2 dt$

$= \frac{1}{2}\left(\frac{1}{3}\right)(t-1)^3\Big|_0^2 = \frac{1}{6}(1+1) = \frac{1}{3}$

$I_{rms} = \underline{\sqrt{\frac{1}{3}}\text{ mA}}$

10.12 (a) $I_{rms}^2 = \frac{1}{T}\int_0^T i^2 dt$

$= 60\int_0^{\frac{1}{60}}[I(4-3\cos 377t)]^2 dt$

$= 60I^2\int_0^{\frac{1}{60}}(16-24\cos 377t + 9\cos^2 377t)dt$

$= 60I^2\left(\frac{41}{2}\right)\left(\frac{1}{60}\right) = \frac{41}{2}I^2$, $\underline{I_{rms} = \sqrt{\frac{41}{2}}\,I\text{ A}}$

(b) $V_{rms}^2 = \frac{w}{2\pi}\int_0^{\frac{2\pi}{w}}[12\sin wt + 5\cos(wt+45^\circ)]^2 dt$

$= \frac{w}{2\pi}\int_0^{\frac{2\pi}{w}}[144\sin^2 wt + 120\sin wt\cos(wt+45^\circ) + 25\cos^2(wt+45^\circ)]dt$

10.12 Cont.

$= \frac{w}{2\pi}\int_0^{\frac{2\pi}{w}}\left\{\left[\frac{144-144\cos 2wt}{2} + \frac{120\sin(2wt+45^\circ)+120\sin(-45^\circ)}{2} + \frac{25+25\cos(2wt+90^\circ)}{2}\right]\right\}dt$

$= \frac{144}{2} + \frac{120(-\frac{1}{\sqrt{2}})}{2} + \frac{25}{2} = 84.5 - 42.4 = 42.1$

$V_{rms} = \sqrt{42.1} = \underline{6.5\text{ V}}$

(c) $V_{rms}^2 = \frac{w}{2\pi}\int_0^{\frac{2\pi}{w}}[60\cos wt + 120\sin(2wt+30^\circ)]^2 dt$

$= \frac{w}{2\pi}\int_0^{\frac{2\pi}{w}}[3600\cos^2 wt + 14400\cos wt\sin(2wt+30^\circ) + 14400\sin^2(2wt+30^\circ)]dt$

$= 3600\frac{w}{2\pi}\int_0^{\frac{2\pi}{w}}\left[\frac{1+\cos 2wt}{2} + (4)\frac{\sin(3wt+30^\circ)+\sin(wt+30^\circ)}{2} + (4)\frac{1-\cos(4wt+60^\circ)}{2}\right]dt$

$= 3600\,\frac{5}{2} = 9000$

$V_{rms} = \sqrt{9000} = \underline{94.9\text{ V}}$

10.13 $I_{rms}^2 = \frac{1}{3}\left[\int_0^1 (\sqrt{t})^2 dt + \int_1^3 0\,dt\right] = \frac{1}{3}\left(\frac{1}{2}\right) = \frac{1}{6}$

$I_{rms} = 1/\sqrt{6} = \underline{0.354\text{ A}}$

10.14 $V_1 = -I(10)$, then KVL gives

$I(10+j100) + 11V_1 = 100$ or

$I = \frac{100}{-100+j100} = \frac{1}{\sqrt{2}}\angle -135^\circ\text{ A}$

$I_{rms} = \frac{1/\sqrt{2}}{\sqrt{2}} = \underline{1/2\text{ A}}$

10.15 In rms form, the circuit diagram is

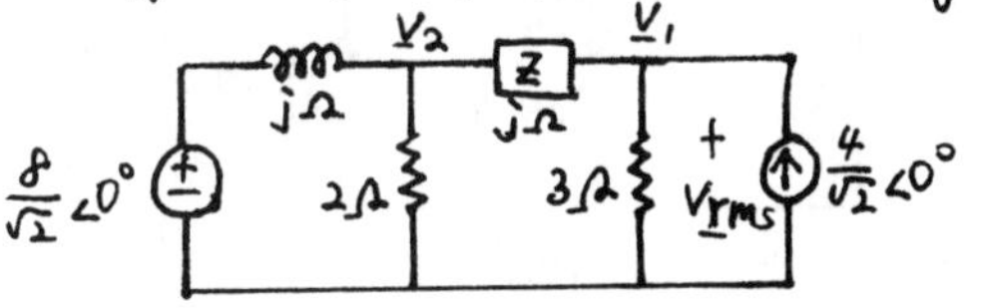

$\frac{8}{\sqrt{2}}\angle 0^\circ$ V_2 $j\Omega$ 2Ω Z $j\Omega$ V_1 3Ω $+$ V_{rms} $\frac{4}{\sqrt{2}}\angle 0^\circ$

Nodal analysis gives

$\begin{cases}\frac{V_2 - \frac{8}{\sqrt{2}}}{j} + \frac{V_2}{2} + \frac{V_2 - V_1}{j} = 0 \\ \frac{V_1 - V_2}{j} + \frac{V_1}{3} = \frac{4}{\sqrt{2}}\end{cases}$ or $\begin{cases}2V_1 - (4+j)V_2 = -8\sqrt{2} \\ V_1(3+j) - 3V_2 = j\frac{12}{\sqrt{2}}\end{cases}$

$V_1 = \begin{vmatrix}-8\sqrt{2} & -(4+j) \\ j6\sqrt{2} & -3\end{vmatrix} \Big/ \begin{vmatrix}2 & -(4+j) \\ 3+j & -3\end{vmatrix} = 4.93 - j0.1146 = 4.93\angle -1.3^\circ$

$V_{rms} = V_1 = \underline{4.93\cos(3t - 1.3^\circ)\text{ V}}$

10.16 $\theta = \cos^{-1}(0.707) = 45°$; $= P\tan\theta$

$Q = 40\tan 45° = 40\ VAR$

$\underline{S} = P + jQ = 40 + j40 = \underline{40\sqrt{2}\angle 45°\ VA}$

10.17 $\underline{S} = \underline{S}_1 + \underline{S}_2 + \underline{S}_3$

$= 2 + j3 + 3 - j + 1 + j6$

$= 6 + j8 = 10\angle 53.13°$

$PF = \cos(53.13°) = \underline{0.6}$ (lagging)

$\underline{S} = \underline{V}_{rms}\underline{I}^*_{rms} \Rightarrow 10\angle 53° = 20\,\underline{I}^*_{rms}$

$\underline{I}_{rms} = \underline{0.5\angle -53.13°A}$ rms

10.18 $\underline{S}_1 = 2 + j0\ VA$;

$\theta_2 = \cos^{-1}(0.6) = 53.13°$

$\underline{S}_2 = 10\angle 53.13° = 6 + j8\ VA$

$\underline{S} = \underline{S}_1 + \underline{S}_2 = 2 + 6 + j8 = 8\sqrt{2}\angle 45°\ VA$

$\underline{S} = \underline{V}_{rms}\underline{I}^*_{rms} \Rightarrow \underline{V}_{rms} = \dfrac{8\sqrt{2}\angle 45°}{2\sqrt{2}\angle -15°}$

$= \underline{4\angle 60°\ V}$

10.19 KCL yields

$$\frac{10 - 5\underline{I}}{j2} - \underline{I} + \frac{10 - 5\underline{I} - 2\underline{I}}{-j4} = 0$$

$\underline{I} = 2\angle -53.1°A$ rms

$\underline{S} = \underline{V}_{rms}\underline{I}^*_{rms} = 10\ (2\angle 53.1°) = 20\angle 53.1° VA$

$PF = \cos(53.1°) = \underline{0.6}$ (lagging)

$\underline{S} = 20\angle 53.1° = \underline{12 + j16\ VA}$

10.20 In rms form, the circuit diagram is

V_2, V_1, $j\Omega$, $j3\Omega$, $-j2\Omega$, 2Ω, 3Ω, $\frac{8}{\sqrt{2}}\angle 0° V$, $\frac{4}{\sqrt{2}}\angle 0° A$

From the solution of Prob. 10.15

$\underline{V}_1 = 4.93\angle -1.3° \cong 4.93$

By KCL

$$\frac{V_2 - \frac{8}{\sqrt{2}}}{j} + \frac{V_2}{2} + \frac{V_2 - V_1}{j} = 0 \text{ or } V_2 = \frac{8\sqrt{2} + 2V_1}{4 + j}$$

$$V_2 \cong \frac{8\sqrt{2} + 9.9}{4 + j} = \frac{21.2}{4.1\angle 14°} = 5.2\angle -14° = 5.0 + j(-1.3)$$

The current flows out of the positive terminal of the voltage source is

$\underline{I} = \left[\frac{8}{\sqrt{2}} - (5.0 - j1.3)\right]/j = (0.7 + j1.3)/j = -1.3 + j0.7$

10.20 Cont.

$\underline{I} = 1.5\angle -28.3°$

Complex power delivered by the voltage source:

$P = \frac{8}{\sqrt{2}}(1.5\angle -28.3°)^* = \frac{8}{\sqrt{2}}(1.5\angle 28.3°) = \underline{8.5\angle 28.3° VA}$

The average power delivered by the voltage source is $\text{Re}\{8.5\angle 28.3°\} = \underline{7.5 W}$

10.21 $V_{grms} = \dfrac{7.07}{\sqrt{2}} = 5\angle 0° = 5\ V_{rms}$

Let the voltage on top of R_2 be $\underline{V}_1$ and the voltage at the inverting input terminal of op amp be $\underline{V}_0$. Nodal analysis yields:

$$\begin{cases} \dfrac{V_1 - 5}{R_1} + \dfrac{V_1}{R_2} + \dfrac{V_1}{1/j\omega C_1} + \dfrac{V_1 - V_0}{1/j\omega C_2} = 0 \\ \dfrac{0 - V_1}{1/j\omega C_1} + \dfrac{0 - V_0}{R_3} = 0 \end{cases} \quad \text{and } R_1 = R_2,\ C_1 = C_2$$

$$\text{or} \begin{cases} \underline{V}_1\left(\frac{2}{R_1} + j2\omega C_1\right) - \underline{V}_0(j\omega C_1) = \frac{5}{R_1} \\ V_1(j\omega C_1) + V_0\frac{1}{R_1} = 0 \end{cases}$$

$j\omega C_1 = j2\pi \times 10^4 \times 10^{-8} = j0.000628$

$\frac{1}{R_1} = \frac{1}{22\times 10^3} = 0.000045$

$$\Rightarrow \begin{cases} \underline{V}_1(0.000090 + j0.001256) - \underline{V}_0(j0.000628) = 0.000225 \\ \underline{V}_1(j0.000628) + 0.000045\,\underline{V}_0 = 0 \end{cases}$$

$$\underline{V}_1 = \begin{vmatrix} 0.000225 & -j0.000628 \\ 0 & 0.000045 \end{vmatrix} \Bigg/ \begin{vmatrix} 0.000090 + j0.001256 & -j0.000628 \\ j0.000628 & 0.000045 \end{vmatrix}$$

$= -0.0254 - j0.0037$

$P = \underline{V}_g\underline{I}_g = 5\left[\dfrac{5 - (-0.0254 - j0.0037)}{22\times 10^3}\right] \cong \underline{1.14\angle 0° mVA}$

(b) Both resistor and capacitor are linear elements. The negative feedback of op-amp is approximated linear. Hence any phase delay (advance) introduced at voltage source V_g will cause same phase delay (advance) in network voltage and current.
∴ The complex power delivered by the voltage source will depend on phase angle ϕ but the magnitude is unchanged.

10.22 $\mathbf{I} = (\mathbf{V}_{g1} - \mathbf{V}_{g2})/R = 5+j5/1\text{k} = 5\sqrt{2}\angle 45^\circ$ mA

$P = \frac{I_m^2}{2}(1\text{k}\Omega) = \underline{25\text{mW}}$

10.23 v = voltage across R; $i = v/R = v$

With only the voltage source active $v = v_1$ and KCL yields

$$\frac{V_1-18}{3+j2} + \frac{V_1}{1} + \frac{V_1}{1+j-j2} = 0 \Rightarrow V_1 = 2\sqrt{2}\angle -45^\circ \text{ V}$$

$\therefore I_1 = \frac{V_1}{1} = 2\sqrt{2}\angle -45^\circ$ A

With only the current source active $i = i_2$ and by current division

$i_2 = \frac{3}{3+1}\,8 = 6\text{A (dc)}$

$P = i_2^2 R + \frac{1}{2}|I_1|^2 R = (6)^2 + \frac{1}{2}(2\sqrt{2})^2 = \underline{40\text{W}}$

10.24 i = current down in the 1Ω; with only the current source active, $i = i_1$ and by current div.

$$I_1 = \frac{j2-j2+2}{j2-j2+2+1}(3) = 2\angle 0^\circ \text{ A}$$

With only the voltage source active $i = i_2$ and KVL yields

$$I_2 = \frac{6}{1+j4-j+2} = \sqrt{2}\angle -45^\circ \text{A}$$

$P = P_1 + P_2 = \frac{1}{2}|I_1|^2(1) + \frac{1}{2}|I_2|^2(1)$

$= \frac{1}{2}[(2)^2 + (\sqrt{2})^2] = \underline{3\text{W}}$

10.25 (a) Consider a subcircuit with terminal variables i and v in a circuit containing two sinusoidal sources of frequencies ω_1 and ω_2; $\omega_1 \neq \omega_2$.

By superposition $i_1 + i_2 = i$, $v_1 + v_2 = v$.

Where $i_1 = i_{11} + j\,i_{12}$

$i_2 = i_{21} + j\,i_{22}$

$v_1 = v_{11} + j\,v_{12}$

$v_2 = v_{21} + j\,v_{22}$

i_1, v_1 and i_2, v_2 are currents and voltages produced by source 1 and source 2, respectively.

The average reactive power is

10.25 Cont.

$Q = \frac{1}{T}\int_0^T \text{Im}(i\,v)\,dt$, $\text{Im}(\cdot)$ is the imaginary part of $i\cdot v$

$= \frac{1}{T}\int_0^T \text{Im}(i_1 v_1 + i_2 v_2 + i_1 v_2 + i_2 v_1)\,dt$

$= Q_1 + Q_2 + \frac{1}{T}\int_0^T \text{Im}(i_1 v_2 + i_2 v_1)\,dt$

$\because \frac{1}{T}\int_0^T \text{Im}(i_1 v_2)\,dt$

$= \frac{1}{T}\int_0^T \text{Im}\{(i_{11}v_{21} - i_{12}v_{22}) + j(i_{12}v_{21} + i_{11}v_{22})\}\,dt$

$= \frac{1}{T}\int_0^T (i_{12}v_{21} + i_{11}v_{22})\,dt$

Due to sinusoidal excitations, i_{11}, i_{12}, v_{21} and v_{22} are of the forms

$i_{11} = I_{11}\cos(\omega_1 t + \phi_{i_{11}})$

$i_{12} = I_{12}\cos(\omega_1 t + \phi_{i_{12}})$

$v_{21} = V_{21}\cos(\omega_2 t + \phi_{v_{21}})$

$v_{22} = V_{22}\cos(\omega_2 t + \phi_{v_{22}})$

$\because \omega_1 \neq \omega_2 \Rightarrow \frac{1}{T}\int_0^T (i_{12}v_{21} + i_{11}v_{22})\,dt = 0$

$\Rightarrow \frac{1}{T}\int_0^T \text{Im}(i_1 v_2)\,dt = 0$

Similarly, $\frac{1}{T}\int_0^T \text{Im}(i_2 v_1)\,dt = 0$

$\Rightarrow Q = Q_1 + Q_2 \Rightarrow$ <u>superposition holds for reactive power if $\omega_1 \neq \omega_2$</u>

(b) A similar derivation leads to $P = P_1 + P_2$ (superposition holds for average power)

$\because S = P + Q = (P_1 + P_2) + (Q_1 + Q_2)$

$= (P_1 + Q_1) + (P_2 + Q_2)$

$= S_1 + S_2$

$\Rightarrow$ <u>superposition holds for complex power</u>

10.26 (a) The impedance seen by Z_L is

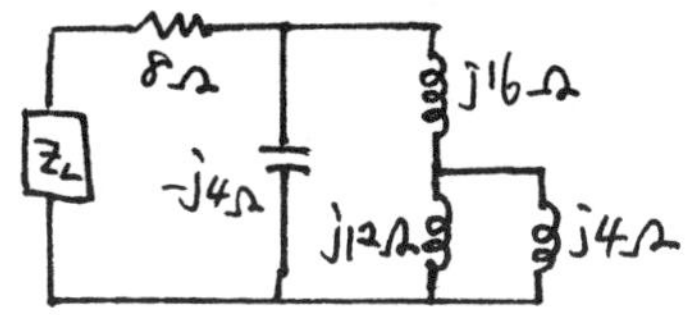

$R_{TH} = 8 + \{-j4 \,//\, [j16 + (j12 // j4)]\}$

$= 8 + \{-j4 \,//\, [j16 + j3]\} = 8 + \{-j4 // j19\}$

$= 8 + \frac{76}{j15} = 8 - j\frac{76}{15}$

10.26 Cont.

To absorb maximum power

$$Z_L = \left(8 - j\frac{76}{15}\right)^* = 8 + j\frac{76}{15}\ \Omega$$

(b)

$-j4\,\Omega$, V_1, V_2, $8\,\Omega$, V_3, $j12\,\Omega$, $j16\,\Omega$, $j4\,\Omega$, $40\angle 0^\circ$ V, Z_L $8+j\frac{76}{15}$

$$\begin{cases} V_1\left(\frac{1}{j12}+\frac{1}{j4}+\frac{1}{j16}\right) - V_2\left(\frac{1}{j16}\right) = \frac{40}{j12} \\ V_1\left(-\frac{1}{j16}\right) + V_2\left(\frac{1}{j16}+\frac{1}{8}+\frac{1}{-j4}\right) - V_3\left(\frac{1}{8}\right) = \frac{40}{-j4} \\ 0 - V_2\left(\frac{1}{8}\right) + V_3\left(\frac{1}{8}+\frac{1}{8+j\frac{76}{15}}\right) = 0 \end{cases}$$

or

$$\begin{cases} -j0.396\,V_1 + j0.0625\,V_2 = -j3.333 \\ j0.625\,V_1 + (0.125+j0.187)\,V_2 - 0.125\,V_3 = j10 \\ 0\,V_1 - 0.125\,V_2 + (0.214-j0.057)\,V_3 = 0 \end{cases}$$

$$V_3 = \begin{vmatrix} -j0.396 & j0.0625 & -j3.333 \\ j0.625 & 0.125+j0.187 & j10 \\ 0 & -0.125 & 0 \end{vmatrix} \Big/ |\Delta|$$

where

$$|\Delta| = \begin{vmatrix} -j0.396 & j0.0625 & 0 \\ j0.625 & 0.125+j0.187 & -0.125 \\ 0 & -0.125 & 0.214-j0.057 \end{vmatrix}$$

$$V_3 = \frac{0.235}{0.0214 - j0.0109} = 8.724 + j4.427 = 9.78\angle 26.9^\circ \text{ V}$$

$$P_{max} = \frac{|V_3|^2}{2\,Re\{Z_L\}} = \frac{9.78^2}{2(8)} = 5.97 \cong 6\text{ W}$$

10.27

V_1, $-j\frac{1}{2}\,\Omega$, I, $6\angle 0^\circ$ A, $7\,\Omega$, rI

$$\frac{V_1}{7} + \frac{V_1 - rI}{-j\frac{1}{2}} = 6 \quad \text{and} \quad I = \frac{V_1}{7}$$

$$\Rightarrow V_1\left(-j\tfrac{1}{2} + 7 - r\right) = 6$$

$$V_1 = \frac{6}{(7-r) - j\frac{1}{2}}$$

Output impedance of the independent source

$$Z = \frac{V_1}{6} = \frac{6}{(7-r)-j\frac{1}{2}} \Big/ 6 = \frac{1}{(7-r)-j\frac{1}{2}} = \frac{(7-r)+j\frac{1}{2}}{(7-r)^2+\frac{1}{4}}$$

To have maximum power delivered by the independent source, $Z_{load} = Z^* = \frac{(7-r)-j\frac{1}{2}}{(7-r)^2+\frac{1}{4}}$

$$P = \frac{1}{2}(6)^2 Re\{Z\} = 18\frac{7-r}{(7-r)^2+\frac{1}{4}}$$

10.27 Cont.

To maximize P

$$\frac{\partial P}{\partial r} = 0 \Rightarrow \frac{\partial\left\{(7-r)\left[(7-r)^2+\frac{1}{4}\right]^{-1}\right\}}{\partial r} = 0$$

$$\Rightarrow -\left[(7-r)^2+\tfrac{1}{4}\right]^{-1} + (-1)(7-r)\left[(7-r)^2+\tfrac{1}{4}\right]^{-2} = 0$$

$$\Rightarrow -\left[(7-r)^2+\tfrac{1}{4}\right] + (r-7) = 0$$

$$\Rightarrow r^2 - 14r + 49 + \tfrac{1}{4} - r + 7 = 0$$

$$\Rightarrow \left(r - \tfrac{15}{2}\right)^2 = 0 \Rightarrow r = \tfrac{15}{2}$$

10.28 The impedance Z_{eq} seen by the load Z_L

$j\frac{7}{2}\,\Omega$, $-j3\,\Omega$, $5\,\Omega$, Z_{eq}

$$Z_{eq} = \left(j\tfrac{7}{2}+5\right)//(-j3) = \frac{(5+j\frac{7}{2})(-j3)}{5+j\frac{7}{2}-j3} = \frac{\frac{21}{2}-j15}{5+j\frac{1}{2}}$$

$$= 1.78 - j3.2$$

For maximum power transfer $Z_L = Z_{eq}^*$

$$\Rightarrow Z_L = 1.78 + j3.2$$

Z_L = $1.78\,\Omega$, 0.457 H at $\omega = 7$

10.29 (a) The Thevenin equivalent seen by Z_L

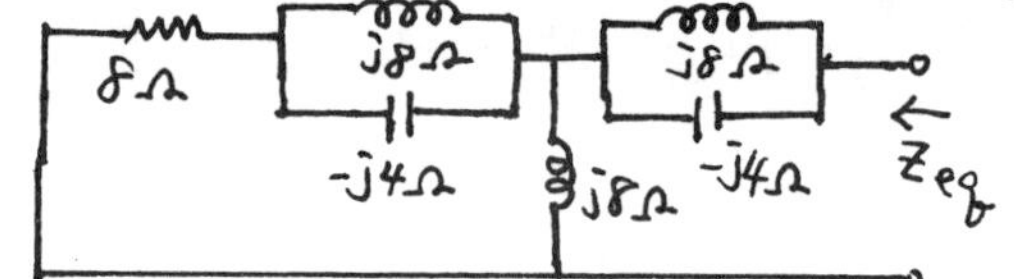

$$R_{eq} = (j8//-j4) + \{[(j8//-j4)+8]//j8\}$$

$$= (-j8) + \{(-j8+8)//j8\}$$

$$= -j8 + \frac{(-j8+8)(j8)}{8-j8+j8} = -j8 + (8-j8)$$

$$= 8 - j16\ \Omega$$

$$V_{TH} = A \times \left(\frac{j8}{8-j8+j8}\right) = jA = A\angle 90^\circ$$

$8-j16$, $A\angle 90^\circ$, Z_L

$$I = \frac{A\angle 90^\circ}{(8-j16)+8} = \frac{A\angle 90^\circ}{16-j16} = \frac{A\angle 90^\circ}{16\sqrt{2}\angle 45^\circ} = \frac{A}{16\sqrt{2}}\angle 45^\circ$$

$$P = \frac{1}{2} I_m^2 Re\{Z\} = \frac{1}{2}\frac{A^2}{16^2 \times 2} \times 8 = 10 \Rightarrow A = 8\sqrt{20}\text{ V}$$

(b) $Z_L = R_{eq}^* = 8 + j16\ \Omega$

10.30 

RMS phasor circuit diagram

$163/\sqrt{2} \cong 115$

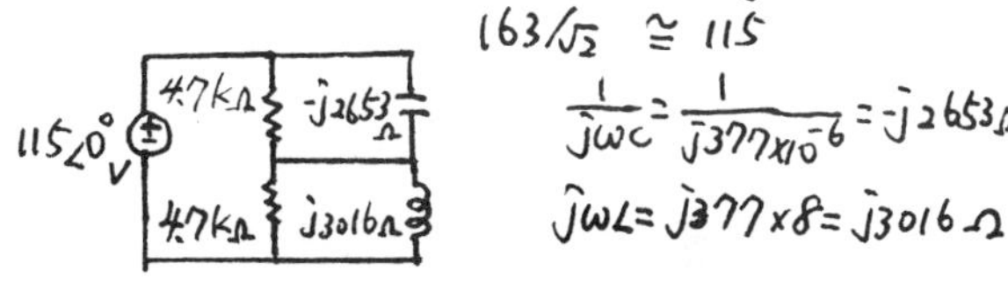

$\frac{1}{j\omega C} = \frac{1}{j377\times10^{-6}} = -j2653\,\Omega$

$j\omega L = j377\times 8 = j3016\,\Omega$

$$Z = \frac{(4700)(-j2653)}{4700 - j2653} + \frac{(4700)(j3016)}{4700 + j3016}$$

$= 2506.6 + j124.36 = 2509.7\angle 2.84°$

(a) Apparent power $= V_{rms}\cdot I_{rms} = \frac{V_{rms}^2}{|Z|}$

$= \frac{115^2}{2509.7} = \underline{5.27\,VA}$

(b) The power factor $PF = \cos(2.84°)$

$= \underline{0.999 \text{ (leading)}}$

10.31 At $\omega = 3$, the phasor diagram

$$\frac{V_1}{2} + \frac{V_1 - 4}{j4} = 0 \Rightarrow V_1(j4+2) = 8 \Rightarrow V_1 = \frac{8}{j4+2}$$

$$I_1 = \frac{V_1}{2} = \frac{\frac{8}{j4+2}}{2} = \frac{2}{j2+1} = \frac{2(1-j2)}{5}$$

(a) The reactive power absorbed by the inductor $Q = I_m\left[\frac{1}{2}\left(\frac{2-j4}{5}\right)(j4)^2\right]$

$= I_m\left(\frac{-32+j64}{10}\right) = \underline{6.4\,VA}$

(b) at $\omega = 0$, the inductor is shorted

$Q = \underline{0\,VA}$

10.32 Phasor circuit diagram

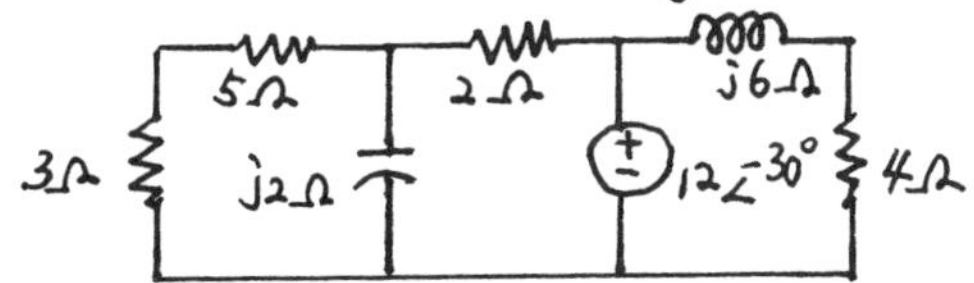

Impedance seen by the voltage source

$Z = [(8 // j2) + 2] // (4 + j6)$

$= \left(\frac{j16}{8+j2} + 2\right) // (4+j6) = \frac{16+j20}{8+j2} // (4+j6)$

$= (2.47 + j1.88) // (4 + j6) = \frac{(2.47+j1.88)(4+j6)}{6.47 + j7.88}$

10.32 Cont.

$Z = 1.61 + j1.50 = 2.2\angle 43°$

$I = \frac{V}{Z} = \frac{12\angle -30°}{2.2\angle 43°} = 5.5\angle -73°$

$S = V_{rms} I_{rms}^{*} = (12\angle -30°)(5.5\angle 73°)\left(\frac{1}{2}\right)$

$= 33\angle 43° = 24.1 + j22.5$

Hence the reactive power absorbed by the inductor and capacitor is $\underline{22.5\,VA}$

10.33

(a) $Z = 2506.6 + j124.36$

$Y = \frac{1}{Z} \cong 0.0004 - j0.00002 = 0.0004\angle -2.86°$

The angle of the admittance is the angle by which the current leads the voltage

To achieve 0.95 leading

$\cos^{-1} 0.95 = 18.2°$. One has to increase the leading angle from $-2.86°$ to $18.2°$; that is the new admittance

$Y_T = 0.0004 + jX$, where

$\tan^{-1}\frac{X}{0.0004} = 18.2°$ or $X = 0.0004 \tan 18.2°$

$= 0.00013$

Hence $Y_T = 0.0004 + j0.00013$

or $Y_1 = Y_T - Y = j(0.00013 + 0.00002)$

$= j0.00015$

$Z_1 = 1/Y_1 = -j6666.7\,\Omega$

At frequency $\omega = 377$, this requires an capacitance of

$$\frac{1}{j(377)C} = -j6666.7 \Rightarrow C = \underline{0.398\,\mu F}$$

(b) To achieve 0.95 lagging, $Y = 0.0004 + jX$

where $\tan^{-1}\frac{X}{0.0004} = -18.2$

$X = 0.0004\tan(-18.2°) = -0.00013$

$Y_1 = (-j0.00013 - 0.00002) = -j0.00015$

$\Rightarrow Z_1 = 1/Y_1 = j6666.7\,\Omega$

$\Rightarrow j\omega L = j(377)(L) = j6666.7 \Rightarrow L = \underline{17.7\,H}$

10.34 $X_1 = \frac{R^2 + X^2}{R\tan(\cos^{-1}PF) - X} = \frac{(200)^2 + (100)^2}{200\tan(\cos^{-1}0.95) - 100}$

$= -1459$

Since negative $C = -1/\omega X_1 = \underline{0.685\,\mu F}$

10.35 $X_1 = \dfrac{R^2+X^2}{-R\tan(\cos^{-1}PF)-X}$

$= -301.7$ since negative

$C = -1/\omega X_1 = \underline{3.315\ \mu F}$

10.36

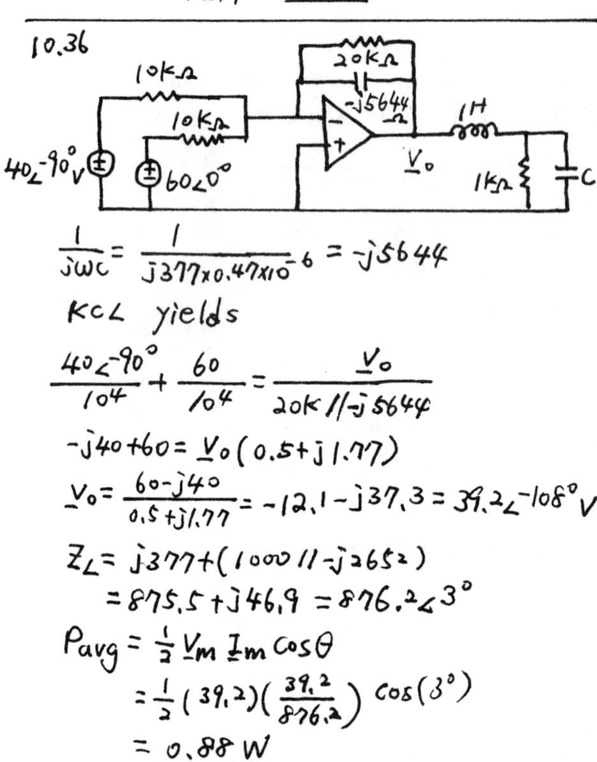

$\dfrac{1}{j\omega C} = \dfrac{1}{j377\times0.47\times10^{-6}} = -j5644$

KCL yields

$\dfrac{40\angle-90^\circ}{10^4} + \dfrac{60}{10^4} = \dfrac{V_o}{20k // -j5644}$

$-j40+60 = V_o(0.5+j1.77)$

$V_o = \dfrac{60-j40}{0.5+j1.77} = -12.1-j37.3 = 39.2\angle-108^\circ V$

$Z_L = j377 + (1000 // -j2652)$

$= 875.5 + j46.9 = 876.2\angle3^\circ$

$P_{avg} = \frac{1}{2} V_m I_m \cos\theta$

$= \frac{1}{2}(39.2)\left(\frac{39.2}{876.2}\right)\cos(3^\circ)$

$= \underline{0.88\ W}$

10.37 The output impedance of the op-amp circuit seen by the load Z_L is

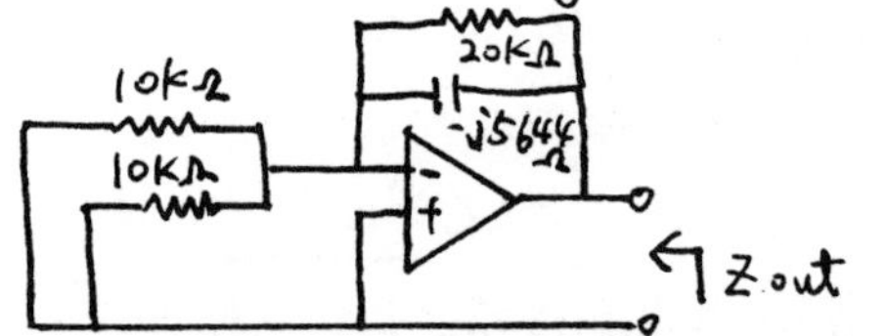

Replace op-amp by practical op-amp model

1MΩ, 20kΩ, -j5644Ω, 10kΩ, 10kΩ, 30Ω, z out

$z_{out} \cong 30\ \Omega$

∵ Z_{out} is approximated resistive, to have maximum power transfer, Z_L has to

10.37 Cont.

be resistive too.

$Z_L = j\omega + (1000 // \frac{1}{j\omega C})$, where $\omega = 377$

$= j\omega + \dfrac{1000\times\frac{1}{j\omega C}}{1000+\frac{1}{j\omega C}} = j\omega + \dfrac{10^3}{j\omega C\times10^3+1}$

$= \dfrac{10^3 + j[\omega - 10^6(1-\omega^2 C)\omega C]}{1+(\omega C\times10^3)^2}$

To be purely resistive,

$\omega - 10^6(1-\omega^2 C)\omega C = 0 \Rightarrow C = 5.8\ \mu F$

or $C = 1.2\ \mu F$

For $C = 5.8\ \mu F \Rightarrow Z_L = 173\ \Omega$

$C = 1.2\ \mu F \Rightarrow Z_L = 830\ \Omega$

To have maximum power transfer to Z_L, the tuning capacitor is $\underline{1.2\ \mu F}$

10.38

```
X_U1    0 1 2 OPAMP
R_R1    1 2 20K
R_R2    1 4 10K
R_R3    1 5 10K
R_R4    3 0 1K
L_L1    2 3 1
C_C1    1 2 0.47U
C_C2    3 0 C_LOAD 1U
V_V1    4 0 SIN (0 40 377 0 0 -90)
V_V2    5 0 SIN (0 60 377 0 0  0)

*Control statement
.INC   C:\PS\OPAMP.LIB
.MODEL C_LOAD CAP(C=1)
*Output V(3)
.TRAN 10MS 100MS
.PRINT TRAN V(3)
.END
```

100V
C=1UF
-20V
0s 50ms 100ms
▫ V(3)

Chapter 11
Three-Phase Circuit

11.1 Single-phase, three-wire systems

11.1 Find the currents I_{aA}, I_{bB} and I_{nN} in the single phase system.

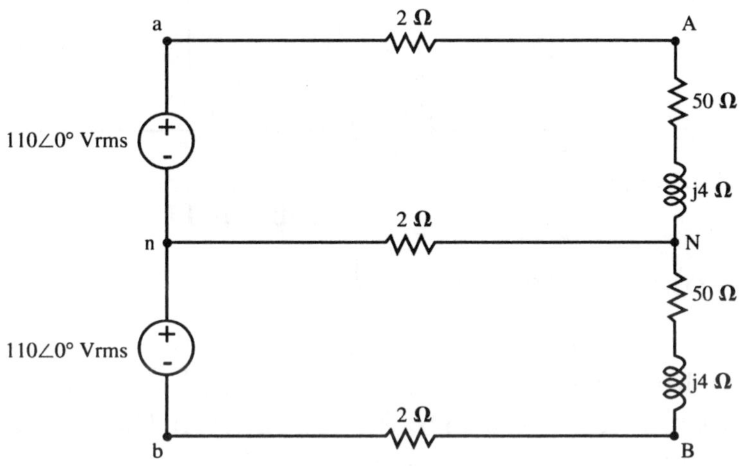

Problem 11.1

11.2 If the 2Ω resistors in Prob.11.1 are to be modelled as wire impedances, find the power delivered by each of two voltage sources and the total power lost in the lines.

11.3 Suppose in the circuit of Prob.11.1 the voltage source connected to points n-b is out of phase of the other voltage source by $\theta°$ leading. Find the currents I_{aA}, I_{bB} and I_{nN} in terms of θ.

11.4 For the phase difference, $\theta°$, defined in Prob.11.3, find the value of θ which results in maximum and minimum power lost in the lines.

11.5 Let $v_1 = 15\angle 0° V_{rms}$, $Z_1 = 2 + j\Omega$, $Z_2 = 5\Omega$, $Z_3 = 2\Omega$ and $Z_4 = 8\Omega$. Find the average power absorbed by the loads, lost in the lines, and delivered by the sources. Assume the source frequency is 60Hz.

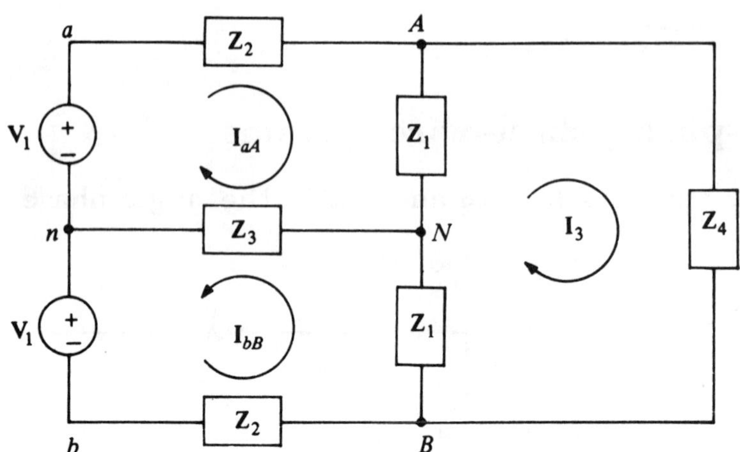

Problem 11.5

11.2 Three-phase wye-wye system

11.6 A balanced Y-Y 60 Hz, 3 wire(no neutral line), positive phase sequence system has line voltage $v_{bc} = 208\angle 45° V_{rms}$ and the line current $I_{aA} = 4\angle 174° A_{rms}$. Find the phase impedance Z_P and sketch the set of line voltages and line currents in a phasor diagram.

11.7 A balanced three phase wye-wye system with phase voltage $v_{an} = \frac{100}{\sqrt{3}}\angle 0° V_{rms}$, $Z_L = 0$ and the phase sequence is abc. The load in each phase is a series combnation of a 10Ω resistor and a 20uF capacitor. The frequency is ω=500rad/sec. Find the line currents and the power delivered to the load.

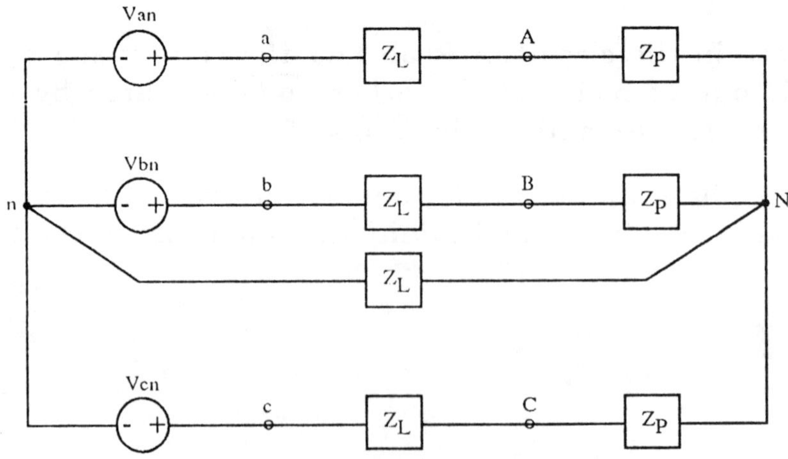

Problem 11.7

11.8 Repeat Prob11.7 with a voltage source of negative sequence and the line voltage $v_{ab} = 100\angle{-30^\circ}V_{rms}$.

11.9 In Prob.11.7 the line currents form a balanced, positive sequence set with $I_{aA} = 20\angle 0^\circ A_{rms}$ and $V_{ab} = 60\angle 60^\circ V_{rms}$. Find Z_P and the power delivered to the three-phase load.

11.10 For the power delivery system in Prob11.7, the lines connect points a-A, b-B and c-C are imperfect conductors with line impedance $Z_l = j3\Omega$, $Z_p = 10 - j23$, $V_{an} = 230\angle 0^\circ$. Compute the three line voltage V_{ab}, V_{bc}, V_{ca} at the source end of the lines, the line voltages V_{AB}, V_{BC}, V_{CA} at the load end and sketch them in a single phasor diagram. Assume positive phase sequence.

11.11 If in the balanced three phase circuit shown in Prob.11.7, the phase sources are $4400V_{rms}$ of magnitude. The lines connect points aA, bB and cC are modelled as inductive with inductance 0.1 mH/km. Find an equation relating the line length in km to rms line current.

11.12 Repeat Prob.11.11 for the line length in km vs. total power delivered to the three phase load.

11.13 A balanced Y-connected source, $V_{an} = 100\angle 0^\circ V_{rms}$, negative sequence, is connected by four perfect conductors (having zero impedance) to an unbalanced Y-connected load, $Z_{AN} = 5\Omega$, $Z_{BN} = j15\Omega$, and $Z_{CN} = -j10\Omega$. Find the four line currents.

11.14 A balanced three-phase Y-connected load draws 1KW at a power factor of 0.707 leading. A balanced Y of capacitors is to be placed in parallel with the load so that the power factor of the combination is 1.0. If the frequency is 60Hz and the line voltages are a balanced $200V_{rms}$ set, find the capacitances required.

11.15 If $Z_1 = 3 + j3\Omega$, $Z_2 = 3 - j3\Omega$, and the line voltage is $V_L = 300V_{rms}$, find the current I_L in each line.

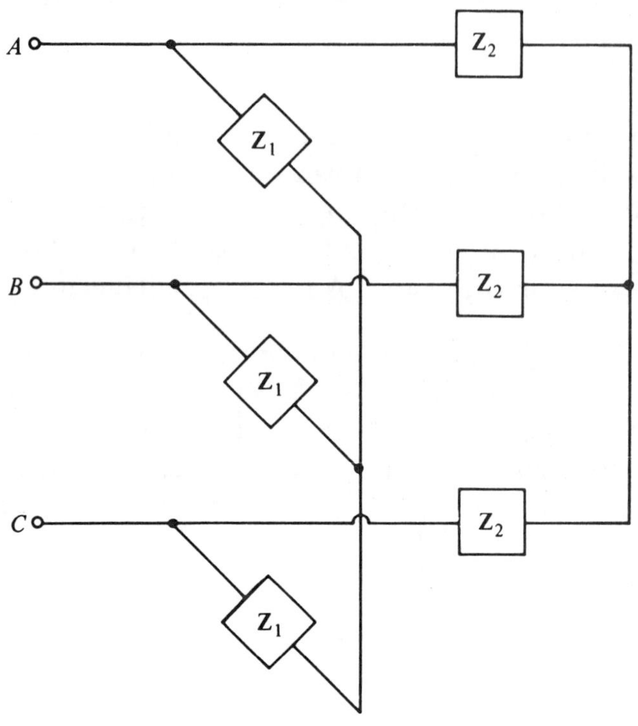

Problem 11.15

11.16 In Prob.11.7 if the source is balanced, with positive phase sequence, and $V_{an} = 20\angle 0^{o} V_{rms}$, and if the source provides 9KW at pf=0.707 lagging, find Z_p.

11.17 For a three phase wye-wye system shown in Prob.11.7 has no line impedance and $Z_P = 3 + j3\Omega$ for each phase. Initally the source is balanced with a line voltage magnitude of $208V_{rms}$. If a power failure occurs that one of the three phase voltages drop to a magnitude of $10V_{rms}$ while all else(including phase voltage angles) remain the same. Compute the power delivered to the load (a) before the voltage drop, (b) after the voltage drop and (c) after the voltage drop assuming the neutral line is removed.

11.18 For a three phase wye-wye system shown in Prob.11.7 with phase voltage $120\angle 0^{o} V_{rms}$, if a "phase fault" occurs such that the load of phase C is short-circuited, find the line voltage and phase current for the load of phase A of the system. Assume $Z_L = 1\Omega$ and $Z_P = 2 + j3\Omega$. Assume positive sequence.

11.3 Single phase vs. three phase power delivery

11.19 For the circuit shown with $V = 120\angle 0° V_{rms}$, $Z_L = Z_N = 1\Omega$ and $Z_3 = 2 + j3\Omega$, for what load impedance Z_1 will the currents in the single and three phase circuit be equal? Compute the power delivered to the load for the single phase circuit and the power delivered to the three phase load at this line current. Assume equal power factor of the single and three phase loads.

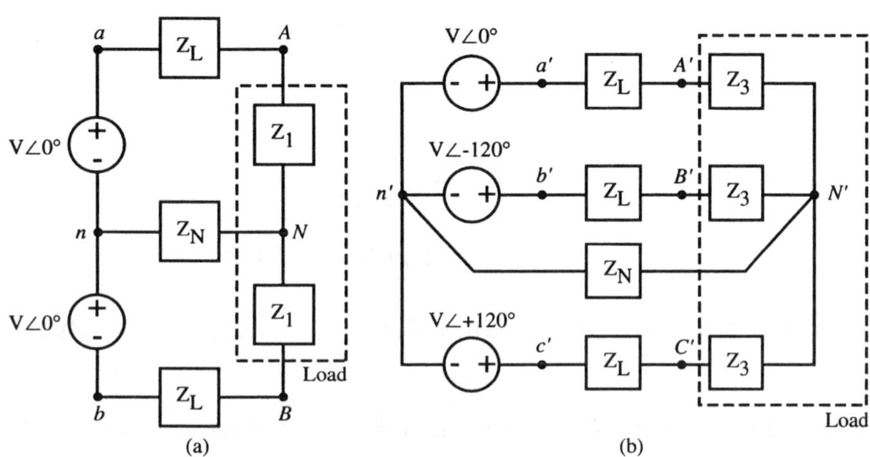

Problem 11.19

11.20 For the Prob.11.19, if equal power is delivered to the single phase load and three phase load, compute the total line losses in the single phase circuit and three phase circuit. Assume $V = 120\angle 0° V_{rms}$, $Z_L = Z_N = 1\Omega$ and $Z_3 = 2 + j3\Omega$ and equal power factor of the single phase load and three phase load.

11.4 The Delta Connection

11.21 Solve Prob.11.7 if the source and load are unchanged except that the load is Δ-connected.

11.22 A balanced Δ-connected load has a line voltage of $V_L = 100V_{rms}$ at the load terminals and absorbs a total power of 4.8KW. If the power factor of the load is 0.8 leading, find the phase impedance.

11.23 A balanced three-phase, positive-sequence source with $V_{ab} = 200\angle 0° V_{rms}$ is supplying a Δ-connected load, $Z_{AB} = 3 - j4\Omega$, $Z_{BC} = 20\angle 60°\Omega$ and $Z_{CA} = 50\angle 30°\Omega$. Find the phasor line current. (Assume perfectly conducting lines)

11.5 Wye-delta transformations

11.24 Find the equivalent delta and wye connected loads if $Z_1 = 3 - j6\Omega$ and $Z_2 = \frac{5}{4} + j4\frac{5}{4}\Omega$.

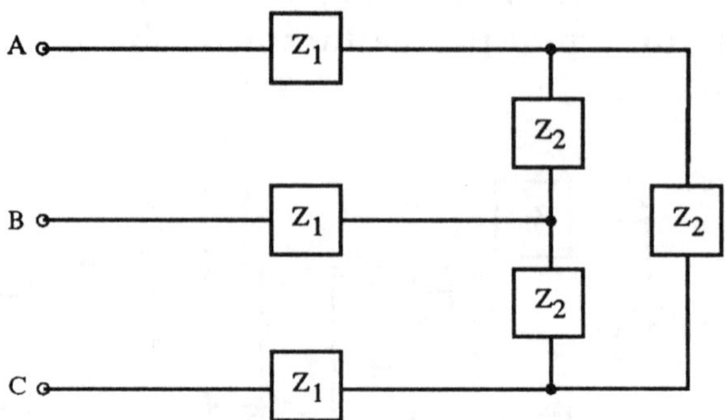

Problem 11.24

11.25 Determine the Thevenin equivalent of the circuit of Prob.11.5 at the terminals A-B.

11.26 In Prob.11.25 if a capacitor connected between points A and B causes the complex power delivered by the source to be purely real, what is the value of the capacitance?

11.27 A balanced three-phase source with $V_L = 100V_{rms}$ is delivering power to a balanced Y-connected load with phase impedance $Z_1 = 5 + j12\Omega$ in parallel with a balanced Δ-connected load with phase impedance $Z_2 = 15\Omega$. Find the power delivered by the source.

11.28 A balanced three-phase positive sequence source with $V_{ab} = 100\angle 0° V_{rms}$ is supplying a parallel combination of a Y-connected load and a Δ-connected load. If the Y and Δ loads are balanced with phase impedance of $3 - j3\Omega$ and $9 + j9\Omega$, respectively, find the line current I_L and the power supplied by the source, assuming perfectly conducting lines.

11.29 Use the result of Exer.11.1.3 to find an equivalent for the three phase load shown in a balanced wye-wye circuit.

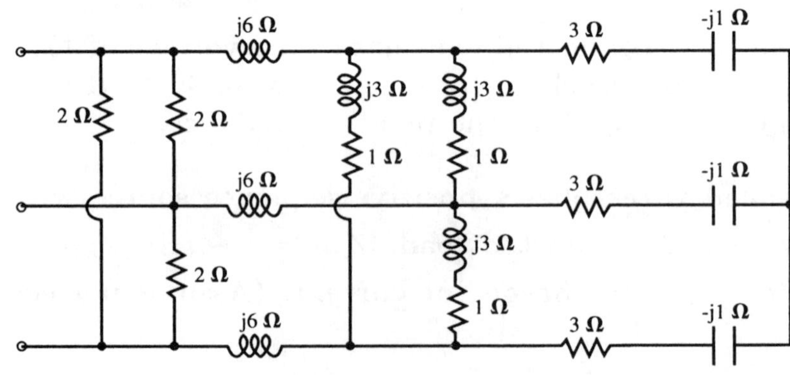

Problem 11.29

11.30 Find the wye-connected equivalent load if $Z_1 = 1\Omega$, $Z_2 = -j1\Omega$.

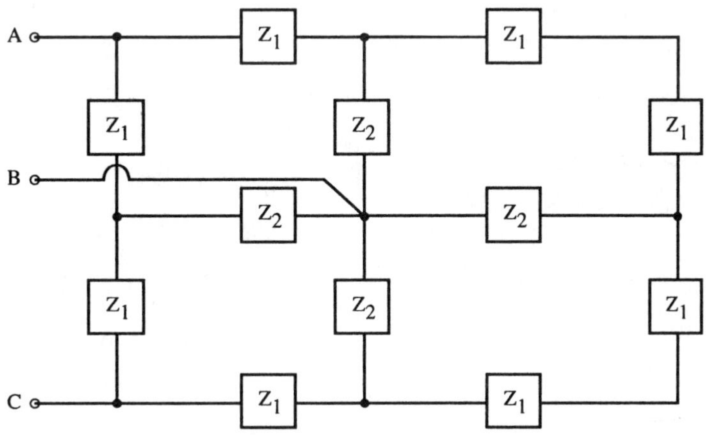

Problem 11.30

11.31 For the Prob.11.7, replace the wye-connected three phase load by an equivalent Δ-connected load. Solve the problem use SPICE.

11.32 Solve Prob.11.30 with the help of SPICE.

11.33 Find the average power delivered by the voltage source to rest of the circuit.

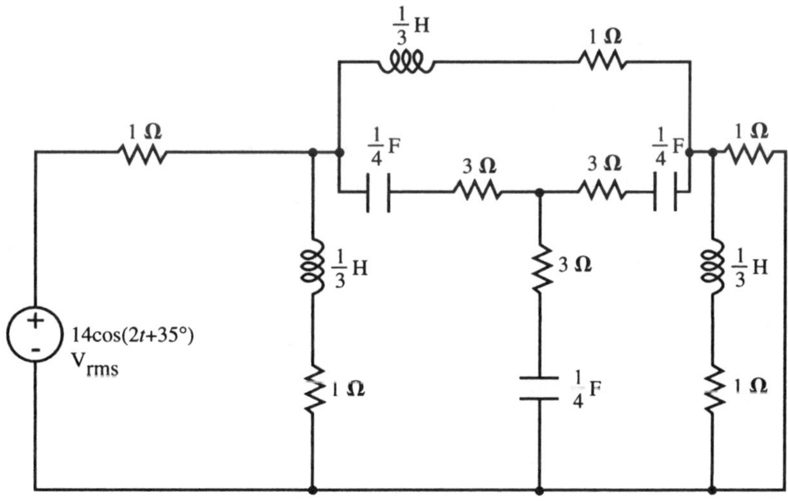

Problem 11.33

11.1 The circuit is symmetric

$I_{aA} = -I_{bB}$. By KVL

$$\begin{cases} I_{aA}(2+50+j4+2) - (-I_{bB})2 = 110 \\ -I_{bB}(2+50+j4+2) - (I_{aA})2 = 110 \end{cases}$$

Add the two equations

$$I_{aA}(52+j4) - I_{bB}(52+j4) = 220$$

$$I_{aA} = \frac{110}{52+j4} = \underline{2.1\angle -4.4^\circ A}$$

$$I_{bB} = -I_{aA} = \underline{-2.1\angle -4.4^\circ A}$$

$$I_{nN} = \underline{0 A}$$

11.2 The power delivered by each of the two voltage sources

$$P = (2.1)^2(52) = \underline{229.3 W}$$

The total power lost in the lines

$$P_{Loss} = (2.1)^2(2+2) = \underline{17.6 W}$$

11.3 $I_{aA}(2+50+j4+2) - (-I_{bB})2 = 110$ (1)

$-I_{bB}(2+50+j4+2) - (I_{aA})2 = 110\angle\theta^\circ$ (2)

(1)+(2), $I_{aA}(52+j4) - I_{bB}(52+j4) = 110(1+\angle\theta^\circ)$

or $I_{aA} - I_{bB} = \frac{110}{52+j4}(1+\angle\theta^\circ)$

Let $1+\angle\theta^\circ = \alpha \Rightarrow I_{bB} = I_{aA} - \frac{110}{52+j4}\alpha$ (3)

substitute (3) into (1)

$$I_{aA}(54+j4) + 2\left(I_{aA} - \frac{110}{52+j4}\alpha\right) = 110$$

$$I_{aA} = \frac{110}{56+j4} + \frac{110}{(52+j4)(56+j4)}(2\alpha)$$

Similarly, $I_{bB} = I_{aA} - \frac{110}{52+j4}\alpha$

$$= \frac{110}{56+j4} + \frac{110}{(52+j4)(56+j4)}(-27-j2)(2\alpha)$$

Hence I_{aA}, I_{bB} can be expressed as

$I_{aA} = a + x\alpha$ where $a = \frac{110}{56+j4} = 1.96\angle -4.09^\circ$

$I_{bB} = a + y\alpha$ $b = \frac{110}{(52+j4)(56+j4)}(2)$

$$Y = \frac{110}{(52+j4)(56+j4)}(-27-j2)(2)$$

11.3 Cont.

From the result of Prob. 11.1 at $\theta = 0^\circ$, $\alpha = 2$ and $I_{aA} = -I_{bB}$

One obtain the general equations for any θ.

$I_{aA} = a + b\alpha$ where $b = x = \frac{110}{(52+j4)(56+j4)}(2) = 0.0752\angle -8.49^\circ$

$I_{bB} = a-(a+b)\alpha$

Hence $I_{aA} = \underline{a + b\alpha\ A}$

$I_{bB} = \underline{a-(a+b)\alpha\ A}$

$I_{nN} = I_{aA} + I_{bB} = \underline{2a - a\alpha\ A}$

where $\alpha = 1+\angle\theta^\circ = 1+\cos\theta + j\sin\theta$

$a = 1.96\angle -4.09^\circ$

$b = 0.0752\angle -8.49^\circ$

11.4 From the solution of Prob. 11.3

$I_{aA} = a + b\alpha$ where $\alpha = 1+\cos\theta + j\sin\theta$

$I_{bB} = a-(a+b)\alpha$ $a = 1.96\angle -4.09^\circ$

$I_{nN} = 2a - a\alpha$ $b = 0.0752\angle -8.49^\circ$

Complex power lost in the lines

$$P_{comp} = R\left[(I_{aA}^*)^2 + (I_{bB}^*)^2 + (I_{nN}^*)^2\right]$$

It is seen that $a \gg b$ in magnitude.

One can approximate

$I_{aA} = a$

$I_{bB} = a - a\alpha$

$I_{nN} = 2a - a\alpha$

$$P_{comp} = 2\left[(a^*)^2 + \{[a(1-\alpha)]^*\}^2 + \{[a(2-\alpha)]^*\}^2\right]$$

$$= 4(a^*)^2\left[(\alpha^*)^2 - 3(\alpha^*) + 3\right]$$

$$P_{avg} = Re\{4(a^*)^2[(\alpha^*)^2 - 3\alpha^* + 3]\}$$

$$(\alpha^*)^2 - 3\alpha^* + 3 = (1+\cos\theta - j\sin\theta)^2 - 3(1+\cos\theta - j\sin\theta) + 3$$

$$= (2\cos\theta - 1)(\cos\theta - j\sin\theta) = (2\cos\theta - 1)(\cos\theta - j\sin\theta)$$

$$4(a^*)^2 = 15.4\angle 8.09^\circ \cong 15.4$$

$$Re\{(\alpha^*)^2 - 3\alpha^* + 3\} = (2\cos\theta - 1)\cos\theta$$

$$\frac{\partial\{(2\cos\theta - 1)\cos\theta\}}{\partial\theta} = 0 \Rightarrow \theta = 0 \text{ or } \pi$$

11.4 Cont.

Thus maximum power lost in the lines occurs at $\theta = \pi \Rightarrow P_{avg} = 15.4(3) = \underline{46.2W}$

minimum power lost in the lines occurs at $\theta = 0 \Rightarrow P_{avg} = 15.4(1) = \underline{15.4W}$

11.5 Since $\underline{I}_{aA} + \underline{I}_{bB} = 0$, the first mesh equation can be written, using $\underline{I}_{aA} = \underline{I}_1$,

$(Z_1 + Z_2)\underline{I}_1 - Z_1 \underline{I}_3 = \underline{V}_1$

The equation for the right mesh can be written

$-2Z_1 \underline{I}_1 + (2Z_1 + Z_4)\underline{I}_3 = 0$

Then $\underline{I}_1 = \dfrac{(2Z_1 + Z_4)\underline{V}_1}{(2Z_1 + Z_4)Z_2 + Z_1 Z_4}$

For $Z_1 = 2+j$, $Z_2 = 5$, $Z_3 = 2$ and $Z_4 = 8$

$\underline{I}_1 = \underline{I}_{aA} = -\underline{I}_{bB} = \dfrac{(4+2j+8)(15)}{(12+j2)5 + 16+j8}$

$= 2.336\angle -3.86°$ Arms

$\underline{I}_3 = \dfrac{2Z_1}{2Z_1 + Z_4}\underline{I}_1 = \dfrac{2(2+j)}{12+j2}(2.34\angle -3.9°)$

$= 0.429\angle 13.24°$ Arms

$P_{Z_4} = 8|\underline{I}_3|^2 = 1.475W$

$P_{Z_1} = Re\{(2+j)|\underline{I}_3 - \underline{I}_1|^2\}$

$\underline{I}_3 - \underline{I}_1 = (0.429\angle 13.24° - 2.336\angle -3.86°)$

$= 1.93\angle 172.4°$ Arms

$P_{Z_1} = (2)(1.93)^2 = 7.451W$,

$P_{aA} = 5|\underline{I}_{aA}|^2 = 27.30W$

$P_{bB} = 5|\underline{I}_{aA}|^2 = 27.30W$

$P_{Top\ source} = |\underline{V}_1| \cdot |\underline{I}_{aA}| \cos[\text{ang}\,\underline{V}_1 - \text{ang}\,\underline{I}_{aA}]$

$= 15(2.34)\cos(3.86°) = 34.97W$

$P_{Bottom\ source} = |\underline{V}_1||-\underline{I}_{bB}|\cos(3.86°)$

$= (15)(2.34)\cos(3.86°)$

$= 34.97W$

$P_{Load} = 2P_{Z_1} + P_{Z_4} = \underline{16.38W}$

$P_{Loss} = P_{aA} + P_{bB} = \underline{54.59W}$

$P_{delivered} = P_{Load} + P_{Loss} = \underline{70.97W}$

11.6 $\underline{V}_{bc} = 208\angle 45°$, $\underline{I}_{aA} = 4\angle 174°$

$\underline{V}_{ab} = 208\angle 165°$, $\underline{I}_{bB} = 4\angle 54°$

$\underline{V}_{ca} = 208\angle -75°$, $\underline{I}_{cC} = 4\angle -66°$

$\underline{V}_{an} = \dfrac{208}{\sqrt{3}}\angle 45° - 30° = 120\angle 15°$ Vrms

$Z_p = \dfrac{\underline{V}_{an}}{\underline{I}_{aA}} = \dfrac{120\angle 15°}{4\angle 174°} = \underline{30\angle -159°\ \Omega}$

11.7 $Z_p = 10 - j\dfrac{1}{(500)(20\times10^{-6})} = 100.5\angle -84.3°\ \Omega$

$\underline{I}_{aA} = \dfrac{\underline{V}_{an}}{Z_p} = \dfrac{57.74\angle 0°}{100.5\angle -84.3°} = \underline{0.574\angle 84.3°}$ Arms

$\underline{I}_{bB} = 0.574\angle 84.3° - 120° = \underline{0.574\angle -35.71°}$ Arms

$\underline{I}_{cC} = \underline{0.574\angle -155.7°}$ Arms

$P = 3V_p I_p \cos\theta$

$= 3(57.74)(0.574)\cos(-84.3°)$

$= \underline{9.90W}$

11.8 For negative sequence

$\underline{V}_{an} = V_p\angle 0°$, $\underline{V}_{bn} = V_p\angle 120°$, $\underline{V}_{cn} = V_p\angle -120°$

and $\underline{V}_{ab} = \underline{V}_{an} - \underline{V}_{nb} = V_p\angle 0° - V_p\angle 120°$

$= \sqrt{3}\,V_p\angle -30°$

$\underline{V}_{an} = V_p\angle 0° = \dfrac{100}{\sqrt{3}}\angle -30° + 30° = 57.74\angle 0°$ Vrms

$\underline{V}_{bn} = 57.74\angle 120°$, $\underline{V}_{cn} = 57.74\angle -120°$

$Z_p = 10 - j\dfrac{1}{(500)(20\times10^{-6})} = 100.5\angle -84.3°\ \Omega$

$\underline{I}_{aA} = \dfrac{\underline{V}_{an}}{Z_p} = \dfrac{57.74\angle 0°}{100.5\angle -84.3°} = \underline{0.574\angle 84.3°}$ Arms

$\underline{I}_{bB} = \underline{0.574\angle -155.7°}$ Arms

$\underline{I}_{cC} = \underline{0.574\angle -35.7°}$ Arms

$P_{delivered} = 3V_p I_p \cos\theta = \underline{9.9W}$

11.9 $\underline{V}_{an} = \frac{60\angle 30^\circ}{\sqrt{3}}$ Vrms

$\underline{I}_{aA} = 20\angle 0^\circ$ Arms

$\underline{Z}_p = \underline{V}_{an}/\underline{I}_{aA} = \frac{60}{20\sqrt{3}}\angle 30^\circ = \sqrt{3}\angle 30^\circ\,\Omega$

$P = 3V_p I_p \cos\theta = 3\left(\frac{60}{\sqrt{3}}\right)20\cos 30^\circ = 1800\,W$

11.10 At the source end

$\underline{V}_{ab} = \underline{V}_{an} + \underline{V}_{nb} = \sqrt{3}\,V_p\angle 30^\circ = 230\sqrt{3}\angle 30^\circ$ Vrms

$\underline{V}_{bc} = 230\sqrt{3}\angle -90^\circ$ Vrms

$\underline{V}_{ca} = 230\sqrt{3}\angle -210^\circ$ Vrms

$\frac{10-j23}{10-j23+j3} = \frac{10-j23}{10-j20} = 1.1\angle -3.1^\circ$

At the load end

$\underline{V}_{AB} = \underline{V}_{AN} + \underline{V}_{NB} = (230\angle 0^\circ - 230\angle -120^\circ)(1.1\angle -3.1^\circ)$

$= 438.2\angle 26.9^\circ$ Vrms

$\underline{V}_{BC} = 438.2\angle -93.1^\circ$ Vrms

$\underline{V}_{CA} = 438.2\angle -213.1^\circ$ Vrms

11.11 $\frac{1}{j\omega C} = \frac{1}{j500\times 20\times 10^{-6}} = -j100$

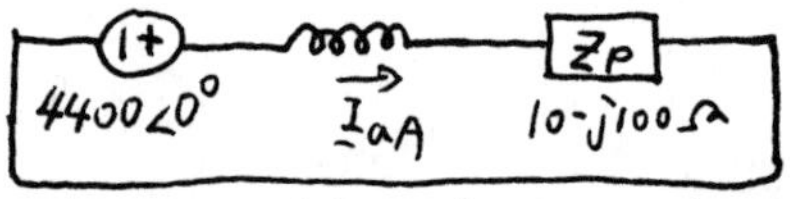

Let the length of line be x km

$\underline{I}_{aA} = \frac{4400\angle 0^\circ}{j(500)(0.0001)(x) + 10 - j100} = \frac{4400}{\sqrt{\frac{1}{4}x^2 - 100x + 10100}}\angle\theta^\circ$ Arms

where $\theta = \tan^{-1} 0.05x - 10$

$\underline{I}_{bB} = \frac{4400}{\sqrt{\frac{1}{4}x^2 - 100x + 10100}}\angle -120-\theta^\circ$ Arms

$\underline{I}_{cC} = \frac{4400}{\sqrt{\frac{1}{4}x^2 - 100x + 10100}}\angle 120-\theta^\circ$ Arms

---

11.12 $P = \underline{V}_p \underline{I}_p \cos\theta$

$= (4400\angle\theta^\circ)\left(\frac{4400}{\sqrt{\frac{1}{4}x^2 - 100x + 10100}}\right)\cos\theta$

$= \frac{19.36\times 10^6}{\sqrt{\frac{1}{4}x^2 - 100x + 10100}}\cos\theta\ W$

where $\theta = \tan^{-1} 0.05x - 10$.

11.13 $\underline{I}_{AN} = \frac{\underline{V}_{an}}{\underline{Z}_{AN}} = \frac{100}{5} = 200$ Arms

$\underline{I}_{BN} = \frac{\underline{V}_{bn}}{\underline{Z}_{BN}} = \frac{100\angle 120^\circ}{j15} = 6.67\angle 30^\circ$ Arms

$\underline{I}_{CN} = \frac{\underline{V}_{cn}}{\underline{Z}_{CN}} = \frac{100\angle -120^\circ}{-j10} = 10\angle -30^\circ$ Arms

$\underline{I}_{nN} = -(\underline{I}_{aN} + \underline{I}_{bN} + \underline{I}_{cN})$

$= -(20 + 5.77 + j3.33 + 8.66 - j5)$

$= 34.47\angle 177.2^\circ$ Arms

11.14 $P_1 = \frac{1}{3}$ kW, $Q_1 = P_1\tan(\cos^{-1}0.707)$

$= P_1 = \frac{1}{3}$ kW.

$Q_T = Q_1 + Q_2 = P_T\tan(\cos^{-1}1) = 0$

$Q_2 = -Q_1 = -\frac{1}{3}$ kVAR

$C = \frac{-Q_2}{2\pi f\,V_p^2} = \frac{\frac{1}{3}\times 10^3}{377(200/\sqrt{3})^2} = 66.3\,\mu F$

11.15 $\underline{Z}_p = \frac{\underline{Z}_1\underline{Z}_2}{\underline{Z}_1 + \underline{Z}_2} = \frac{(3+j3)(3-j3)}{3+j3+3-j3} = 3\,\Omega$

$V_p = \frac{V_L}{\sqrt{3}} = \frac{300}{\sqrt{3}} = 173.2$ Vrms

$I_L = I_p = \frac{V_p}{|Z_p|} = \frac{173.2}{3} = 57.74$ Arms

---

11.16 $P = 3V_p I_p\cos\theta = 3\frac{V_p^2}{|Z_p|}\cos\theta$

$|Z_p| = \frac{3V_p^2\cos\theta}{P} = \frac{3(200)^2(0.707)}{9000} = 9.43\,\Omega$

$\underline{Z}_p = 9.43\angle\cos(0.707) = 9.43\angle 45^\circ\,\Omega$

11.17 $|\underline{V}_{aA}| = |\underline{V}_{bB}| = |\underline{V}_{cC}| = 208$ Vrms

$|\underline{V}_{an}| = |\underline{V}_{bn}| = |\underline{V}_{cn}| = \frac{208}{\sqrt{3}} = 120$ Vrms

(a) Before the voltage drop

$\underline{V}_{an} = 120\angle 0^\circ$ Vrms

$\underline{Z}_p = 3 + j3 = 4.24\angle 45^\circ\,\Omega$

$\underline{I}_{aA} = \frac{120\angle 0^\circ}{4.24\angle 45^\circ} = 28.3\angle -45^\circ$ Arms

$P_{delivered} = 3(28.3)^2(3) = 7028$ W

(b) After the voltage drop

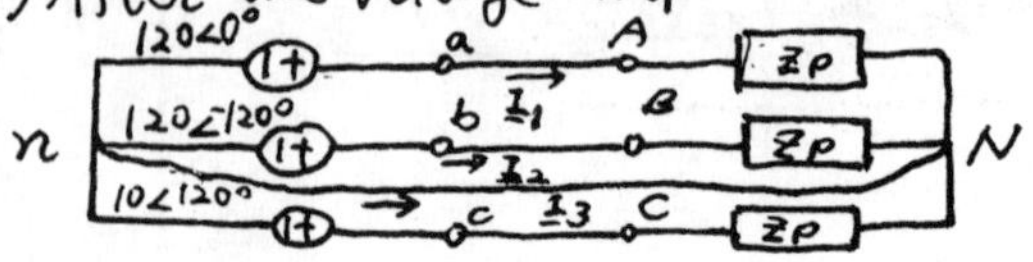

11.17 Cont.

$$\begin{cases} \underline{I}_1(3+j3+3+j3)-\underline{I}_2(3+j3)=120\angle 0^\circ-120\angle -120^\circ \\ -\underline{I}_1(3+j3)+\underline{I}_2(3+j3)=120\angle -120^\circ \\ -\underline{I}_3(3+j3)=10\angle 120^\circ \end{cases}$$

or
$$\begin{cases} \underline{I}_1(6+j6)-\underline{I}_2(3+j3)=120\sqrt{3}\angle 30^\circ=180+j103.9 \\ -\underline{I}_1(3+j3)+\underline{I}_2(3+j3)=-60-j103.9 \\ -\underline{I}_3(3+j3)=-5+j8.7 \end{cases}$$

$$\underline{I}_1(3+j3)=120 \Rightarrow \underline{I}_1=\frac{120}{3+j3}=20-j20$$

$$\underline{I}_2=[(-60-j103.9)+120]/3+j3=-7.3-j27.3$$

$$\underline{I}_3=\frac{5-j8.7}{3+j3}=-0.62-j2.28$$

$$P_{delivered}=|\underline{I}_1|^2(3)+|\underline{I}_1-\underline{I}_2|^2(3)+|\underline{I}_3|^2(3)$$
$$=(28.3)^2 3+|20-j20+7.3+j27.3|^2(3)+(2.4)^2(3)$$
$$=2402.7+2395.7+17.3=\underline{4815.7\,W}$$

(c) With the neutral line removed

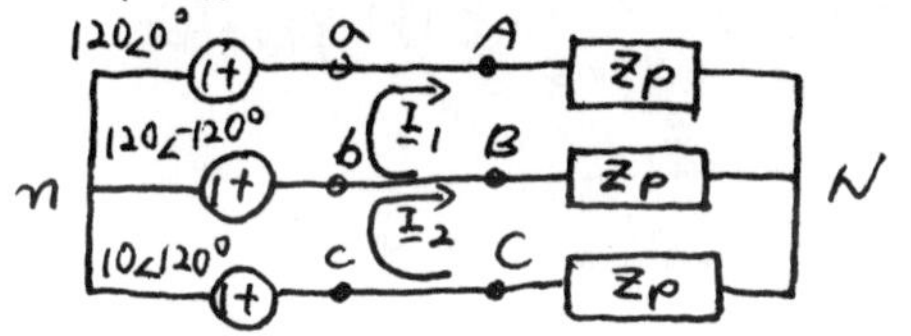

$$\begin{cases} \underline{I}_1(6+j6)-\underline{I}_2(3+j3)=120\angle 0^\circ-120\angle -120^\circ \\ -\underline{I}_1(3+j3)+\underline{I}_2(6+j6)=120\angle -120^\circ-10\angle 120^\circ \end{cases}$$

$$\begin{cases} \underline{I}_1(6+j6)-\underline{I}_2(3+j3)=180+j103.9 \\ -\underline{I}_1(3+j3)+\underline{I}_2(6+j6)=-55-j112.6 \end{cases}$$

$$\underline{I}_1=\begin{vmatrix} 180+j103.9 & -3-j3 \\ -55-j112.6 & 6+j6 \end{vmatrix} \Big/ \begin{vmatrix} 6+j6 & -3-j3 \\ -3-j3 & 6+j6 \end{vmatrix}$$
$$=(629.4+j1200.6)/j54=22.2-j11.7$$

$$\underline{I}_2=\begin{vmatrix} 6+j6 & 180+j103.9 \\ -3-j3 & -55-j112.6 \end{vmatrix} \Big/ \begin{vmatrix} 6+j6 & -3-j3 \\ -3-j3 & 6+j6 \end{vmatrix}$$
$$=(573.9-j153.9)/j54=-2.85-j10.62$$

$$P=|\underline{I}_1|^2(3)+|\underline{I}_1+\underline{I}_2|^2(3)+|\underline{I}_2|^2(3)$$
$$=(629.7)(3)+(872.6)(3)+(120.9)(3)$$
$$=\underline{4869.6\,W}$$

11.18

120∠0°, a, 1Ω, A, 2+j3; 120∠-120°, b, 1Ω, I_{aA}, B, 2+j3; n, N; 1Ω, I_{bB}; 120∠120°, c, 1Ω, I_{cC}, C

11.18 Cont.

$$\begin{cases} \underline{I}_{aA}(6+j6)-\underline{I}_{bB}(3+j3)=120\angle 0^\circ-120\angle -120^\circ \\ -\underline{I}_{aA}(3+j3)+\underline{I}_{bB}(4+j3)-(-\underline{I}_{cC})=120\angle -120^\circ \\ -\underline{I}_{bB}+(-\underline{I}_{cC})(2)=-120\angle 120^\circ \end{cases}$$

$$\begin{cases} \underline{I}_{aA}(6+j6)-\underline{I}_{bB}(3+j3)=180+j104 \\ -\underline{I}_{aA}(3+j3)+\underline{I}_{bB}(4+j3)+\underline{I}_{cC}=-60-j104 \\ -\underline{I}_{bB}-\underline{I}_{cC}(2)=60-j104 \end{cases}$$

$$\underline{I}_{aA}=\begin{vmatrix} 180+j104 & -3-j3 & 0 \\ -60-j104 & 4+j3 & 1 \\ 60-j104 & -1 & -2 \end{vmatrix} \Big/ \begin{vmatrix} 6+j6 & -3-j3 & 0 \\ -3-j3 & 4+j3 & 1 \\ 0 & -1 & -2 \end{vmatrix}$$
$$=(-1392-j692)/(-6-j42)$$
$$=20.8-j30.2=\underline{36.6\angle -55.4^\circ\ A\,rms}$$

$$\underline{I}_{bB}=\begin{vmatrix} 6+j6 & 180+j104 & 0 \\ -3-j3 & -60-j104 & 1 \\ 0 & 60-j104 & -2 \end{vmatrix} \Big/ \begin{vmatrix} 6+j6 & -3-j3 & 0 \\ -3-j3 & 4+j3 & 1 \\ 0 & -1 & -2 \end{vmatrix}$$
$$=(-1968+j528)/(-6-j42)$$
$$=-5.8-j47.7=48\angle -96.9^\circ\ A\,rms$$

$$\underline{V}_{AB}=\underline{I}_{aA}(4+j6)-\underline{I}_{bB}(2+j3)$$
$$=(20.8-j30.2)(4+j6)-(-5.8-j47.7)(2+j3)$$
$$=132.9+j116.8=\underline{176.9\angle 41.3^\circ}\ V\,rms$$

11.19 $\underline{I}_{aN}=\frac{120\angle 0^\circ}{3+j3}=\frac{120\angle 0^\circ}{1+Z_1}\Rightarrow Z_1=2+j3=\sqrt{13}\angle 56.3^\circ\Omega$

$$P_{aN}=\frac{V^2\cos\theta}{|3+j3|}=\frac{120^2\cos(56.3^\circ)}{\sqrt{13}}=2216\,W$$

Power delivered to the single phase circuit $P_1=P_{aN}+P_{bN}=2P_{aN}=2(2216)$
$=\underline{4432\,W}$

Power delivered to the three phase circuit $P_3=3P_{a'N'}=3(2216)=\underline{6648\,W}$

11.20 $P_{single}=2P_{aN}=2\frac{(120)^2}{|1+Z_1|}\cos\theta\times\left(\frac{2+j3}{3+j3}\right)$

$$P_{Three}=3P_{a'N'}=3\frac{(120)^2}{|3+j3|}\cos\theta\times\left(\frac{2+j3}{3+j3}\right)$$

$$P_{single}=P_{Three}\Rightarrow\frac{|1+Z_1|}{|3+j3|}=\frac{2}{3}$$

and $|\underline{I}_{a'N'}|=\frac{120}{|3+j3|}=28.3\ A\,rms$

$$\frac{|\underline{I}_{aN}|}{|\underline{I}_{a'N'}|}=\frac{V}{|1+Z_1|}\Big/\frac{V}{|3+j3|}=\frac{|3+j3|}{|1+Z_1|}=3/2$$

11.20 Cont.

Power loss in the single phase circuit

$P_L = 2|I_{aN}|^2 Re\{Z_L\} = 2(\frac{9}{4})|I_{a'N'}|^2 \cdot 1$

$= 3604\ W$

Power loss in the three phase circuit

$P_L' = 3|I_{a'N'}|^2 Re\{Z_L\} = 3(28.3)^2$

$= 2402.7\ W$

11.21 $V_{ab} = 100\angle 30°$ Vrms

$Z_p = 10 - j100\ \Omega = 100.5\angle -84.3°\ \Omega$

$I_{AB} = \frac{V_{ab}}{Z_p} = \frac{100\angle 30°}{100.5\angle -84.3°} = 0.995\angle 114.3°$ Arms

$I_{aA} = 0.995\sqrt{3}\angle 114.3° - 30°$

$= 1.723\angle 84.3°$ A rms

$I_{bB} = 1.723\angle -35.7°$ A rms

$I_{cC} = 1.723\angle -155.7°$ A rms

11.22 $|Z_p| = \frac{3V_p^2 \cos\theta}{P} = \frac{3(100)^2(0.8)}{4800}$

$= 5\ \Omega$

$Z_p = 5\angle -\cos(0.8) = 5\angle -36.9°\ \Omega$

$= 4 - j3\ \Omega$

11.23 $I_{AB} = \frac{V_{ab}}{Z_{AB}} = \frac{200}{3-j4} = 20\angle 53.1°$ Arms

$I_{BC} = \frac{V_{bc}}{Z_{BC}} = \frac{200\angle -120°}{20\angle 60°} = -10$ A rms

$I_{CA} = \frac{V_{ca}}{Z_{CA}} = \frac{200\angle 120°}{50\angle 30°} = -j4$ Arms

$I_{aA} = I_{AB} - I_{CA} = 12 + j12$ Arms

$I_{bB} = I_{BC} - I_{AB} = -22 - j16$ Arms

$I_{cC} = I_{CA} - I_{BC} = 10 + j4$ Arms

11.24 By delta-wye transformation

$Z_Y = \frac{Z_2^2}{3Z_2} = \frac{Z_2}{3} = \frac{5}{12} + j\frac{5}{12}\ \Omega$

Equivalent wye connected load

$Z_A = Z_B = Z_C = Z_1 + Z_Y = (3 - j6) + (\frac{5}{12} + j\frac{5}{12})$

$= \frac{41}{12} - j\frac{67}{12}\ \Omega$

$Y_A = Y_B = Y_C = 1/(\frac{41}{12} - j\frac{67}{12}) = 0.080 + j0.13$

11.24 Cont.

Equivalent delta connected load

$Y_{AB} = Y_{BC} = Y_{CA} = \frac{(Y_A)^2}{3Y_A} = Y_A = 0.0266 + j0.0434$

$Z_{AB} = Z_{BC} = Z_{CA} = 1/(0.0266 - j0.0434)$

$= 10.25 - j16.75\ \Omega$

11.25

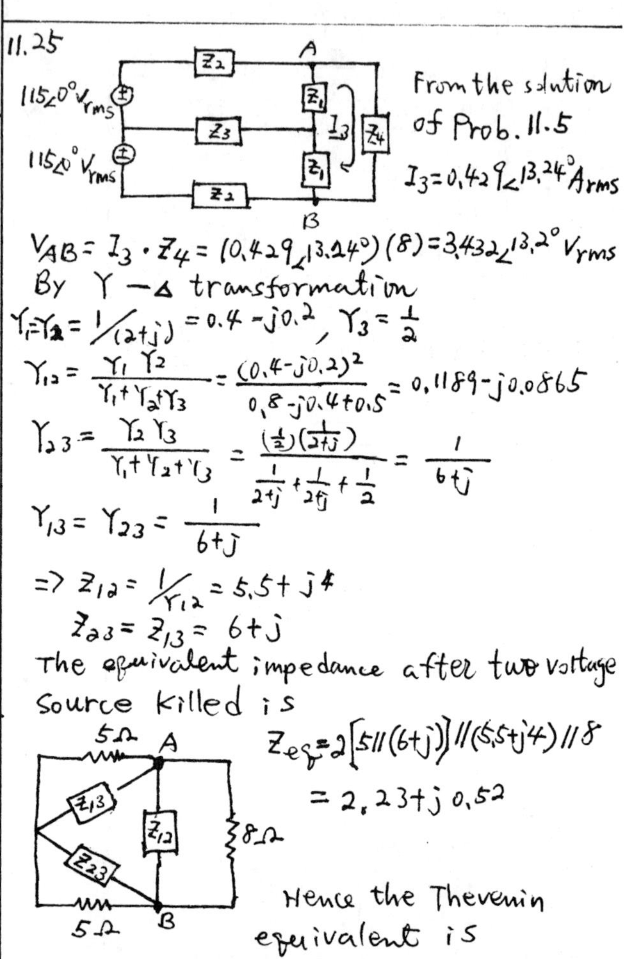

From the solution of Prob. 11.5

$I_3 = 0.429\angle 13.24°$ Arms

$V_{AB} = I_3 \cdot Z_4 = (0.429\angle 13.24°)(8) = 3.432\angle 13.2°$ Vrms

By Y – Δ transformation

$Y_1 = Y_2 = 1/(2+j) = 0.4 - j0.2,\ Y_3 = \frac{1}{2}$

$Y_{12} = \frac{Y_1 Y_2}{Y_1 + Y_2 + Y_3} = \frac{(0.4 - j0.2)^2}{0.8 - j0.4 + 0.5} = 0.1189 - j0.0865$

$Y_{23} = \frac{Y_2 Y_3}{Y_1 + Y_2 + Y_3} = \frac{(\frac{1}{2})(\frac{1}{2+j})}{\frac{1}{2+j} + \frac{1}{2+j} + \frac{1}{2}} = \frac{1}{6+j}$

$Y_{13} = Y_{23} = \frac{1}{6+j}$

$\Rightarrow Z_{12} = 1/Y_{12} = 5.5 + j4$

$Z_{23} = Z_{13} = 6 + j$

The equivalent impedance after two voltage source killed is

$Z_{eq} = 2[5 \| (6+j)] \| (5.5 + j4) \| 8$

$= 2.23 + j0.52$

Hence the Thevenin equivalent is

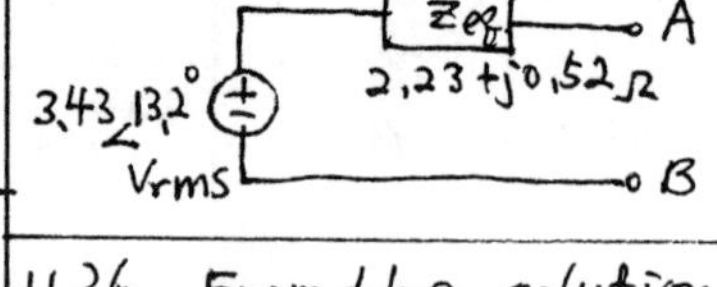

11.26 From the solution of Prob. 11.25

$P = \frac{(3.43\angle 13.2°)^2}{2.23 + j0.56 - j\frac{1}{\omega C}} = \frac{11.8\angle 26.4°}{2.23 + j(0.56 - \frac{1}{377C})}$

To have purely real power,

$\tan(26.4°) = (0.56 - \frac{1}{377C})/2.23 \Rightarrow C = 0.00494F$

11.27 $Z_{2Y} = \frac{1}{3}Z_2 = \frac{1}{3}(15) = 5\,\Omega$

$Z_P = \frac{Z_1 Z_{2Y}}{Z_1 + Z_{2Y}} = \frac{(5+j12)(5)}{5+j12+5} = 4.16\angle 17.2^\circ\,\Omega$

$I_P = \frac{V_P}{|Z_P|} = \frac{100/\sqrt{3}}{4.16} = 13.87$ Arms

$P = \sqrt{3} V_L I_L \cos\theta = \sqrt{3}(100)(13.87)\cos 17.2^\circ$
$= 2,296\,W$

11.28 By Δ-Y transformation

$Z_Y = \frac{1}{3}Z_\Delta = \frac{1}{3}(9+j9) = 3+j3\,\Omega$

$Z_{eq} = \frac{(3-j3)(3+j3)}{3-j3+3+j3} = 3\,\Omega$

$I_L = \frac{V_P}{|Z_{eq}|} = \frac{100/\sqrt{3}}{3} = 19.25$ Arms

$P = 3I_L^2\,\mathrm{Re}\{Z_{eq}\} = 3(19.25)^2(3) = \frac{10}{3}\,kW$

11.29 By Y-Δ transformation from right to left.

$Y_1 = \frac{(1/(3-j))^2}{3(1/(3-j))} = 0.1+j0.0333$

$Z_1 = 1/Y_1 = 9-j3$

$Z_2 = (9-j3)//(1+j3) = \frac{(9-j3)(1+j3)}{9-j3+1+j3} = 1.8+j2.4$

By Δ-Y transformation

$Z_3 = \frac{Z_2}{3} = 0.6+j0.8$

$Z_4 = Z_3 + j6 = 0.6+j6.8$

By Y-Δ transformation

$Y_5 = \frac{1/(0.6+j6.8)}{3} \Rightarrow Z_5 = 1.8+j20.4$

$Z_6 = Z_5//2 = \frac{(1.8+j20.4)(2)}{1.8+j20.4} = 1.96+j0.19$

By Δ-Y transformation

$Z_7 = \frac{Z_6}{3} = 0.65+j0.0632\,\Omega$

11.30 $Z_3 = Z_1 + Z_1 = 2\,\Omega \Rightarrow Y_3 = 1/2\,S$

By Y-Δ transformation

$Y_4 = \frac{(1/2)^2}{3(\frac{1}{2})} = \frac{1}{6} \Rightarrow Z_4 = 6\,\Omega$

$Z_5 = Z_2//Z_4 = \frac{2\times 6}{2+6} = 2\,\Omega$

11.30 Cont.

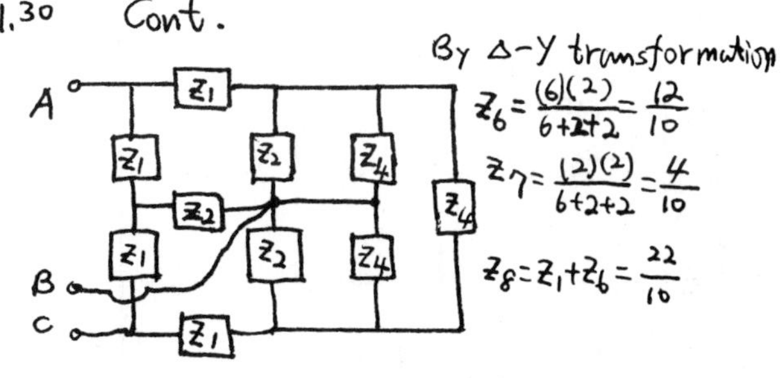

By Δ-Y transformation

$Z_6 = \frac{(6)(2)}{6+2+2} = \frac{12}{10}$

$Z_7 = \frac{(2)(2)}{6+2+2} = \frac{4}{10}$

$Z_8 = Z_1 + Z_6 = \frac{22}{10}$

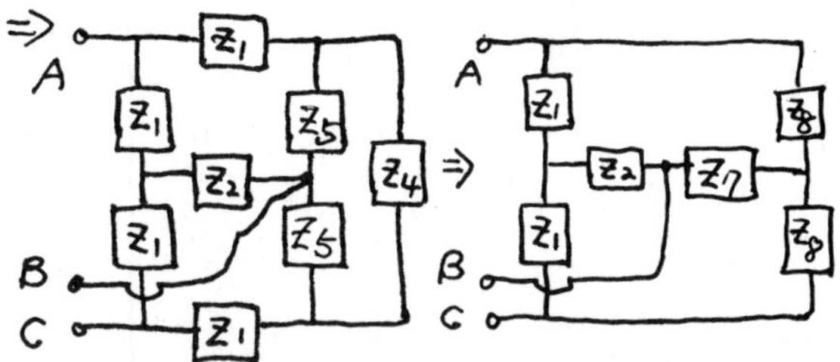

By Y-Δ transformation

$Y_9 = \frac{(1)(\frac{1}{2})}{\frac{1}{1}+\frac{1}{1}+\frac{1}{2}} = \frac{\frac{1}{2}}{\frac{5}{2}} = \frac{1}{5} \Rightarrow Z_9 = 5\,\Omega$

$Y_{10} = \frac{1}{\frac{5}{2}} = \frac{2}{5} \Rightarrow Z_{10} = \frac{5}{2}\,\Omega$

$Y_{11} = \frac{(\frac{10}{4})(\frac{22}{10})}{\frac{10}{4}+\frac{2\times 10}{22}} = \frac{\frac{22}{4}}{\frac{300}{88}} = \frac{484}{300} \Rightarrow Z_{11} = \frac{75}{121} = 0.62\,\Omega$

$Y_{12} = \frac{(\frac{22}{10})^2}{\frac{10}{4}+\frac{2\times 10}{22}} = \frac{\frac{484}{100}}{\frac{300}{88}} = 1.42 \Rightarrow Z_{12} = 0.70\,\Omega$

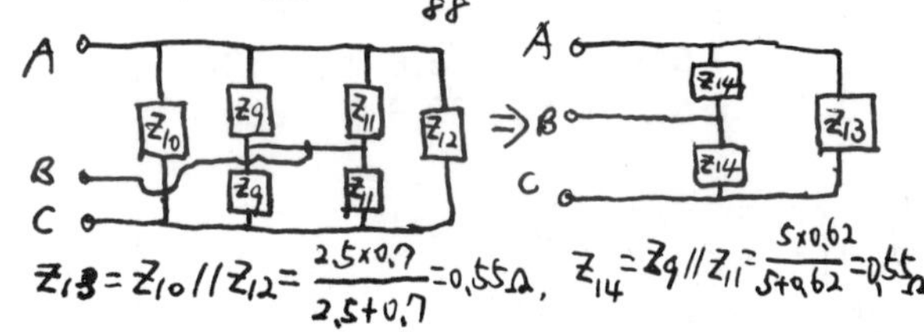

$Z_{13} = Z_{10}//Z_{12} = \frac{2.5\times 0.7}{2.5+0.7} = 0.55\,\Omega$, $Z_{14} = Z_9//Z_{11} = \frac{5\times 0.62}{5+0.62} = 0.55\,\Omega$

11.33

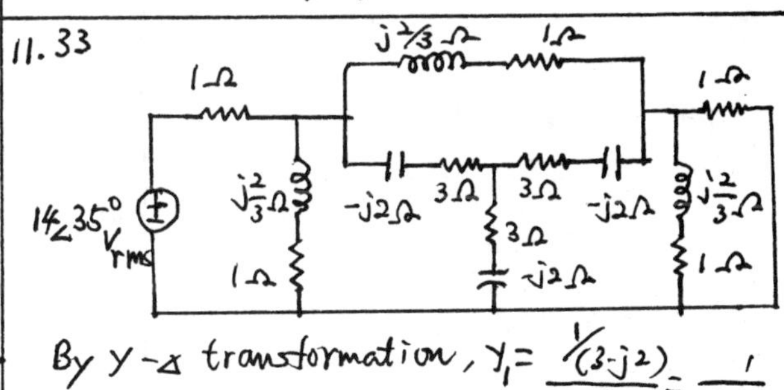

By Y-Δ transformation, $Y_1 = \frac{1/(3-j2)}{3} = \frac{1}{9-j6}$

$\Rightarrow Z_1 = 9-j6\,\Omega$

$Z_2 = Z_1//(1+j\frac{2}{3}) = \frac{(9-j6)(1+j\frac{2}{3})}{9-j6+1+j\frac{2}{3}} = 1.01+j0.54$

By Δ-Y transformation

$Z_3 = \frac{Z_2}{3} = 0.34+j0.18$

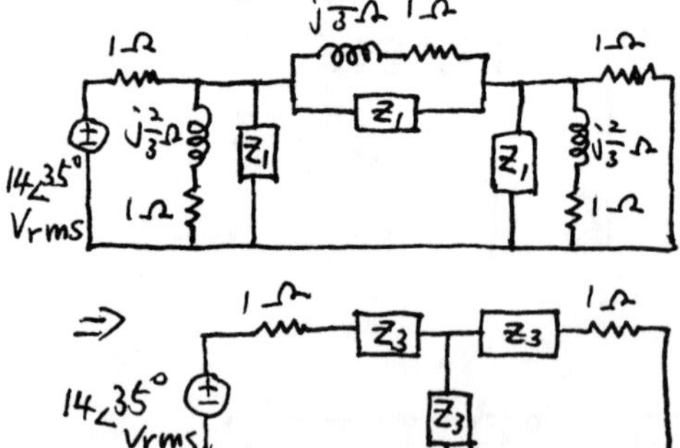

Impedance seen by the voltage source

$Z = 1.34 + j0.18 + (1.34 + j0.18) // (0.34 + j0.18)$

$= 1.62 + j0.40$

$|Z| = 1.67$

$$P_{avg} = V_m^2 \frac{Re\{Z\}}{|Z|^2} = (14)^2 \frac{(1.62)}{(1.67)^2} = 114.2\,W$$

11.31

```
V_V1    1 0 SIN (0 100 79.6 0 0 0)
R_R1    2 1 10
C_C1    3 2 20uF
R_R2    4 3 10
C_C2    5 4 20uF
R_R3    6 5 10
C_C3    1 6 20uF
V_V2    3 0 SIN (0 100 79.6 0 0 -120)
V_V3    5 0 SIN (0 100 79.6 0 0 120)

*Control statement
.TRAN 10MS 100MS
.PRINT TRAN I(R_R1)
.END
```

TIME	I(R_R1)
0.000E+00	0.000E+00
1.000E-02	-1.069E+00
2.000E-02	9.115E-01
3.000E-02	1.702E+00
4.000E-02	-5.571E-03
5.000E-02	-1.619E+00
6.000E-02	-9.750E-01
7.000E-02	1.086E+00
8.000E-02	1.623E+00
9.000E-02	-2.068E-01
1.000E-01	-1.740E+00

11.32

```
V_V1    1 0 SIN (0 1 1 0 0 0)
R_R1    2 1 1
R_R2    3 2 1
R_R3    4 3 1
R_R4    5 4 1
R_R5    6 5 1
R_R6    7 6 1
R_R7    8 7 1
R_R8    1 8 1
C_C1    2 0 1
C_C2    4 0 1
C_C3    6 0 1
C_C4    8 0 1

*Control statement
.TRAN 1S 10S
.PRINT TRAN I(V_V1)
.END
```

TIME	I(V_V1)
0.000E+00	0.000E+00
1.000E+00	-1.911E-01
2.000E+00	-2.565E-01
3.000E+00	-2.629E-01
4.000E+00	-2.693E-01
5.000E+00	-2.731E-01
6.000E+00	-2.798E-01
7.000E+00	-2.684E-01
8.000E+00	-2.465E-01
9.000E+00	-2.577E-01
1.000E+01	-2.827E-01

Chapter 12
The Laplace Transform

12.1 The S-domain

12.1 Find the Laplace transforms of functions
(a) $f(t) = 7\cos(8t)u(t)$
(b) $f(t) = 2\sin(3t)u(t)$
(c) $f(t) = [6\cos(4t) + 7\sin(4t)]u(t)$

12.2 Find the Laplace transforms of functions
(a) $f(t) = 12e^{6t}u(t)$
(b) $f(t) = (1 + t + e^{-2t} + te^{-3t} + t\cos 7t)u(t)$

12.3 Find the Laplace transforms of functions
(a) $\cos(4t)\cos(2t)u(t)$
(b) $\sinh^2(kt)u(t)$

12.4 Find the Laplace transforms of functions

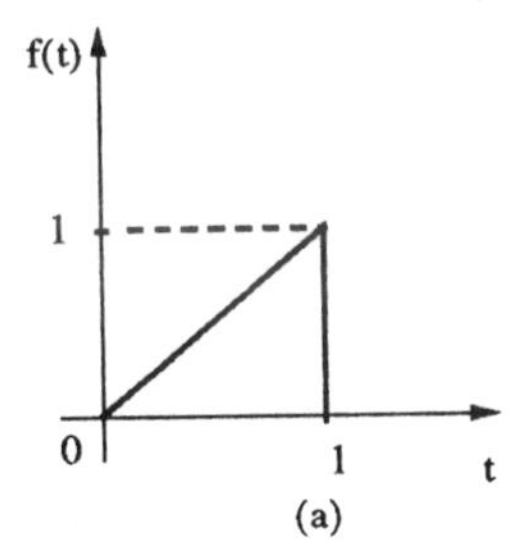

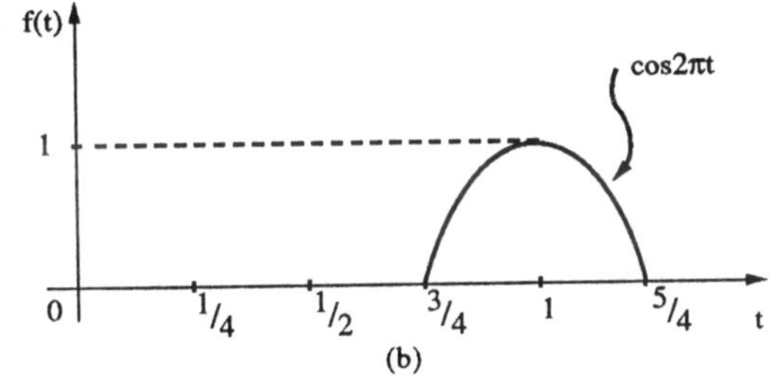

Problem 12.4

12.5 Find the inverse Laplace transforms of
$(a)\frac{1}{s} - \frac{3}{s^2+8}$
$(b)\frac{4s+7}{s^2+4}$

12.6 Find the inverse Laplace transforms of
$(a)\frac{6}{s+3} + \frac{1}{s^3}$
$(b)\frac{s}{(s+2)^2}$

12.2 Singularity functions

12.7 Evaluate $\int_a^{\infty} e^{-3t}\delta(t-2)dt$ for (a) a=3 and (b) a=1.

12.8 Find the Laplace transform of function $f(t) = \sum_{n=0}^{\infty} \delta(t-n)$.

12.9 Find the Laplace transform of function $f(t) = 4u(t) - 7u(t-2) + 2u(t-6) + tu(t-7)$.

12.10 Find the Laplace transform of function

$$\begin{aligned} f(t) &= 0 & -\infty < t < 2 \\ &= t^2 - 4t + 4 & 2 \leq t < 3 \\ &= -3t + 8 & 3 \leq t < 4 \\ &= 2 & 3 \leq t < 4 \end{aligned}$$

12.3 Other transform properties and pairs

12.11 Use the result of Prob.12.4a and linearity to find $L\{f_1(t)\}$ for $f_1(t)$ shown.

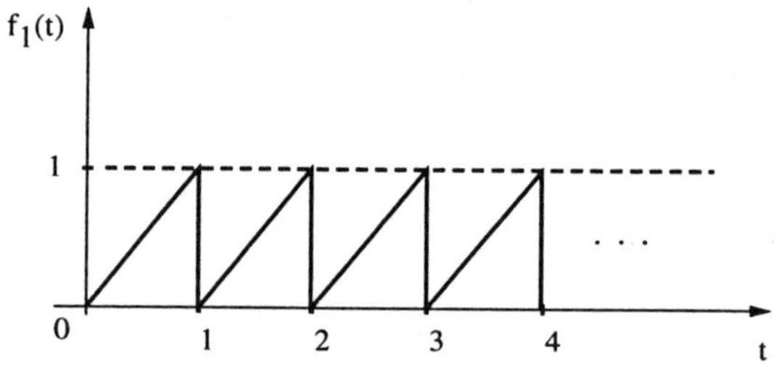

Problem 12.11

12.12 Find the Laplace transform of $\frac{df(t)}{dt}$ if $f(t) = e^{-t}\sin t$.

12.13 Find the Laplace transform of $\frac{df(t)}{dt}$ if $f(t) = (5-t)u(t) - (5-t)u(t-5)$.

12.14 Find the Laplace transform of $f(t) = t^3 + t^2 + t + 1$.

12.15 Find the Laplace transform of function $f(t) = \cos^2 t$.

12.16 Find the inverse Laplace transforms of
$(a) F(s) = \frac{1}{s} + \frac{2}{s+3} + \frac{3}{(s+5)^2+4}$
$(b) F(s) = \frac{1}{(s-3)^4} + \frac{1}{(s-2)^3}$

12.17 Find the Laplace transform of $f(t) = e^{5t}(t + t^2)u(t)$

12.18 Find the Laplace transform of $f(t) = e^{-3t}\cos[(n+1)\pi t + \phi]$

12.4 Partial fraction expansion

12.19 Find the inverse Laplace transform of

$$F(s) = \frac{s^3 + 11s^2 + 7s + 4}{s^2 + 5s + 6}$$

12.20 Find the inverse Laplace transform of

$$F(s) = \frac{s + 2}{s^3 + 4s^2 + 3s}$$

12.21 Find the inverse Laplace transforms of

$(a)\frac{3s+3}{3s^2+9}$
$(b)\frac{s+2}{3s^2+4s+6}$

12.22 Find the inverse Laplace transform of

$$F(s) = \frac{s}{(s+2)(s^2 + 6s + 12)}$$

12.23 Find the inverse Laplace transform of

$$F(s) = \frac{3s^3 + 2s^2 + 4s + 5}{(s+1)^2(s+2)^2}$$

12.24 Find the inverse Laplace transform of

$$F(s) = \frac{6s - 4}{(s^2 - 2s + 5)^2}$$

12.5 Solving integro-differential equations

12.25 Use the Laplace transform to solve for $y(t)$ **for** $t > 0$

$$y'' + 2y' + y = 3te^{-t}$$

$$y(0) = 4, \quad y'(0) = 2$$

12.26 Use the Laplace transform to solve for $y(t)$ **for** $t > 0$

$$y'' - 4y' + 4y = 4\cos(2t)$$

$$y(0) = 2, \qquad y'(0) = 5$$

12.27 Use the Laplace transform to solve $y(t)$ **for** $t > 0$

$$\int_0^t y(\tau)\sin(t-\tau)d\tau = y(t) + sin(t) - cos(t)$$

12.28 Solve for $x(t)$ for $t > 0$

$$\begin{aligned} x'' + x &= f(t) \\ f(t) &= 1 \qquad 0 \le t < \frac{\pi}{2} \\ &= 0 \qquad t \ge \frac{\pi}{2} \\ x(0) &= 0 \qquad x'(0) = 1 \end{aligned}$$

12.29 Use the Laplace transform to solve $x(t)$ for $t > 0$

$$\begin{aligned} x' &= t + \int_0^t x(t-\tau)\cos(\tau)d\tau \\ x(0) &= 4 \end{aligned}$$

12.30 Find the forced response voltage v. Assume the voltage source $V_{in} = f_1(t)$ that $F_1(t) = -\frac{e^{-s}}{s(1-e^{-s})} + \frac{1}{s}$ is given in Prob.12.11. Initial conditions $v_c(0^-) = 0$, $i(0^-) = 0$.

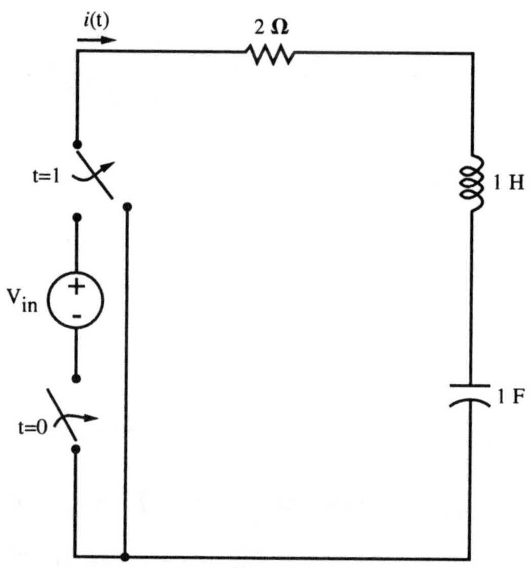

Problem 12.30

12.31 Use the describing equations and Laplace transforms to find i for $t > 0$ if the circuit is in steady-state at $t = 0^-$.

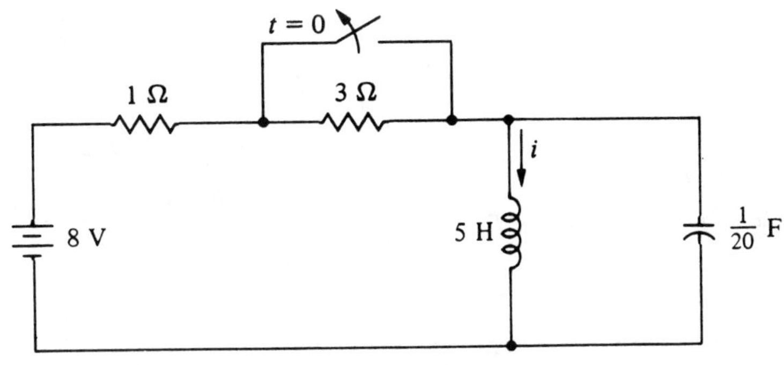

Problem 12.31

12.32 Use the describing equations and Laplace transforms to find the current in the inductor for $t > 0$ if L=3H and the circuit is in steady-state at $t = 0^-$.

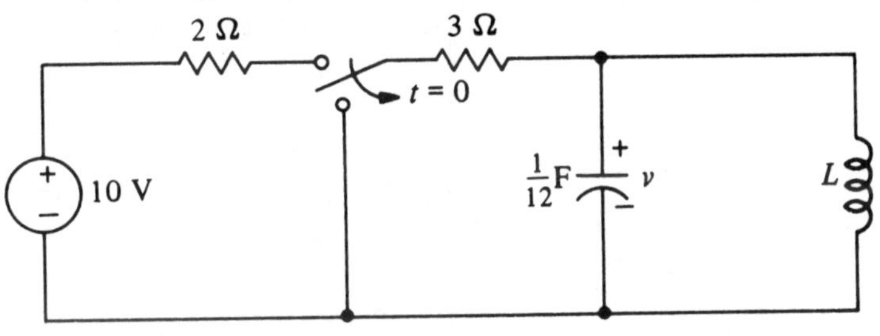

Problem 12.32

12.1 (a) $\mathcal{L}\{7\cos 8t\,u(t)\}$

$=\int_0^\infty 7\cos 8t\, e^{-st}dt$

$=\frac{7s}{s^2+64}$

(b) $\mathcal{L}\{2\sin 3t\,u(t)\}=\int_0^\infty 2\sin 3t\, e^{-st}dt$

$=\frac{6}{s^2+9}$

(c) $\mathcal{L}\{[6\cos 4t+7\sin 4t]u(t)\}$

$=\int_0^\infty (6\cos 4t+7\sin 4t)e^{-st}dt$

$=\frac{6s}{s^2+16}+\frac{28}{s^2+16}$

12.2 (a) $\mathcal{L}\{12e^{6t}u(t)\}=\int_0^\infty 12e^{6t}e^{-st}dt$

$=\frac{12}{s-6}$

(b) $\mathcal{L}\{t\cos 7t\}=-\frac{d\{\mathcal{L}[\cos 7t]\}}{ds}$

$=-\frac{d\left(\frac{s}{s^2+7^2}\right)}{ds}=-\left[\frac{1}{s^2+7^2}-\frac{s}{(s^2+7^2)^2}\right]$

$=\frac{-s^2+s+7}{(s^2+7^2)^2}$

Hence $\mathcal{L}\{f(t)\}=\frac{1}{s}+\frac{1}{s^2}+\frac{1}{s+2}+\frac{1}{(s+3)^2}-\frac{s^2-s+7^2}{(s^2+7^2)^2}$

12.3 (a) $\cos 4t\cos 2t\,u(t)=\frac{1}{2}[\cos 6t+\cos 2t]u(t)$

$\mathcal{L}\{f(t)\}=\int_0^\infty \frac{1}{2}[\cos 6t+\cos 2t]e^{-st}dt$

$=\frac{1}{2}\left(\frac{s}{s^2+6^2}+\frac{s}{s^2+2^2}\right)$

(b) $\sinh^2 kt\,u(t)=\frac{1}{2}[\sinh 2kt-1]u(t)$

$=\frac{1}{2}\left[\frac{1}{2}(e^{2kt}+e^{-2kt})-1\right]u(t)$

$=\left[\frac{1}{4}(e^{2kt}+e^{-2kt})-\frac{1}{2}\right]u(t)$

$\mathcal{L}\{\sinh^2 kt\,u(t)\}=\frac{1}{4}\left(\frac{1}{s-2k}+\frac{1}{s+2k}\right)-\frac{1}{2s}$

12.4 (a) $\mathcal{L}\{f(t)\}=\int_0^\infty f(t)e^{-st}dt$

$=\int_0^1 te^{-st}dt=-\frac{1}{s}e^{-st}t\Big|_0^1-\int_0^1\left(-\frac{1}{s}\right)e^{-st}dt$

$=-\frac{1}{s}e^{-s}-\left(\frac{1}{s^2}\right)e^{-st}\Big|_0^1=-\frac{1}{s}e^{-s}-\frac{e^{-s}}{s^2}+\frac{1}{s^2}$

(b) $f(t)=\cos 2\pi\left(t-\frac{3}{4}\right)u\left(t-\frac{3}{4}\right)-\cos 2\pi\left(t-\frac{5}{4}\right)u\left(t-\frac{5}{4}\right)$

$\mathcal{L}\{f(t)\}=\frac{s}{s^2+4\pi^2}e^{-\frac{3}{4}s}-\frac{s}{s^2+4\pi^2}e^{-\frac{5}{4}s}$

$=\frac{s}{s^2+4\pi^2}\left(e^{-\frac{3}{4}s}-e^{-\frac{5}{4}s}\right)$

12.5 (a) $\mathcal{L}^{-1}\left\{\frac{1}{s}-\frac{3}{s^2+8}\right\}=u(t)-\frac{3}{2\sqrt{2}}\sin 2\sqrt{2}t\,u(t)$

$=\left(1-\frac{3}{2\sqrt{2}}\sin 2\sqrt{2}t\right)u(t)$

(b) $\frac{4s+7}{s^2+4}=\frac{4s}{s^2+4}+\frac{7}{s^2+4}$

$f(t)=\mathcal{L}^{-1}\left\{\frac{4s}{s^2+4}+\frac{7}{s^2+4}\right\}$

$=4\cos 2t\,u(t)+7/2\sin 2t\,u(t)$

$=\left(4\cos 2t+\frac{7}{2}\sin 2t\right)u(t)$

12.6 (a) $f(t)=\mathcal{L}^{-1}\left\{\frac{6}{s+3}+\frac{1}{s^3}\right\}$

$=6e^{-3t}u(t)+\frac{1}{2}t^2u(t)$

(b) $\because \mathcal{L}\left\{\frac{df(t)}{dt}\right\}=sF(s)-f(0)$

Let $sF(s)=\frac{s}{(s+2)^2}\Rightarrow F(s)=\frac{1}{(s+2)^2}$

$\Rightarrow f(t)=te^{-2t}u(t)$, and $f(0)=0$

Hence $g(t)=\frac{df(t)}{dt}=[e^{-2t}-2te^{-2t}]u(t)$

12.7 (a) Since 2 is not between 3 and ∞

then $\int_3^\infty e^{-3t}\delta(t-2)dt=0$

(b) Since 2 is between 1 and ∞

$\int_1^\infty e^{-3t}\delta(t-2)dt=e^{-3(2)}=e^{-6}$

12.8 $\mathcal{L}\{t-n\}=e^{-sn}$

$\mathcal{L}\left\{\sum_{n=0}^{\infty}\delta(t-n)\right\}=\sum_{n=0}^{\infty}e^{-sn}$

12.9 $f(t)=4u(t)-7u(t-2)+2u(t-6)+tu(t-7)$

$=4u(t)-7u(t-2)+2u(t-6)+(t-7)u(t-7)+7u(t-7)$

$\mathcal{L}\{f(t)\}=\frac{4}{s}-\frac{7e^{-2s}}{s}+\frac{2e^{-6s}}{s}+\frac{e^{-7s}}{s^2}+\frac{7e^{-7s}}{s}$

12.10 $f(t)=(t-2)^2[u(t-2)-u(t-3)]+(-3t+8)[u(t-3)-u(t-4)]$

$+2u(t-4)$

$=(t-2)^2u(t-2)-(t-3)^2u(t-3)-(2t-5)u(t-3)+(-3t+8)[u(t-3)-$

$u(t-4)]+2u(t-4)$

$=(t-2)^2u(t-2)-(t-3)^2u(t-3)+(-5t+13)u(t-3)-(-3t+8)u(t-4)$

$+2u(t-4)$

$=(t-2)^2u(t-2)-(t-3)^2u(t-3)-5(t-3)u(t-3)-2u(t-3)$

$+3(t-4)u(t-4)+4u(t-4)+2u(t-4)$

$=(t-2)^2u(t-2)-(t-3)^2u(t-3)-5(t-3)u(t-3)-2u(t-3)$

$+3(t-4)u(t-4)+6u(t-4)$

$F(s)=\mathcal{L}\{f(t)\}=\frac{2e^{-2s}}{s^3}-\frac{2e^{-3s}}{s^3}-\frac{5e^{-3s}}{s^2}-\frac{2e^{-3s}}{s}+\frac{3e^{-4s}}{s^2}+\frac{6e^{-4s}}{s}$

12.11 $f_1(t) = f(t) + f(t-1) + f(t-2) + \cdots$
where $f(t) = tu(t) - tu(t-1)$ is the function of Prob. 12.4 (a)

$\mathcal{L}\{f_1(t)\} = \mathcal{L}\{f(t) + f(t-1) + f(t-2) + \cdots\}$
$= F(s) + e^{-s}F(s) + e^{-2s}F(s) + \cdots$
$= F(s)\{1 + e^{-s} + e^{-2s} + \cdots\}$
$= F(s)\dfrac{1}{1-e^{-s}}$

From the solution of Prob. 12.4 (a)
$F(s) = -\dfrac{e^{-s}}{s} + \dfrac{1}{s^2}(1-e^{-s})$

$\mathcal{L}\{f_1(t)\} = \left[-\dfrac{e^{-s}}{s} + \dfrac{1}{s^2}(1-e^{-s})\right]\left(\dfrac{1}{1-e^{-s}}\right)$
$= -\dfrac{e^{-s}}{s(1-e^{-s})} + \dfrac{1}{s^2}$

12.12 $\mathcal{L}\left\{\dfrac{df(t)}{dt}\right\} = sF(s) - f(0)$

$F(s) = \mathcal{L}\{e^{-t}\sin t\} = \dfrac{1}{(s+1)^2+1}$

$f(0) = e^{-0}\sin 0 = 0$

Hence $\mathcal{L}\left\{\dfrac{df(t)}{dt}\right\} = \dfrac{s}{(s+1)^2+1} = \dfrac{s}{s^2+2s+2}$

12.13 $\mathcal{L}[f(t)] = \int_0^5 (5-t)e^{-st}\,dt$
$= \dfrac{5}{s} + \dfrac{e^{-5s}-1}{s^2}$

$\mathcal{L}\left[\dfrac{df(t)}{dt}\right] = s\mathcal{L}\{f(t)\} - f(0)$
$= s\left(\dfrac{5}{s} + \dfrac{e^{-5s}-1}{s^2}\right) - 5$
$= \dfrac{e^{-5s}-1}{s}$

12.14 $\mathcal{L}\{f(t)\} = \mathcal{L}\{t^3+t^2+t+1\}$
$= \mathcal{L}\{t^3\} + \mathcal{L}\{t^2\} + \dfrac{1}{s^2} + \dfrac{1}{s}$

Let $x(t) = t^3$, $x(0) = 0$, $x'(0) = 3t^2|_0 = 0$
$x''(t) = 6t|_0 = 0$, $x'''(\ ) = 6$

$\mathcal{L}\{x'''(t)\} = s^3\mathcal{L}\{t^3\} - s^2x(0) - sx(0) - x''(0)$
$\Rightarrow \mathcal{L}\{6\} = \dfrac{6}{s} = s^3\mathcal{L}\{t^3\} \Rightarrow \mathcal{L}\{t^3\} = \dfrac{6}{s^4}$

Similarly, $\mathcal{L}\{t^2\} = \dfrac{2}{s^3}$

$\mathcal{L}\{f(t)\} = \dfrac{6}{s^4} + \dfrac{2}{s^3} + \dfrac{1}{s^2} + \dfrac{1}{s}$

12.15 $f(t) = \cos^2(t) \Rightarrow f(0) = 1$
$f'(t) = 2\cos(-\sin t) = -2\sin t\cos t = -\sin 2t$
$\mathcal{L}\{f'(t)\} = s\mathcal{L}\{f(t)\} - f(0)$

12.15 Cont.
$\mathcal{L}\{-\sin 2t\} = -\dfrac{2}{s^2+4} = s\mathcal{L}\{\cos^2 t\} - 1$
$\Rightarrow \mathcal{L}\{\cos^2 t\} = \dfrac{1}{s}\left[1 - \dfrac{2}{s^2+4}\right]$
$= \dfrac{1}{s} - \dfrac{2}{s(s^2+4)}$

12.16 (a) $F(s) = \dfrac{1}{s} + \dfrac{2}{s+3} + \dfrac{3}{(s+5)^2+4}$
$f(t) = (1 + 2e^{-3t} + \frac{3}{2}\sin 2t\, e^{-5t})u(t)$

(b) $f(t) = \dfrac{1}{3!}t^3e^{3t} + \dfrac{1}{2!}t^2e^{2t}$

12.17 $e^{5t}(t+t^2)u(t) = e^{5t}tu(t) + 2\frac{1}{2!}e^{5t}t^2u(t)$
$F(s) = \dfrac{1}{(s-5)^2} + \dfrac{2}{(s-5)^3}$

12.18 $\mathcal{L}\{e^{-3t}\cos[(n+1)\pi t + \phi]\}$
$= \mathcal{L}\left\{e^{-3t}\cos\left[(n+1)\pi\left(t + \dfrac{\phi}{(n+1)\pi}\right)\right]\right\}$
$= e^{\frac{\phi}{(n+1)\pi}s}\dfrac{s+3}{(s+3)^2+(n+1)^2\pi^2}$

12.19 $F(s) = \dfrac{s^3+11s^2+75+4}{s^2+5s+6}$
$= s+6 - \dfrac{29s+32}{(s+2)(s+3)}$
$= s+6 - \dfrac{26}{s+2} + \dfrac{55}{s+3}$
$\mathcal{L}^{-1}\{F(s)\} = \delta'(t) + 6\delta(t) - 26e^{-2t} + 55e^{-3t}$

12.20 $F(s) = \dfrac{s+2}{s(s+3)(s+1)} = \dfrac{A}{s} + \dfrac{B}{s+3} + \dfrac{C}{s+1}$
$A = sF(s)|_{s=0} = \dfrac{2}{3}$
$B = (s+3)F(s)|_{s=-3} = -\dfrac{1}{6}$
$C = (s+1)F(s)|_{s=-1} = \dfrac{-2}{-2} = 1$
$F(s) = \dfrac{2}{3}\dfrac{1}{s} - \dfrac{1}{6}\dfrac{1}{s+3} + \dfrac{1}{s+1}$
$f(t) = \left[\dfrac{2}{3} - \dfrac{1}{6}e^{-3t} + e^{-t}\right]u(t)$

12.21 (a) $\dfrac{3s+3}{s^2+9} = \dfrac{3s+3}{(s+j3)(s-j3)} = \dfrac{A}{s+j3} + \dfrac{A^*}{s-j3}$
$A = \dfrac{3(-j3)+3}{-j3-j3} = \dfrac{3-j9}{-j6} = \dfrac{3}{2} + j\dfrac{1}{2} = \dfrac{\sqrt{10}}{2}\angle 73.3°$
$A^* = \dfrac{\sqrt{10}}{2}\angle -73.3°$
$f(t) = \dfrac{\sqrt{10}}{2}\cos(3t+73.3°) + \dfrac{\sqrt{10}}{2}\cos(3t-73.3°)$
$= \sqrt{10}\cos(3t+73.3°)$

(b) $\dfrac{s+2}{s^2+4s+6} = \dfrac{s+2}{(s+2)^2+2} \Rightarrow f(t) = e^{-2t}\cos\sqrt{2}t$

12.22 $F(s)=\frac{s}{(s+2)(s^2+6s+12)}=\frac{A}{s+2}+\frac{Bs+C}{(s+3)^2+3}$

$A=\frac{s}{(s+3)^2+3}\Big|_{s=-2}=\frac{-2}{1+3}=-\frac{1}{2}$

$-\frac{1}{2}(s^2+6s+12)+(Bs+C)(s+2)=s \Rightarrow B=\frac{1}{2}$

and $2c=6 \Rightarrow C=3$

$F(s)=\frac{-\frac{1}{2}}{s+2}+\frac{\frac{1}{2}s+3}{(s+3)^2+3}=\frac{-\frac{1}{2}}{s+2}+\frac{1}{2}\frac{s+3}{(s+3)^2+3}+\frac{\sqrt{3}}{2}\frac{\sqrt{3}}{(s+3)^2+3}$

$f(t)=-\frac{1}{2}e^{-2t}+\frac{1}{2}e^{-3t}\cos\sqrt{3}t+\frac{\sqrt{3}}{2}e^{-3t}\sin\sqrt{3}t$

$=-\frac{1}{2}e^{-2t}+e^{-3t}(\sin 30^\circ\cos\sqrt{3}t+\cos 30^\circ\sin\sqrt{3}t)$

$=-\frac{1}{2}e^{-2t}+e^{-3t}\sin(\sqrt{3}t+30^\circ)$

12.23 $F(s)=\frac{s^3+2s^2+4s+5}{(s+1)^2(s+2)^2}$

$=\frac{A}{(s+1)^2}+\frac{B}{s+1}+\frac{C}{(s+2)^2}+\frac{D}{s+2}$

$s^3+2s^2+4s+5=(B+D)s^3+(A+5B+C+4D)$
$+(4A+8B+2C+5D)s+4A+4B+C+2D$

equate coefficients

$1=B+D$
$2=A+5B+C+4D$
$4=4A+8B+2C+5D$
$5=4A+4B+C+2D$

Solve simult. equations
$A=2,\ B=-1$
$C=-3,\ D=2.$

$F(s)=\frac{2}{(s+1)^2}-\frac{1}{s+1}-\frac{3}{(s+2)^2}+\frac{2}{s+2}$

$\mathcal{L}^{-1}[F(s)]=2te^{-t}-e^{-t}-3te^{-2t}+2e^{-2t}$

$=[(2t-1)e^{-t}+(2-3t)e^{-2t}]u(t).$

12.24 $F(s)=\frac{6-s}{[(s-1)^2+2^2]}=\frac{A}{(s-p)^2}+\frac{B}{(s-p)}+\frac{A^*}{(s-p^*)^2}+\frac{B^*}{(s-p^*)}$

where $p=1+j\sqrt{2}$.

$A=\frac{6s-4}{(s-p^*)^2}\Big|_{s=p}=\frac{6(1+j\sqrt{2})-4}{(p-p^*)^2}=\frac{2+j6\sqrt{2}}{(j2\sqrt{2})^2}=\frac{2+j6\sqrt{2}}{-8}$

$=-\frac{1}{4}-j\frac{3\sqrt{2}}{4}=1.09\angle-103.3^\circ$

$B=\frac{d}{ds}\left[\frac{6s-4}{(s-p^*)^2}\right]\Big|_{s=p}=\frac{d}{ds}[(6s-4)(s-p^*)^{-2}]$

$=6(s-p^*)^{-2}+(-2)\frac{(6s-4)}{(s-p^*)^3}\Big|_{s=p}$

$=\frac{6}{(j2\sqrt{2})^2}-\frac{2(6+j6\sqrt{2}-4)}{(j2\sqrt{2})^3}=-\frac{6}{8}+\frac{2(2+j6\sqrt{2})}{j16\sqrt{2}}$

$=-\frac{3}{4}+\frac{2+j6\sqrt{2}}{j8\sqrt{2}}=-\frac{3}{4}-\frac{1}{4\sqrt{2}}+\frac{3}{4}=-j\frac{1}{4\sqrt{2}}=0.177\angle-90^\circ$

12.24 Cont.

$f(t)=2\,\mathrm{Re}\{Ate^{pt}+Be^{pt}\}$

$=2\,\mathrm{Re}\{(1.09\angle-103.3^\circ)te^{t}e^{j\sqrt{2}t}+(0.177\angle-90^\circ)e^{t}e^{j\sqrt{2}t}\}$

$=2.18\,te^{t}\,\mathrm{Re}\{e^{j\sqrt{2}t-103.3^\circ}\}+0.177e^{t}\,\mathrm{Re}\{e^{j\sqrt{2}t-90^\circ}\}$

$=2.18te^{t}\cos(\sqrt{2}t-103.3^\circ)+0.177e^{t}\cos(\sqrt{2}t-90^\circ)$

12.25 $f^{(n)}(t)\Longleftrightarrow s^n\mathcal{L}\{f\}-s^{n-1}f(0^+)$
$-s^{n-2}f'(0^+)\cdots f^{(n-1)}(0)$

$(s^2Y(s)-s4-2)+2(sY(s)-4)+Y(s)=3/(s+1)^2$

$(s^2+2s+1)Y(s)=\frac{3}{(s+1)^2}+4s+10$

$Y(s)=\frac{3}{(s+1)^4}+\frac{4s}{(s+1)^2}+\frac{10}{(s+1)^2}$

$=\frac{3}{(s+1)^4}+\frac{4(s+1)}{(s+1)^2}-\frac{4}{(s+1)^2}+\frac{10}{(s+1)^2}$

$y(t)=(\frac{t^3e^{-t}}{2}+4e^{-t}+6te^{-t})u(t)$

$=(4+6t+t^3/2)e^{-t}u(t)$

12.26 $(s^2Y(s)-2s-5)-4(sY(s)-2)+4Y(s)$
$=4s/(s^2+4)$;

$Y(s)(s^2-4s+4)=\frac{4s}{s^2+4}+2s-3$

$Y(s)=\frac{4s}{(s^2+4)(s-2)^2}+\frac{2s}{(s-2)^2}-\frac{3}{(s-2)^2}$

$=\frac{2s^3-3s^2+12s-12}{(s^2+4)(s-2)^2}$

$=\frac{As+B}{s^2+4}+\frac{C}{(s-2)^2}+\frac{D}{s-2}$

equating coefficients

$2=A+D$
$-3=-4A+B+C-2D$
$12=4A-4B+4D$
$-12=4B+4C-8D$

Solve equations
$A=0,\ B=-1$
$C=2,\ D=2$

$Y(s)=\frac{-1}{s^2+4}+\frac{2}{(s-2)^2}+\frac{2}{s-2}$

$y(t)=[2e^{2t}(t+1)-\frac{1}{2}\sin 2t]u(t).$

12.27 $Y(s)(\frac{1}{s^2+1})=Y(s)+\frac{1}{s^2+1}-\frac{s}{s^2+1}$

$Y(s)=Y(s)(s^2+1)+1-s$

$Y(s)=\frac{s-1}{s^2}=\frac{1}{s}-\frac{1}{s^2}$

$y(t)=(1-t)u(t)$

12.28 $x'' + x = u(t) - u(t - \pi/2)$

$s^2 X(s) - 1 + X(s) = 1/s - \dfrac{e^{-\pi s/2}}{s}$

$X(s)(s^2+1) = -\dfrac{e^{-\pi s/2}}{s} + \dfrac{1}{s} + 1$

$X(s) = \dfrac{-e^{-\pi s/2} + 1 + s}{s(s^2+1)} = \dfrac{A}{s} + \dfrac{Bs + C}{s^2+1}$

$s + 1 - e^{-\pi} = (A+B)s^2 + Cs + A$

$0 = A + B, \quad 1 = C, \quad 1 - e^{-\pi s/2} = A$

$B = e^{-\pi s/2} - 1;$

$X(s) = \dfrac{1 - e^{-\pi s/2}}{s} + \dfrac{(e^{-\pi s/2} - 1)s + 1}{s^2+1}$

$= \dfrac{1}{s} + \dfrac{1}{s^2+1} - \dfrac{s}{s^2+1} - \dfrac{e^{-\pi s/2}}{s} + \dfrac{se^{-\pi s/2}}{s^2+1}$

$x(t) = (1 + \sin t - \cos t)u(t)$
$\quad + (-1 + \cos(t - \pi/2))\,u(t - \pi/2)$

$= \underline{(1 + \sin t - \cos t)u(t) + (\sin t - 1)u(t - \pi/2)}$

12.29 $X(s)s - 4 = \dfrac{1}{s^2} + X(s)\left(\dfrac{s}{s^2+1}\right)$

$X(s)\, s^3(s^2+1) - 4s^2(s^2+1) = s^2 + 1 + X(s)s^3$

$X(s)(s^5 + s^3 - s^3) = 4s^4 + 5s^2 + 1$

$X(s) = \dfrac{4s^4 + 5s^2 + 1}{s^5} = \dfrac{4}{s} + \dfrac{5}{s^3} + \dfrac{1}{s^5}$

$x(t) = \underline{\left(4 + \frac{5}{2}t^2 + \frac{t^4}{24}\right)u(t)}.$

12.30 Describing equation

$2i + \dfrac{di}{dt} + \int_0^t i\,dt + v_C(0^-) = f_1(t)$

$2s + sI + \dfrac{I}{s} = -\dfrac{e^{-s}}{s} + \dfrac{1 - e^{-s}}{s^2}$

$I(s^2 + 2s + 1) = -\dfrac{e^{-s}}{s} + \dfrac{1 - e^{-s}}{s^2}$

$I = -\dfrac{e^{-s}}{s(s+1)^2} + \dfrac{1 - e^{-s}}{s^2(s+1)^2} = (-e^{-s})\left[\dfrac{A_1}{s} + \dfrac{A_2}{(s+1)^2} + \dfrac{A_3}{s+1}\right]$

$+ (1 - e^{-s})\left[\dfrac{B_1}{s^2} + \dfrac{B_2}{s} + \dfrac{B_3}{(s+1)^2} + \dfrac{B_4}{(s+1)}\right]$

$A_1 = \dfrac{1}{(s+1)^2}\Big|_{s=0} = 1, \quad A_2 = \dfrac{1}{s}\Big|_{s=-1} = -1$

$A_3 = \dfrac{d}{ds}\left[\dfrac{1}{s}\right]\Big|_{s=-1} = (-1)\dfrac{1}{s^2}\Big|_{s=-1} = -1$

$B_1 = \dfrac{1}{(s+1)^2}\Big|_{s=0} = 1, \quad B_2 = \dfrac{d}{ds}\left[\dfrac{1}{(s+1)^2}\right]\Big|_{s=0} = (-2)\dfrac{1}{(s+1)^3}\Big|_{s=0} = -2$

$B_3 = \dfrac{1}{s^2}\Big|_{s=-1} = 1, \quad B_4 = \dfrac{d}{ds}\left[\dfrac{1}{s^2}\right]\Big|_{s=-1} = (-2)\dfrac{1}{s^3}\Big|_{s=-1} = 2$

$I = -\dfrac{e^{-s}}{s} - \dfrac{-e^{-s}}{(s+1)^2} - \dfrac{-e^{-s}}{s+1} + (1 - e^{-s})\left[\dfrac{1}{s^2} - \dfrac{2}{s} + \dfrac{1}{(s+1)^2} + \dfrac{2}{s+1}\right]$

$= \dfrac{1 - e^{-s}}{s^2} - \dfrac{2(1 - e^{-s})}{s} + \dfrac{2(1 - e^{-s})}{s+1}$

$= \dfrac{1}{s^2} + \dfrac{e^{-s}}{s} + \dfrac{2}{s+1} - \dfrac{e^{-s}}{s+1} + \dfrac{1}{s^2} - \dfrac{e^{-s}}{s^2}$

$i(t) = \underline{-2u(t) + u(t-1) + 2e^{-t} - e^{-(t-1)} + tu(t) - (t-1)u(t-1)}$ A

12.31 At $t = 0^-$; $v_C = 0$, $i = 8$ A

At $t = 0^+$; KVL around right loop $v_C = 5\dfrac{di}{dt}$

KCL yields $\dfrac{1}{20}\dfrac{dv_C}{dt} + \dfrac{v_C}{4} + i = \dfrac{8}{4}.$

By substitution,

$\dfrac{1}{4}\dfrac{d^2 i}{dt} + \dfrac{5}{4}\dfrac{di}{dt} + i = 2$ or $\dfrac{d^2 i}{dt} + 5\dfrac{di}{dt} + 4i = 8$

Since $\dfrac{di(0)}{dt} = \dfrac{v_C(0)}{5} = 0$

$s^2 I - 8s + 5(sI - 8) + 4I = 8/s$

$I = \dfrac{8s^2 + 40s + 8}{s(s^2 + 5s + 4)} = \dfrac{A}{s} + \dfrac{B}{s+1} + \dfrac{C}{s+4}$

$A = \dfrac{8s^2 + 40}{s^2 + 5s + 4}\Big|_{s=0} = \dfrac{8}{4} = 2$

$B = \dfrac{8s^2 + 40s + 8}{s^2 + 4s}\Big|_{s=-1} = \dfrac{8 - 4 + 8}{1 - 4} = -4$

$C = \dfrac{8s^2 + 40s + 8}{s^2 + 5}\Big|_{s=-4} = 10$

$I = \dfrac{2}{s} + \dfrac{4}{s+1} + \dfrac{10}{s+4}$

$i(t) = \underline{[2 - 4e^{-t} + 10e^{-4t}]u(t)}$

12.32 At $t = 0^-$; $v = 0$, $i_L = \dfrac{10}{2+3}$

At $t = 0^+$; KVL of right loop

$v = 3\dfrac{di_L}{dt}$. KCL yields

$\dfrac{v}{3} + \dfrac{1}{12}\dfrac{dv}{dt} + i_L = 0.$ By substitution

$\dfrac{di_L}{dt} + \dfrac{1}{4}\dfrac{d^2 i_L}{dt^2} + i_L = 0$ or

$\dfrac{d^2 i_L}{dt} + 4\dfrac{di_L}{dt} + 4i_L = 0$

Since $\dfrac{di_L(0^+)}{dt} = \dfrac{v(0^+)}{3} = 0$

$s^2 I_L - s2 + 4(sI_L - 2) + 4I_L = 0$

$I_L = \dfrac{2s + 8}{s^2 + 4s + 4} = \dfrac{A}{(s+2)^2} + \dfrac{B}{s+2}$

$2s + 8 = A + B(s+2)$; $A = 4, B = 2$

$I_L = \dfrac{4}{(s+2)^2} + \dfrac{2}{s+2}$

$i_L(t) = (4te^{-2t} + 2e^{-2t})u(t)$

$= \underline{2e^{-2t}(2t+1)u(t)\text{ A}}$

Chapter 13
Circuit Analysis in the S-domain

13.1 Element and Kirchoff's laws

13.1 Write the S-domain describing equation of the circuit shown. Find the voltage $\nu(t)$ by solving the problem in S-domain and inverse Laplace transform. Assume $i(0) = 0\text{A}$, $\nu(0) = 2\text{V}$.

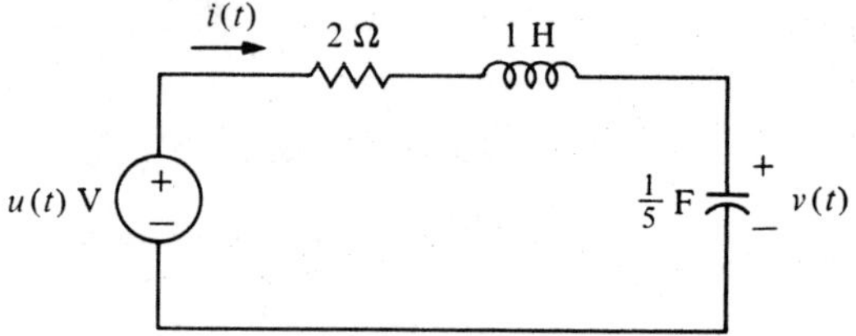

Problem 13.1

13.2 Find the equivalent impedance seen by the voltage source of the circuit shown in s-domain. Assume zero initial conditions.

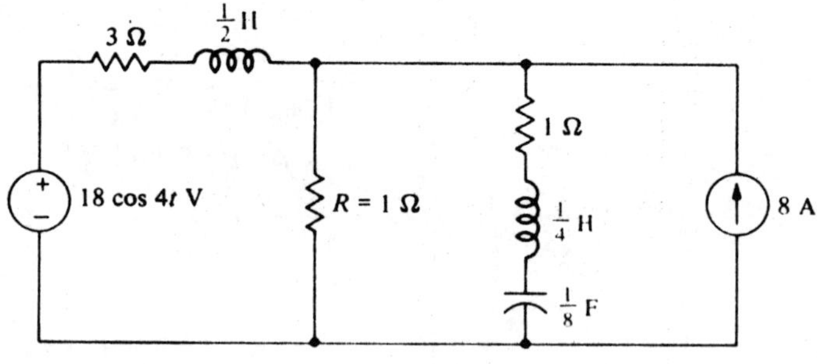

Problem 13.2

13.3 Find the equivalent admittance seen by the current source in s-domain. Assume zero initial conditions.

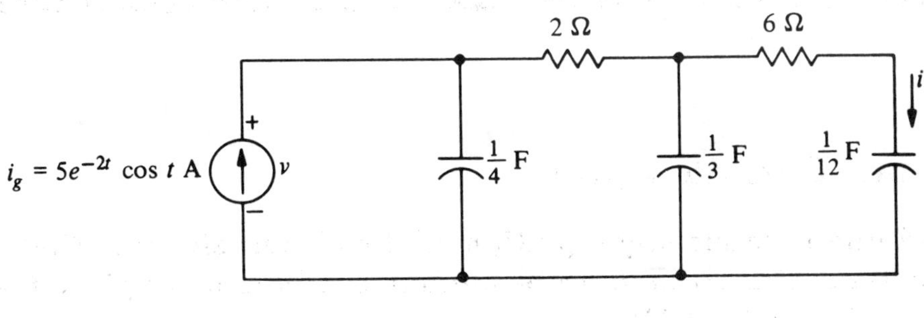

Problem 13.3

13.4 Replace in the corresponding circuit everything except the capacitor by its Thevenin equivalent circuit and use the result to find the forced response i if $v_g = 4e^{-2t}\cos(2t)$V.

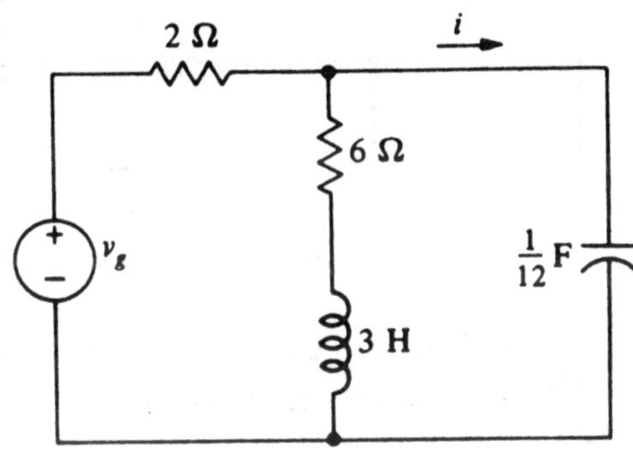

Problem 13.4

13.5 Replace all the circuit except the inductor, by its Norton equivalent and find I_1.

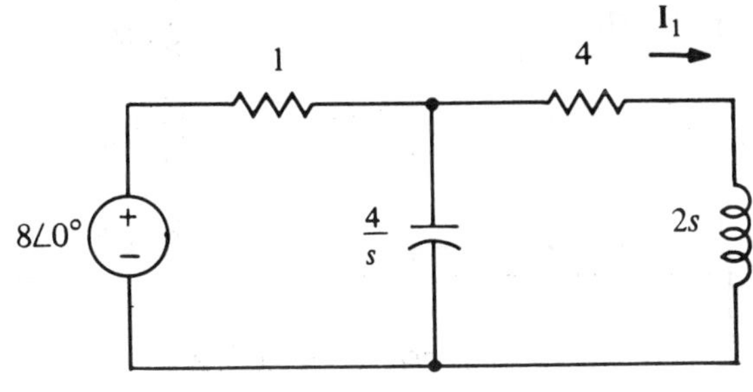

Problem 13.5

13.2 S-domain circuit

13.6 **Find $v(t)$ with initial conditions $i(0) = 1$A, $v(0) = 6$V.**

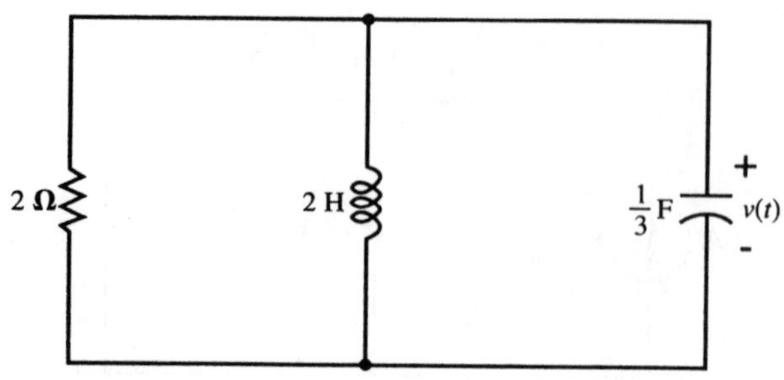

Problem 13.6

13.7 **Find the S-domain Thevenin and Norton equivalents seen by the terminals a - a' for time t > 0. Assume steady state at t < 0.**

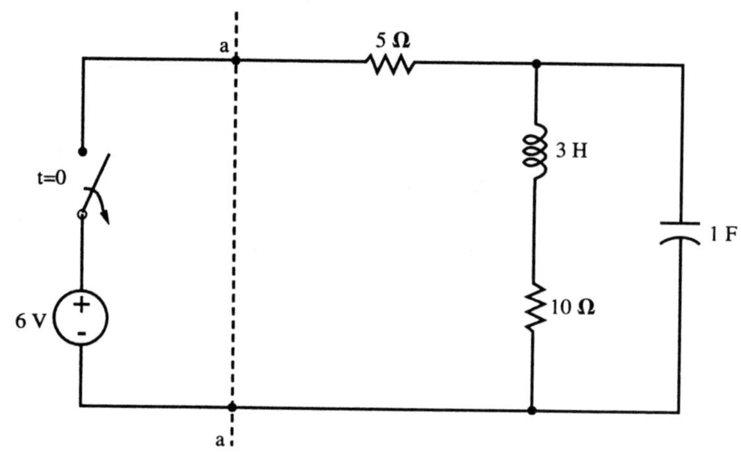

Problem 13.7

13.8 **Find the forced response i if $v_{g1} = 8e^{-t}\cos(t)$V and $i_{g2} = 2e^{-2t}$A.**

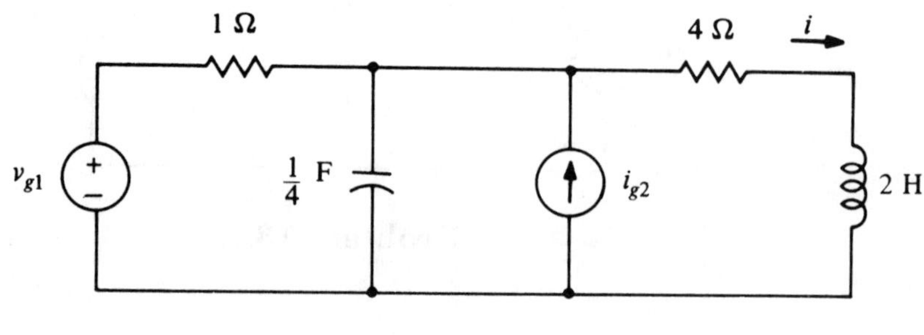

Problem 13.8

13.9 Use the Laplace transformed circuit method to find $i(t)$, t > 0, if $v(0) = 4$V and $i(0) = 2$A.

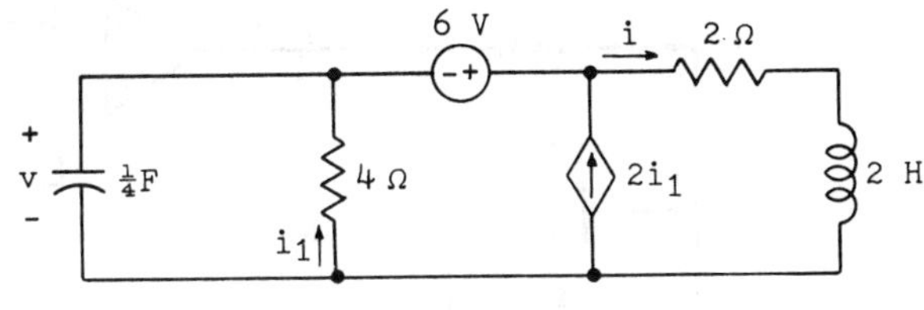

Problem 13.9

13.10 For the circuit shown, replace everything to the left of the terminals a-b by its Thevenin equivalent and use the result to find the forced response i.

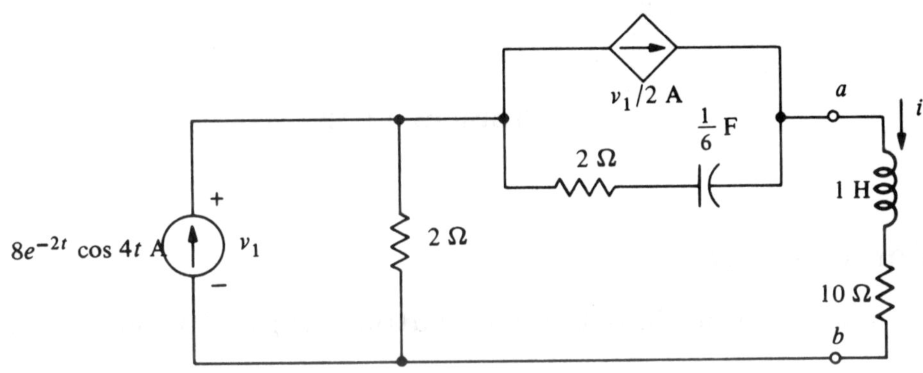

Problem 13.10

13.11 Use the Laplace transformed circuit method to find v, if $v_1(0) = 4$V and $v_2(0) = 2$V.

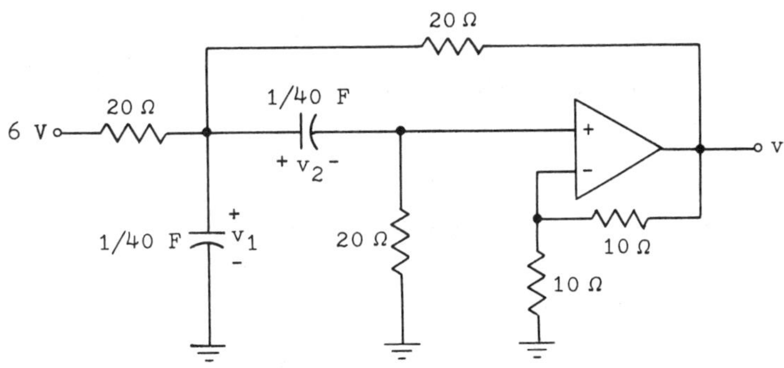

Problem 13.11

13.12 Find the output $\nu_O(t)$ with $\nu_g = te^{-\frac{t}{2}}u(t)$V. Assume no initial energy stored in capacitors.

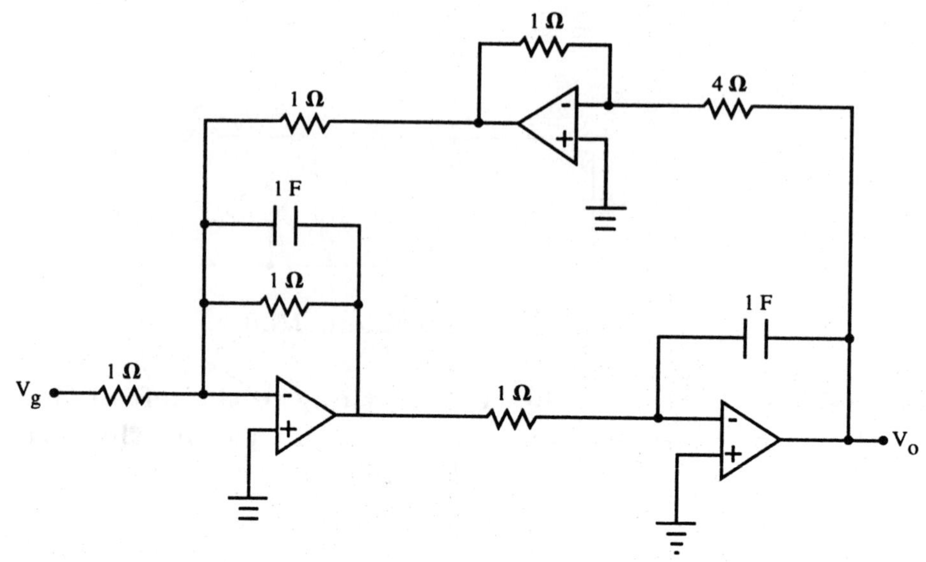

Problem 13.12

13.3 Transfer functions

13.13 Find the transfer functions between output I(s) and inputs V_{s1}(s), I_{s2}(s).

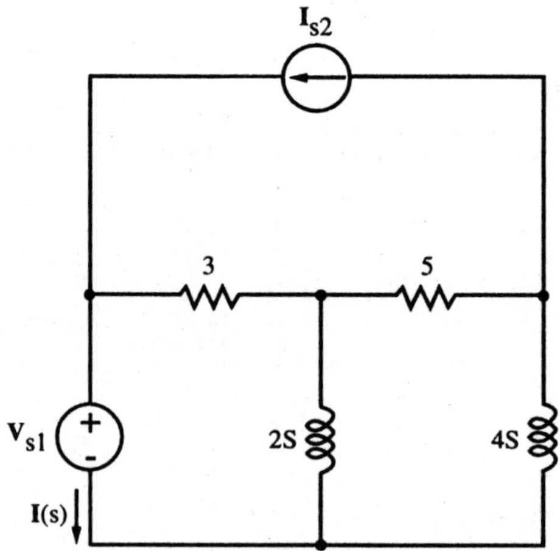

Problem 13.13

13.14 Find $\mathbf{H(s)} = \frac{V(s)}{V_g(s)}$ if the response is ν and $\mathbf{R} = 16\Omega$. Use the result to find the forced response if $\nu_g = 3e^t\mathbf{V}$.

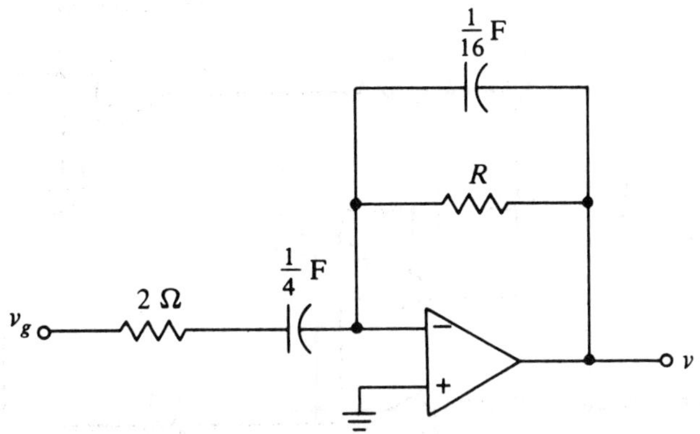

Problem 13.14

13.15 Find $\mathbf{H(s)} = \frac{V(s)}{V_g(s)}$ and use the result to find the forced response ν if $\nu_g = e^{-t}\cos(t)\mathbf{V}$.

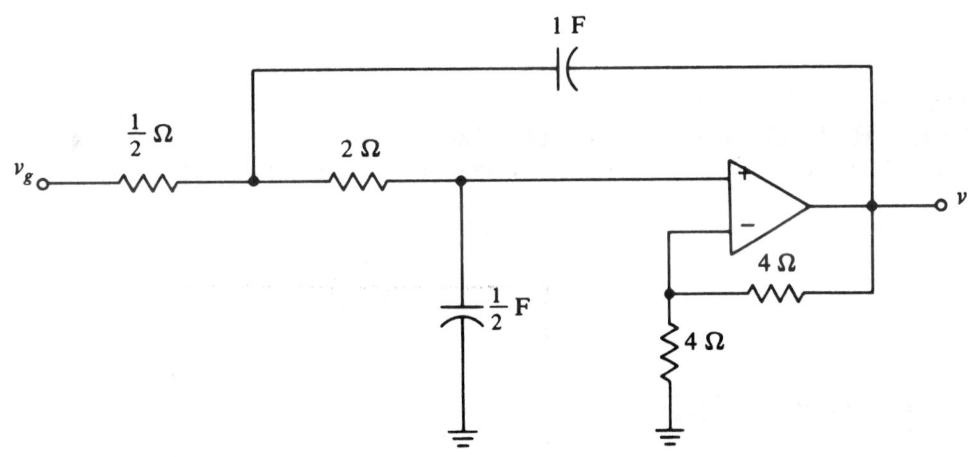

Problem 13.15

13.16 Find the transfer function $H(s) = \frac{V(s)}{V_g(s)}$.

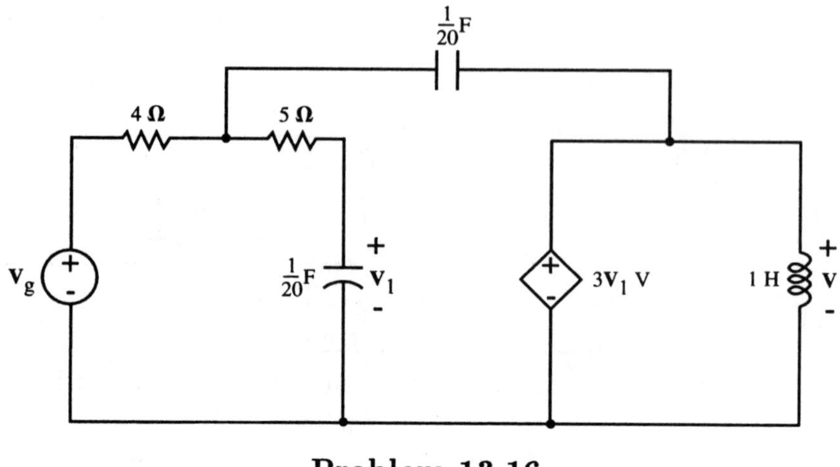

Problem 13.16

13.17 If the zeros of H(s) are s = -4 ± j3, the poles are s = -3, -2 ± j, and H(0) = 5, find H(s).

13.18 Using R, L, C elements, synthesize a two-terminal network with the impedance

$$Z(s) = \frac{s^2 + 4s + 3}{2}$$

13.19 Using R, L, C elements, synthesize a two-terminal network with the impedance

$$Z(s) = \frac{6s^2 + 12s + 8}{3s + 4}$$

13.4 Poles and stability

13.20 Find the poles and zeros for the transfer function T(s). Is this system stable ?

$$T(s) = \frac{-2(s^2 + 6s + 9)}{4s^2 + 1}$$

13.21 Find the poles and zeros for the transfer function T(s). Is this system stable ?

$$T(s) = \frac{2s^2 - 3s + 2}{(s^2 + 4s + 6)^2}$$

13.22 Determine the poles and the ac steady state value of $i_c(t)$ in the circuit shown for the current source $i_s(t) = 6\sin t$A.

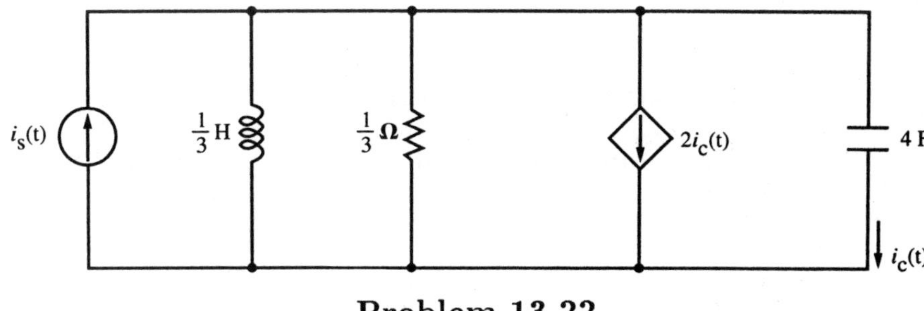

Problem 13.22

13.23 Determine the smallest value of k for which the circuit is not stable.

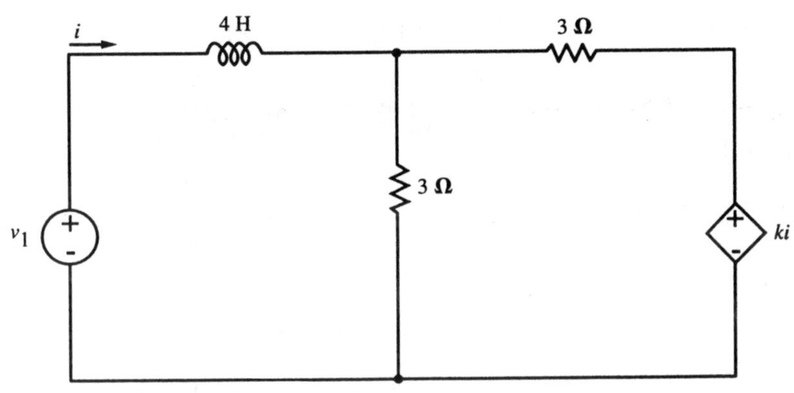

Problem 13.23

13.5 Initial and final value theorems

13.24 Compute the inverse Laplace transform of

$$F(s) = \frac{36}{s^3 + 9s^2 + 24s + 20}$$

and check by applying initial and final value theorems.

13.25 If $i_s(t) =$ u(t) and the initial conditions at $t = 0^-$ are zero in the circuit of Prob.13.22, find $i_c(t)$ at t $= 0^+$ and t $= \infty$.

13.26 A circuit has two independent sources. For the output variable $V_0(t)$, transfer functions with respect to source 1 and source 2 are $H_1(s) = \frac{5s-4}{s^3+3s^2+2s}$ and $H_2(s) = \frac{s^2+2}{s^2+3}$ respectively. Find $V_0(0^+)$ and $V_0(\infty)$ if $\delta'(t)$ is applied to source

1 and u(t) is applied to source 2. Assume all initial conditions at t $= 0^-$ are zero.

13.6 Impulse response and convolution

13.27 Find the transfer function H(s) and unit impulse response h(t) for the input $\nu_1(t)$ and output $\nu_c(t)$.

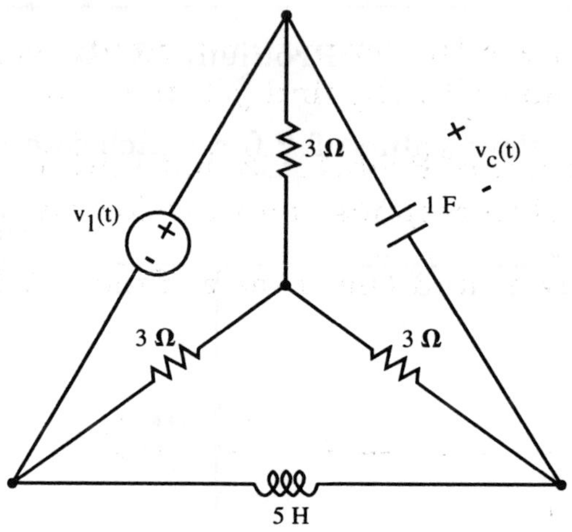

Problem 13.27

13.28 Given the equation $y''(t) + 6y'(t) + 5y(t) = 3x(t)$ where $y(t)$ and $x(t)$ are the output and input, respectively, of a circuit, find the transfer function H(s) and the impulse response h(t).

13.29 An input of $\nu_i(t) = 10\delta(t)$ is applied to a circuit with zero initial conditions and all other independent sources killed. The resulted input is measured to be

$$\nu_0(t) = e^{-4t}\cos(6t)u(t - \frac{\pi}{12})$$

Determine the transfer function H(s) and the impulse response h(t).

13.30 Solve for x and y. Assume $x(t) = y(t) = 0$ for t < 0

$$3x'' + 4x + y' + y = u(t)$$
$$2x'' + 6x + y' + 2y = 0$$

13.31 Find $f * g$ if

$$f(t) = u(t) - u(t - 4)$$
$$g(t) = tu(t) - tu(t - 2)$$

13.32 **Find $f * g$ if**

$$f(t) = 4\cos(4t)u(t)$$
$$g(t) = u(t) - u(t-5)$$

13.33 **Given a circuit with describing equation**

$$y'''(t) + y''(t) + 4y'(t) + 4y(t) + x'(t) + x(t) = \delta(t)$$

where $y(t)$ and $x(t)$ are the output and input, respectively. Find (a) output $y(t)$ in terms of input $x(t)$, (b) find $y(t)$ if $x(t) = \delta(t)$.

13.34 **Given the same circuit of Prob.13.33, but initially relaxed. Find the impulse response $h(t)$, the step response and the forced response to an input e^{-2t}.**

13.35 **For the circuit shown, find the transfer function H(s) with input $\nu_{in}(t)$, and output $\nu_o(t)$.**

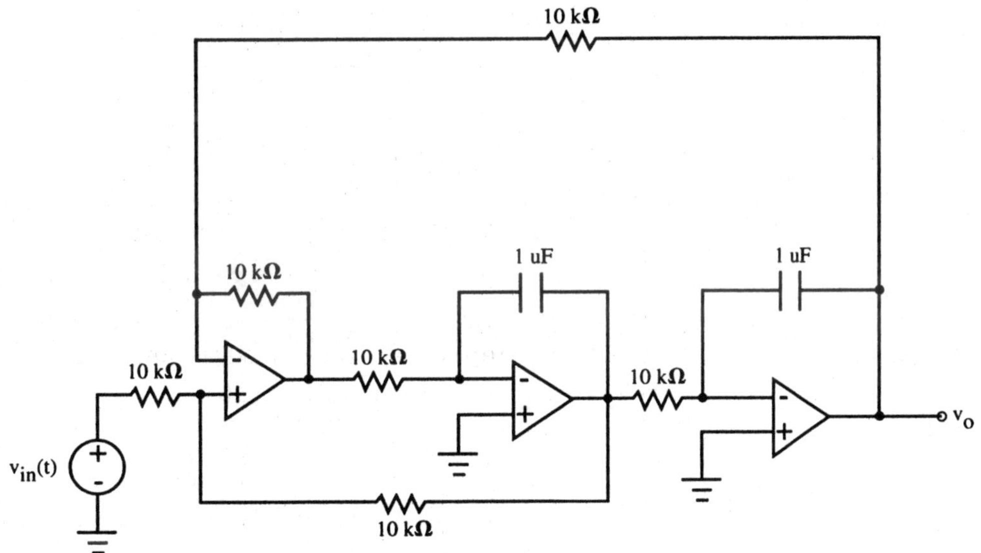

Problem 13.35

13.36 **Find the response $\nu_O(t)$ in Prob.13.35 if the circuit is initially relaxed and $\nu_{in}(t) = u(t) - u(t-1)$.**

13.1

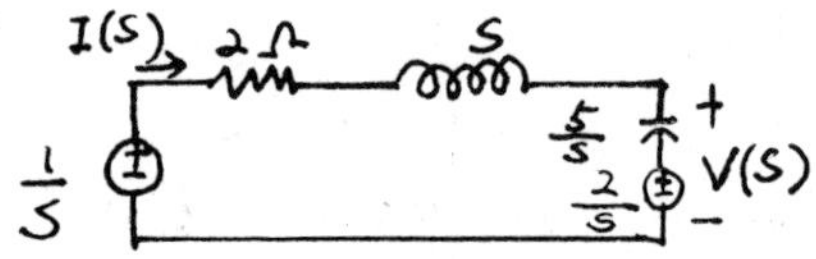

By KVL in S-domain

$$2I(S)+SI(S)+\frac{5}{S}I(S) = \frac{1}{S}$$

$$I(S)=\frac{-\frac{1}{S}}{S+2+\frac{5}{S}}=\frac{-1}{S^2+2S+5}$$

$$V(S)=\frac{5}{S}I(S)+\frac{2}{S}=\frac{-5}{S(S^2+2S+5)}+\frac{2}{S}$$

$$=\frac{2S^2+4S+5}{S(S^2+2S+5)}=\frac{A}{S}+\frac{B}{S^2+2S+5}$$

$$A=\left.\frac{2S^2+4S+5}{S^2+2S+5}\right|_{S=0}=\frac{5}{5}=1$$

$$2S^2+4S+5=S^2+2S+5+BS \Rightarrow S^2+2S=BS$$

$$\Rightarrow B=S+2$$

$$V(S)=\frac{1}{S}+\frac{S+2}{S^2+2S+5}\Rightarrow v(t)=u(t)+e^{-2t}\cos t\,V$$

13.2

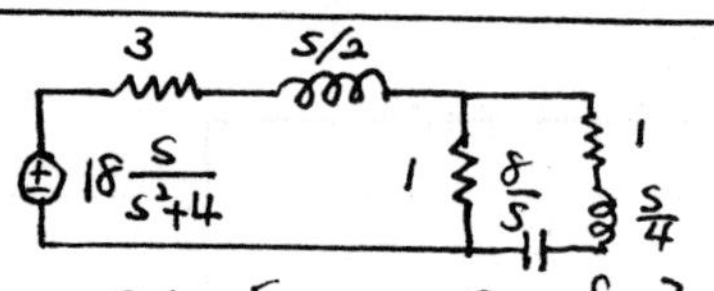

$$R_{eq}=3+S/2+\left[1\,//\left(1+\frac{S}{4}+\frac{8}{S}\right)\right]$$

$$=3+S/2+\frac{1+\frac{S}{4}+\frac{8}{S}}{1+1+\frac{S}{4}+\frac{8}{S}}$$

$$=3+\frac{S}{2}+\frac{S^2+4S+32}{S^2+8S+32}$$

$$=\frac{S^3+16S^2+88S+256}{2(S^2+8S+32)}$$

13.3

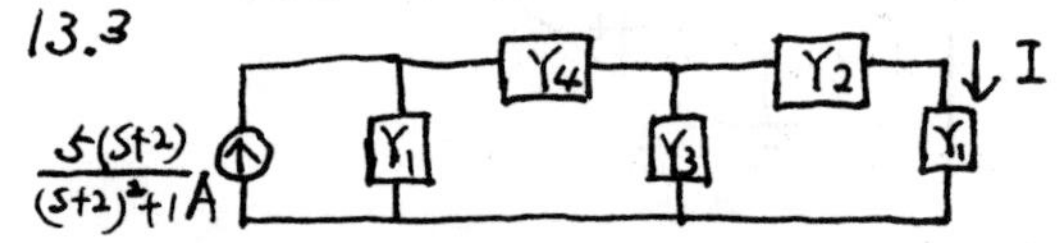

$$Y_1=\frac{1}{12/S}=\frac{S}{12}$$

$$Y_2=\frac{1}{6},\quad Y_{12}=\frac{\frac{1}{6}\times\frac{S}{12}}{\frac{1}{6}+\frac{S}{12}}=\frac{S}{6(2+S)}$$

$$Y_3=\frac{S}{3},\quad Y_{123}=\frac{S}{3}+\frac{S}{6(2+S)}=\frac{3S}{6(2+S)}$$

$$Y_4=\frac{1}{2},\quad Y_{1234}=\frac{\frac{1}{2}\times\frac{3S}{6(S+2)}}{\frac{1}{2}+\frac{3S}{6(S+2)}}=\frac{S}{4(S+1)}$$

$$Y_5=\frac{S}{4}\Rightarrow Y_{12345}=\frac{S}{4(S+1)}+\frac{S}{4}=\frac{S^2+2S}{4(S+1)}$$

13.4 $V_{oc}=\frac{6+3S}{8+3S}V_g$ by voltage division

$$R_{th}=\frac{2(3S+6)}{3S+8}$$

$$I=\frac{V_{oc}}{R_{th}+12/S}=\frac{\frac{4(3S+6)}{8+3S}}{\frac{2(3S+6)}{3S+8}+\frac{12}{S}}=\frac{2S(S+2)}{S^2+8S+16}$$

$$=\frac{2S(S+2)}{(S+4)^2}$$

$$I(-2+j2)=\frac{2(-2+j2)(j2)}{(2+j2)^2}=\sqrt{2}\angle 135^\circ$$

$$i=\sqrt{2}e^{-2t}\cos(2t+135^\circ)A$$

13.5 $V_{oc}=\frac{4/S}{1+4/S}8=\frac{32}{S+4}V$

$$R_{th}=4+\frac{4/S}{1+4/S}=\frac{4(S+5)}{S+4}$$

$$I_1=\frac{V_{oc}}{R_{th}+2S}=\frac{32/S+4}{\frac{4(S+5)}{S+4}+2S}=\frac{16}{S^2+6S+10}$$

13.6

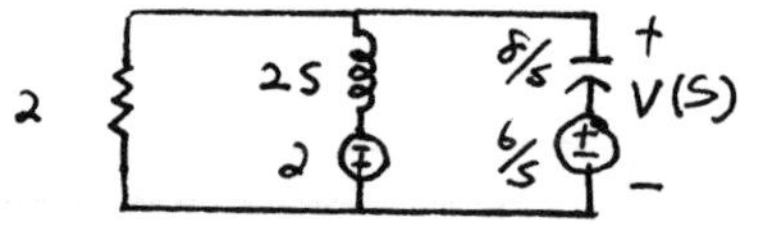

KCL yields

$$\frac{V-(-2)}{2S}+\frac{V-(\frac{6}{S})}{\frac{8}{S}}+\frac{V}{2}=0$$

$$(V+2)\left(\frac{8}{S}\cdot 2\right)+\left(V-\frac{6}{S}\right)(2S\cdot 2)+V\left(2S\cdot\frac{8}{S}\right)=0$$

$$V\left(\frac{16}{S}+4S+16\right)=24-\frac{32}{S}$$

$$V=\frac{6S-8}{S^2+4S+4}=\frac{6}{S+2}-\frac{20}{(S+2)^2}$$

$$v(t)=6e^{-2t}u(t)-20te^{-2t}u(t)\,V$$

13.7 At $t<0$, the inductor is short-circuited, and the capacitor is an open circuit.

$$i_L(0)=\frac{6}{15}=\frac{2}{5}A,$$

$$V_C(0)=\frac{10}{15}\times 6=4V.$$

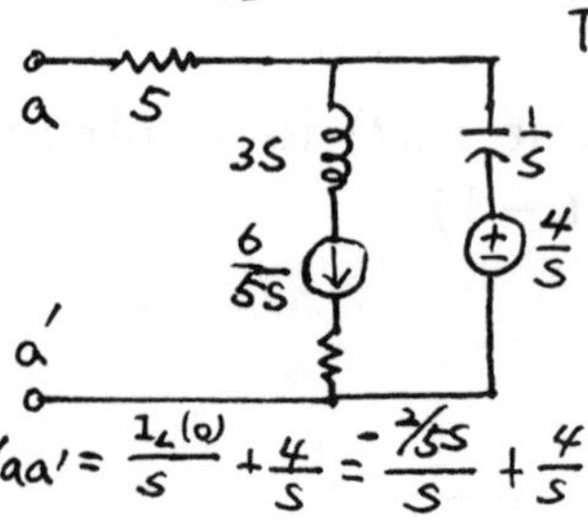

The impedance seen by terminals a-a' is

$$R_{eq}=5+\frac{(3S+10)\frac{1}{S}}{(3S+10)+\frac{1}{S}}$$

$$=5+\frac{3S+10}{3S^2+10S+1}$$

$$V_{aa'}=\frac{i_L(0)}{S}+\frac{4}{S}=\frac{-2/5S}{S}+\frac{4}{S}=\frac{20S-2}{5S^2}$$

13.7 Cont.

Thevenin equivalent

$\frac{20s-2}{5s^2}$, $Z = 5+\frac{3s+10}{3s^2+10s+1}$, a, a'

$$I_{Norton} = \frac{20s-2}{5s^2} \times \frac{15s^2+50s+5+3s+10}{3s^2+10s+1}$$

$$= \frac{20s-2}{5s^2} \times \frac{15s^2+53s+15}{3s^2+10s+1}$$

$\frac{(20s-2)(15s^2+53s+15)}{(5s^2)(3s^2+10s+1)}$, $Z = 5+\frac{3s+10}{3s^2+10s+1}$, a, a'

13.8 $V_{g_1} = 8\angle 0°$, $I_{g_2} = 2\angle 0°$ when $s = -1+j$

With Node voltage V, KCL give

$$V\left(1+\frac{s}{4}+\frac{1}{2s+4}\right) = \frac{V_{g_1}}{1} + I_{g_2} \quad \text{or}$$

$$V = \frac{(2+8)\,2(2s+4)}{s^2+6s+10} \Rightarrow I = \frac{V}{2s+4} = \frac{20}{s^2+6s+10}$$

$$s = -1+j, \quad I = \frac{20}{4+j4} = \frac{5}{\sqrt{2}}\angle -45° A$$

$$i = \frac{5}{\sqrt{2}} e^{-t}\cos(t-45°)\,A$$

13.9 $V = -4I_1$, $6/s$, $I(s)$, $4/s$, $4/s$, 4, $I_1(s)$, $2I_1(s)$, $2s$, 4

$$\text{KCL:}\quad I(s) = 2I_1 + \left(I_1 + \frac{4/s+4I_1}{4/s}\right)$$

$$= (3+s)I_1 + 1 \quad \text{or}$$

$$I(s) - (s+3)I_1 = 1$$

$$I_s = \frac{-4I_1 + 6/s + 4}{2+2s} \quad \text{or}$$

$$I_s(2+2s)s + 4sI_1 = 6+4s$$

$$I(s) = \frac{\begin{vmatrix} 1 & -(s+3) \\ 6+4s & 4s \end{vmatrix}}{\begin{vmatrix} 1 & -(s+3) \\ s(2+2s) & 4s \end{vmatrix}} = \frac{2s^2+11s+9}{s(s^2+4s+5)}$$

$$= \frac{A}{s} + \frac{Bs+C}{s^2+4s+5}$$

equating coefficients

$$\left.\begin{aligned} 2 &= A+B \\ 11 &= 4A+C \\ 9 &= 5A \end{aligned}\right\} \Rightarrow A = 9/5,\ B = 1/5,\ C = 19/5$$

$$I(s) = \frac{9/5}{s} + \frac{s/5}{(s+2)^2+1} + \frac{19/5}{(s+2)^2+1}$$

13.9 Cont.

$$I(s) = \frac{9/5}{s} + \frac{1/5(s+2)}{(s+2)^2+1} + \frac{17/5}{(s+2)^2+1}$$

$$i(t) = \left[9/5 + \frac{1}{5}e^{-2t}(\cos t + 17\sin t)\right]u(t)\,A$$

13.10 With a-b open,

$$V_{oc} = V_1 + \left(\frac{6}{s}+2\right)\left(\frac{V_1}{2}\right) = (2+3/s)V_1$$

where $V_1 = 2\,(8\angle 0°) = 16V$. Therefore

$$V_{oc} = \frac{16(2s+3)}{s}$$

With a-b shorted,

$$I_{sc} = 8 - \frac{V_1}{2} = \frac{V_1}{2} + \frac{V_1}{2+6/s}$$

Therefore $V_1 = \frac{16(s+3)}{3s+6}$ and $I_{sc} = \frac{8(2s+3)}{3(s+2)}$

$$\therefore Z_{th} = \frac{V_{oc}}{I_{sc}} = \frac{16(2s+3)/s}{8(2s+3)/3(s+2)} = \frac{6(s+2)}{s}$$

$$I = \frac{V_{oc}}{Z_{th}+s+10} = \frac{16(2s+3)/s}{\frac{6(s+2)}{s}+s+10} = \frac{16(2s+3)}{s^2+16s+12}$$

$$I(-2+j4) = \frac{16(-1+j8)}{16(-2+j3)} = 2-j = \sqrt{5}\angle -26.6° A$$

$$i = \sqrt{5}e^{-2t}\cos(4t-26.6°)\,A$$

13.11

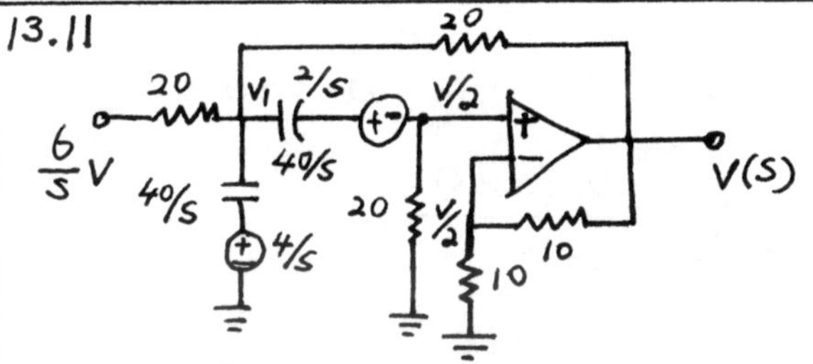

KCL yields:

$$\frac{(V_1 - 6/s)}{20} + \frac{(V_1 - 4/s)s}{40} + \left(V_1 - \frac{V}{2} - \frac{2}{s}\right)\frac{s}{40} + \frac{V_1 - V}{20} = 0$$

$$\text{or } 4s(2+s)V_1 - s(s+4)V = 12(s+2)$$

$$\frac{V/2}{20} + \left(V/2 + 2/s - V_1\right)\frac{s}{40} = 0 \quad \text{or}$$

$$-2sV_1 + (s+2)V = -4$$

$$V = \frac{\begin{vmatrix} 4s(s+2) & 12(s+2) \\ -2s & -4 \end{vmatrix}}{\begin{vmatrix} 4s(s+2) & -s(s+4) \\ -2s & s+2 \end{vmatrix}} = \frac{4(s+2)}{s^2+4s+8}$$

$$= \frac{4(s+2)}{(s+2)^2+4}$$

$$v(t) = 4e^{-2t}\cos 2t\,u(t)\,V$$

13.12

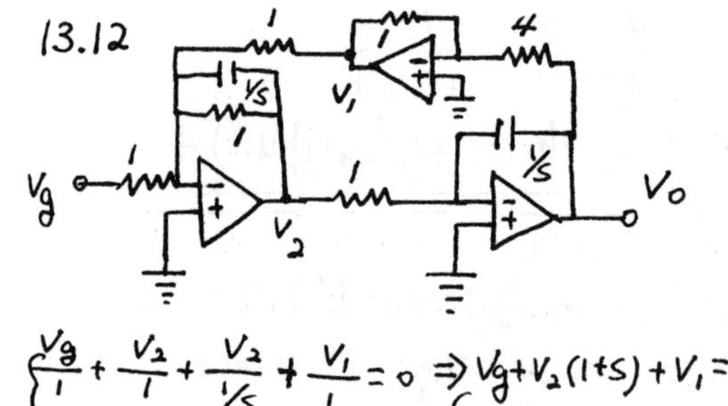

$$\begin{cases} \frac{V_g}{1} + \frac{V_2}{1} + \frac{V_2}{1/s} + \frac{V_1}{1} = 0 \Rightarrow V_g + V_2(1+s) + V_1 = 0 \\ \frac{V_2 - 0}{1} + \frac{V_o}{1/s} = 0 \\ \frac{V_1}{1} + \frac{V_o}{4} = 0 \end{cases} \qquad \begin{cases} sV_o = -V_2 \\ V_o = -4V_1 \end{cases}$$

$$V_g + (-s)(1+s)V_o + \left(-\frac{V_o}{4}\right) = 0$$

or $V_o\left[\frac{1}{4} + s(1+s)\right] = V_g \Rightarrow V_o = \frac{V_g}{s^2+s+\frac{1}{4}} = \frac{4V_g}{(2s+1)^2}$

$v_g = te^{-t/2}u(t) \Rightarrow V_g = \frac{1}{(s+\frac{1}{2})^2}$

Hence $V_o = \frac{V_g}{(s+\frac{1}{2})^2} = \frac{1}{(s+\frac{1}{2})^4}$

$$v_o = \frac{t^3}{3!}e^{-\frac{t}{2}}u(t)$$

13.13

Kill the source current I_{s2}

$I_1(3+2s) - I_2(2s) = V_{s1}$

$-I_1(2s) + I_2(5+6s) = 0$

$$I_1 = \frac{\begin{vmatrix} V_{s1} & -2s \\ 0 & 5+6s \end{vmatrix}}{\begin{vmatrix} 3+2s & -2s \\ -2s & 5+6s \end{vmatrix}} = \frac{V_{s1}(5+6s)}{(3+2s)(5+6s) - 4s^2}$$

$$H_{s1} = \frac{I_1}{V_{s1}} = \frac{5+6s}{8s^2+28s+15}$$

Kill the voltage source V_{s1}

$$\begin{cases} I_1(3+2s) - I_2(2s) - I_3(3) = 0 \\ -I_1(2s) + I_2(5+6s) - I_3(5) = 0 \\ I_3 = -I_{s2} \end{cases}$$

$$\Rightarrow \begin{cases} I_1(3+2s) - I_2(2s) = -3I_{s2} \\ -I_1(2s) + I_2(5+6s) = -5I_{s2} \end{cases}$$

$$I_1 = \frac{\begin{vmatrix} -3I_{s2} & 3+2s \\ -5I_{s2} & -2s \end{vmatrix}}{\begin{vmatrix} 3+2s & -2s \\ -2s & 5+6s \end{vmatrix}} = \frac{I_{s2}(6s+10s+15)}{8s^2+28s+15}$$

$$\Rightarrow H_{s2} = \frac{I_1}{I_{s2}} = \frac{16s+15}{8s^2+28s+15}$$

13.14

KCL at inverting input yields

$$\frac{V_g}{2+(4/s)} + V\left(\frac{1}{16} + \frac{1}{16}s\right) = 0 \Rightarrow$$

$$H(s) = \frac{V}{V_g} = \frac{1}{(2+4/s)(\frac{1}{16}+\frac{1}{16}s)} = \frac{-8s}{(s+2)(s+1)}$$

For $s=1$ and $V_g = 3\angle 0°$

$$V = \frac{-8(1)(3\angle 0°)}{(3)(2)} = -4\angle 0°\ \text{V}$$

$$v = -4e^{t}\ \text{V}$$

13.15

$\frac{V}{2}$ = input voltage for the VCVS

v_1 = voltage at common node of $\frac{1}{2}\Omega$ and 2Ω resistors.

KCL yields

$$\begin{cases} 2(V_1 - V_g) + \frac{1}{2}(V_1 - \frac{V}{2}) + s(V_1 - V) = 0 \\ \frac{1}{2}(\frac{V}{2} - V_1) + \frac{1}{2}s\frac{V}{2} = 0 \end{cases}$$

$$V = \frac{8V_g}{2s^2+3s+4} \Rightarrow H(s) = \frac{V}{V_g} = \frac{8}{2s^2+3s+4}$$

$$V(-1+j) = \frac{8(1)}{2(-1+j)^2 + (-1+j)3 + 4} = 4\sqrt{2}\angle 45°\ \text{V}$$

$$v = 4\sqrt{2}e^{-t}\cos(t+45°)\ \text{V}$$

13.16

$$\begin{cases} \frac{V_2 - V_g}{4} + \frac{V_2 - V_1}{5} + \frac{V_2 - 3V_1}{20/s} = 0 \\ \frac{V_1 - V_2}{5} + \frac{V_1}{20/s} = 0 \end{cases} \Rightarrow \begin{cases} 5(V_2 - V_g) + 4(V_2 - V_1) + s(V_2 - 3V_1) = 0 \\ \text{or } V_1(4+3s) - V_2(s+9) = -5V_g \\ V_1(4+s) - 4V_2 = 0 \end{cases}$$

$$V_1 = \frac{\begin{vmatrix} -5V_g & -(s+9) \\ 0 & -4 \end{vmatrix}}{\begin{vmatrix} 4+3s & -(s+9) \\ 4+s & -4 \end{vmatrix}} = \frac{20V_g}{-16-12s+s^2+13s+36} = \frac{20V_g}{s^2+s+20}$$

$$H(s) = \frac{3V_1}{V_g} = \frac{60}{s^2+s+20}$$

13.17

$$H(s) = \frac{k(s+4+j3)(s+4-j3)}{(s+3)(s+2+j)(s+2-j)}$$

$$H(0) = \frac{k(25)}{3(5)} = 5 \Rightarrow k = 3,$$

$$H(s) = \frac{3(s^2+8s+25)}{(s+3)(s^2+4s+5)}$$

13.18 $Z(s)=\frac{s^2+4s+3}{2} \Rightarrow Y(s)=\frac{2}{s^2+4s+3}$

$Y(s)=\frac{A}{s+3}+\frac{B}{s+1}$

$A=\frac{2}{s+1}\Big|_{-3}=-1$

$B=\frac{2}{s+3}\Big|_{-1}=1$

$Y(s)=\frac{1}{s+3}+\frac{1}{s+1}$. Let $Y_1=\frac{1}{s+3}\Rightarrow Z_1=s+3$

$Y_2=\frac{1}{s+1}\Rightarrow Z_2=s+1$

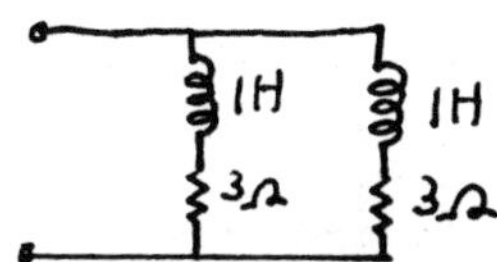

13.19 $Z(s)=\frac{6s^2+12+8}{3s+4}=2s+\frac{4}{3}+\frac{8/3}{3s+4}$

Let $Z_1=2s$, $Z_2=4/3$, $Z_3=\frac{8/3}{3s+4}$

$\Rightarrow Y_3=\frac{3s+4}{8/3}=\frac{9s+12}{8}=\frac{9}{8}s+\frac{3}{2}$

Let $4=\frac{9}{8}s \Rightarrow Z_4=\frac{1}{(9/8)s}$

$Y_5=3/2 \Rightarrow Z_5=2/3$

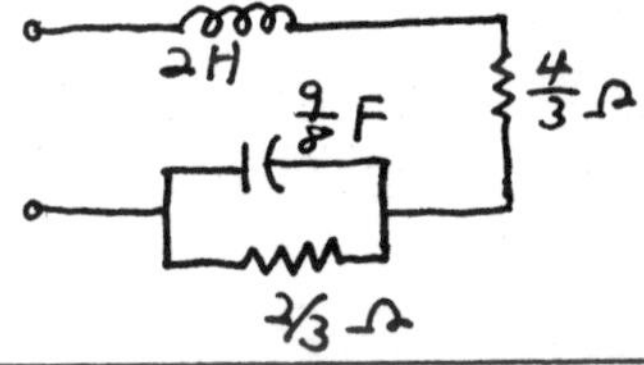

13.20 Zeros: $s^2+6s+9=0$

double zeros at $s=-3$.

poles: $4s^2+1=0$, $s^2=-\frac{1}{4}$, or $s=\pm j\frac{1}{2}$

poles at $s=j\frac{1}{2}, -j\frac{1}{2}$.

The system is not stable.

13.21 Zeros: $s=\frac{3\pm\sqrt{3^2-4^2}}{4}=\frac{3\pm j\sqrt{7}}{4}$

poles: $s^2+4s+6=0 \Rightarrow s=\frac{-4\pm j2\sqrt{2}}{2}=-2\pm j\sqrt{2}$

The system is stable.

13.22

I_s $\quad \frac{s}{3} \quad \frac{1}{3} \quad 2I_c \quad \frac{1}{s} \quad I_c$

KCL yields $\frac{V}{s/3}+\frac{V}{1/3}+\frac{V}{4/s}+2\frac{V}{4/s}=I_s$

$V(3/s+3+3/4s)=I_s$

13.22 Cont.

$\frac{V}{I_s}=\frac{1}{\frac{3}{s}+3+\frac{3s}{4}}=\frac{4s}{3s^2+12s+12}=\frac{4s}{3(s+2)^2}$

$I_c=\frac{V}{4/s}=I_s\frac{4s}{3(s+2)^2}\times\frac{1}{4/s}=I_s\frac{4s^2}{12(s+2)^2}=I_s\frac{s^2}{3(s+2)^2}$

The circuit has repeated poles at $s=-2$

$i_s(t)=6\sin t \Rightarrow I_s=\frac{6}{s^2+1}$

$I_c=\frac{1}{s^2+1}\times\frac{s^2}{3(s+2)^2}=\frac{A}{(s+2)^2}+\frac{B}{s+2}+\frac{C}{s-j}+\frac{C^*}{s+j}$

$C=\frac{-6}{3(j+2)^2(2j)}=\frac{-2}{(-1+j4+4)(2j)}=\frac{-2}{3+j6}$

$i_c(t)=2\left|\frac{4}{\sqrt{3^2+6^2}}\right|\cos(t+\theta)=\frac{8}{3\sqrt{5}}\cos(t+\theta)$

where $\theta=-\arctan 2$

13.23

I $\quad 4s \quad 4i(0^-) \quad 3 \quad V_1 \quad I_1 \quad 3 \quad I_2 \quad kI_1$

$\begin{cases} I_1(4s+3)-I_2(3)=V_1+4i(0^-) \\ -I_1(3)+6I_2=kI_1 \end{cases}$

$\begin{cases} I_1(4s+3)-I_2(3)=V_1+4i(0^-) \\ I_1(-3-k)+6I_2=0 \end{cases}$

$\Delta=\begin{vmatrix}4s+3 & -3\\ -3-k & 6\end{vmatrix}=24s+18-3(3+k)$
$=24s+18-9-3k$
$=24s+3(3-k)$

Pole at $p=\frac{1}{8(k-3)}$

if $k\geq 3$, the circuit is not stable.
$k=3$ is the smallest value such that the circuit is not stable.

13.24 $F(s)=\frac{36}{s^3+9s^2+24s+20}=\frac{36}{(s+2)^2(s+5)}$

$=\frac{A}{(s+2)^2}+\frac{B}{s+2}+\frac{C}{s+5}$

$C=\frac{36}{(s+2)^2}\Big|_{s=-5}=\frac{36}{9}=4$

$A=\frac{36}{s+5}\Big|_{s=-2}=\frac{36}{3}=12$

$\frac{36}{s^3+9s^2+24s+20}=\frac{12(s+5)+B(s+2)(s+5)+4(s+2)^2}{(s+2)^2(s+5)}$

$\Rightarrow s^2(B+4)+s(12+7B+16)+(60+10B+16)=0$

$\Rightarrow B=-4$

$f(t)=12te^{-2t}-4e^{-2t}+4e^{-5t}$

Initial value: $\lim_{s\to\infty} sF(s)=0$, Final value $\lim_{s\to 0} sF(s)=0$

13.25 From the solution of Prob. 13.22

$$I_c = I_s \frac{s^2}{3(s+2)^2}$$

$\because i_s(t) = u(t) \Rightarrow I_s = \frac{1}{s}$

$$I_c = \frac{1}{s}\frac{s^2}{3(s+2)^2}$$

$$i_c(0^+) = \lim_{s\to\infty} s\left[\frac{1}{s}\frac{s^2}{3(s+2)^2}\right] = 0$$

$$i_c(\infty) = \lim_{s\to 0} s\left[\frac{1}{s}\frac{s^2}{3(s+2)^2}\right] = \underline{1}$$

13.26 $V_o(s) = H_1(s)V_1(s) + H_2(s)V_2(s)$

$$= \frac{5s-4}{s^3+3s^2+2s}\cdot s + \frac{s^2+2}{s^2+3}\frac{1}{s}$$

$$V_o(0^+) = \lim_{s\to\infty} s\left[\frac{5s-4}{s^3+3s^2+2s} s + \frac{s^2+2}{s^2+3}\left(\frac{1}{s}\right)\right] = 6$$

$$V_o(\infty) = \lim_{s\to 0} s\left[\frac{5s-4}{s^3+3s^2+2s}\cdot s + \frac{s+2}{s^2+3}\frac{1}{s}\right] = \underline{\frac{2}{3}}$$

13.27 By Y-Δ transformation

$$Y = \frac{1/3}{3} = \frac{1}{9} \Rightarrow Z = 9\ \Omega$$

The s-domain circuit is

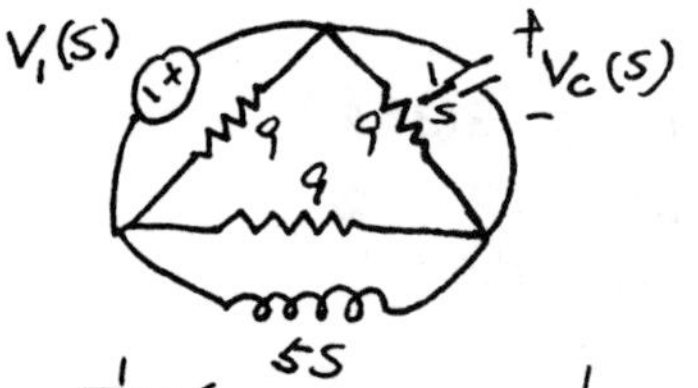

$$V_c = \frac{\frac{1/s}{1+\frac{1}{s}}}{\frac{5s}{1+5s} + \frac{\frac{1}{s}}{1+\frac{1}{s}}} V_1 = \frac{\frac{1}{s+1}}{\frac{5s}{5s+1} + \frac{1}{s+1}} V_1 = \frac{5s+1}{5s(s+1)+(5s+1)}$$

$$= \frac{5s+1}{5s^2+10s+1} V_1$$

$$H(s) = \frac{V_c}{V_1} = \underline{\frac{5s+1}{5s^2+10s+1}}$$

$$5s^2+10s+1 = 0 \Rightarrow s = \frac{-10\pm\sqrt{80}}{10} = \frac{-10\pm 2\sqrt{20}}{10} = -1\pm\frac{2}{\sqrt{5}}$$

$$H(s) = \frac{s+\frac{1}{5}}{s^2+2s+\frac{1}{5}} = \frac{A}{s+(1-\frac{2}{\sqrt{5}})} + \frac{B}{s+(1+\frac{2}{\sqrt{5}})}$$

$$A = \left.\frac{s+\frac{1}{5}}{s^2+2s+\frac{1}{5}}\right|_{s=-1+\frac{2}{\sqrt{5}}} = \frac{-\frac{4}{5}+\frac{2}{\sqrt{5}}}{\frac{4}{\sqrt{5}}} = -\frac{1}{\sqrt{5}}+\frac{1}{2}$$

$$B = \left.\frac{s+\frac{1}{5}}{s+(1-\frac{2}{\sqrt{5}})}\right|_{s=-1-\frac{2}{\sqrt{5}}} = \frac{-1-\frac{2}{\sqrt{5}}+\frac{1}{5}}{\frac{4}{\sqrt{5}}} = \frac{1}{\sqrt{5}}+\frac{1}{2}$$

$$\underline{h(t) = \left(\frac{1}{2}-\frac{1}{\sqrt{5}}\right)e^{-(1-\frac{2}{\sqrt{5}}t)} + \left(\frac{1}{2}+\frac{1}{\sqrt{5}}\right)e^{-(1+\frac{2}{\sqrt{5}}t)}}$$

13.28 $Y(s)[s^2+6s+5] = 3X(s)$

$$H(s) = \frac{Y(s)}{X(s)} = \frac{3}{s^2+6s+5} = \frac{A}{s+1} + \frac{B}{s+5}$$

$$A = \left.\frac{3}{s+5}\right|_{s=-1} = 3/4$$

$$B = \left.\frac{3}{s+1}\right|_{s=-5} = -3/4$$

$$H(s) = \underline{\frac{3/4}{s+1} + \frac{-3/4}{s+5}}$$

The impulse response

$$h(t) = \underline{\tfrac{3}{4}e^{-t} - \tfrac{3}{4}e^{-5t}}$$

13.29 $v_o(t) = e^{-4t}\cos(6t)u(t-\frac{\pi}{12})$

$$= e^{-4(t-\pi/12)}e^{-\pi/3}\cos\left[6(t-\tfrac{\pi}{12})+\tfrac{\pi}{2}\right]u(t-\tfrac{\pi}{12})$$

$$= e^{-\pi/3}e^{-4(t-\frac{\pi}{12})}\left[-\sin 6(t-\tfrac{\pi}{12})\right]u(t-\tfrac{\pi}{12})$$

$$V_o(s) = \frac{-e^{-\pi/3}e^{-\frac{\pi}{12}s}\cdot 6}{(s+4)^2+36}$$

$V_i(s) = 10$

$$\Rightarrow H(s) = \underline{-\frac{3}{5}e^{-\pi/3}\frac{e^{-\pi/12\, s}}{(s+4)^2+36}}$$

The impulse response

$$h(t) = \underline{\tfrac{1}{10}e^{-4t}\cos(6t)u(t-\tfrac{\pi}{12})}$$

13.30
$$\begin{cases} 3x''+4x+y'+y = u(t) \\ 2x''+6x+y'+2y = 0 \end{cases}$$

$$(3s+4)X + (s+1)Y = \frac{1}{s}$$

$$(2s+6)X + (s+2)Y = 0$$

$$X = \frac{\begin{vmatrix} \frac{1}{s} & s+1 \\ 0 & s+2 \end{vmatrix}}{\begin{vmatrix} 3s+4 & s+1 \\ 2s+6 & s+2 \end{vmatrix}} = \frac{\frac{1}{s}(s+2)}{(3s+4)(s+2)-(s+1)(2s+6)}$$

$$= \frac{\frac{1}{s}(s+2)}{3s^2+10s+8-s^2-8s-6} = \frac{s+2}{s(s^2+2s+2)}$$

$$= \frac{A}{s} + \frac{Bs+C}{s^2+2s+2}$$

$$A = \left.\frac{s+2}{s^2+2s+2}\right|_{s=0} = 1 \Rightarrow s^2+2s+2+Bs^2+Cs = 2+s$$

$B = -1,\ C+2 = 1, \Rightarrow C = -1$

$$X = \frac{1}{s} + \frac{-s-1}{s^2+2s+2} = \frac{1}{s} - \frac{(s+1)}{(s+1)^2+1}$$

$$\Rightarrow x(t) = \underline{u(t) - e^{-t}\cos t}.$$

13.30 Cont.

$$Y = \frac{3}{s} + \frac{-3s-8}{(s+1)^2+1} = \frac{3}{s} - \frac{3\,(s+1)}{(s+1)^2+1} - \frac{5}{(s+1)^2+1}$$

$$\Rightarrow y(t) = 3u(t) - 3e^{-t}\cos t - 5e^{-t}\sin t$$

13.31 $F(s) = \frac{1}{s} - \frac{e^{-4s}}{s}$

$G(s) = \frac{1}{s^2} - e^{-2s}\left(\frac{1}{s^2} + \frac{2}{s}\right)$

$$f*g = \mathcal{L}^{-1}[F(s)\,G(s)]$$

$$= \mathcal{L}^{-1}\left[\frac{1}{s^3} - e^{-2s}\left(\frac{1}{s^3} + \frac{2}{s^2}\right) - \frac{e^{-4s}}{s^3} + e^{-6s}\left(\frac{1}{s^3} + \frac{2}{s^2}\right)\right]$$

Since $f(t)u(t-a) \Leftrightarrow e^{-as}\mathcal{L}\{f(t-a)\}$

$t^n \Leftrightarrow \frac{n!}{s^{n+1}}$, $t^2 \Leftrightarrow \frac{2}{s^3}$, $t \Leftrightarrow \frac{1}{s^2}$

then

$$f*g = \frac{t^2}{2}u(t) - \left[\frac{(t-2)^2}{2} + 2(t-2)\right]u(t-2) - \frac{(t-4)^2}{2}u(t-4) + \left[\frac{(t-6)^2}{2} + 2(t-6)\right]u(t-6)$$

$$= \frac{t^2}{2}u(t) - \frac{(t^2+4t-4)}{2}u(t-2) - \frac{(t-4)^2}{2}u(t-4) + \frac{t^2-8t+12}{2}u(t-6)$$

13.32 $F(s) = \frac{4s}{s^2+16}$, $G(s) = \frac{1}{s} - \frac{e^{-5s}}{s}$

$$f*g = \mathcal{L}^{-1}\left[\frac{4}{s^2+16} - \frac{4e^{-5s}}{s^2+16}\right]$$

$$= \sin 4t\,u(t) - \sin 4(t-5)\,u(t-5)$$

13.33 $(s^3+s^2+4s+4)Y + (s+1)X = 1$

$$Y = \frac{1}{(s+1)(s^2+4)} - \frac{s+1}{(s^2+4)(s+1)}X$$

$$= \frac{A}{s+1} + \frac{Bs+C}{s^2+4} - \frac{1}{s^2+4}X$$

$A = \frac{1}{s^2+4}\Big|_{s=-1} = \frac{1}{5}$, $\frac{1/5}{s+1} + \frac{Bs+C}{s^2+4} = \frac{1}{(s+1)(s^2+4)}$

$\Rightarrow B + \frac{1}{5} = 0 \Rightarrow B = -\frac{1}{5}$

$B + C = 0 \Rightarrow C = \frac{1}{5}$

$$\Rightarrow Y = \frac{1/5}{s+1} + \frac{-1/5\,s + 1/5}{s^2+4} - \frac{1}{s^2+4}X$$

$$= \frac{1}{5}\left(\frac{1}{s+1}\right) - \frac{1}{5}\left(\frac{s}{s^2+4}\right) + \frac{1}{10}\left(\frac{2}{s^2+4}\right) - \frac{1}{s^2+4}X$$

(a) $y(t) = \frac{1}{5}e^{-t} - \frac{1}{5}\cos 2t + \frac{1}{10}\sin 2t - \frac{1}{2}\int_{-\infty}^{\infty}(\sin 2\quad)x(t-\tau)\,d\tau$

(b) $y(t) = \frac{1}{5}e^{-t} - \frac{1}{5}\cos 2t - \frac{2}{5}\sin 2t$

13.34 $(s^3+s^2+4s+4)Y + (s+1)X = 0$

$$H(s) = \frac{Y}{X} = -\frac{s+1}{(s+1)(s^2+4)} = -\frac{1}{s^2+4}$$

(a) $h(t) = -\frac{1}{2}\sin 2t$.

(b) the step response

$$Y(t) = -\frac{1}{(s^2+4)s} = \frac{A}{s} + \frac{Bs+C}{s^2+4}$$

$$A = -\frac{1}{s^2+4}\Big|_{s=0} = -\frac{1}{4}$$

$\Rightarrow -\frac{1}{4}(s^2+4) + Bs^2 + Cs = -1 \Rightarrow B - \frac{1}{4} = 0 \Rightarrow B = \frac{1}{4}$

and $C = 0$.

$$Y = \frac{-\frac{1}{4}}{s} + \frac{\frac{1}{4}s}{s^2+4} \Rightarrow y(t) = -\frac{1}{4}u(t) + \frac{1}{4}\cos 2t$$

(c) $Y = -\frac{1}{s^2+4}X = -\frac{1}{(s^2+4)(s+2)} = \frac{A}{s+2} + \frac{Bs+C}{s^2+4}$

equating the coefficients

$A = -\frac{1}{8}$, $B = \frac{1}{8}$, $C = -\frac{1}{4}$

$$Y = -\frac{1/8}{s+2} + \frac{1/8\,s}{s^2+4} + \frac{-\frac{1}{4}}{s^2+4}$$

$$y(t) = -\frac{1}{8}e^{-2t} + \frac{1}{8}\cos 2t - \frac{1}{8}\sin 2t$$

13.35

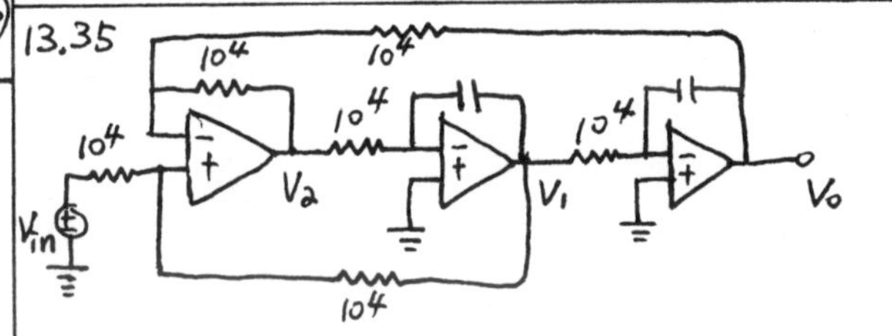

$$\begin{cases} \frac{V_1}{10^4} + \frac{V_o}{10^6/s} = 0 \\ V_3 = \frac{V_1 + V_{in}}{2} \\ V_3 = \frac{V_2 + V_o}{2} \\ \frac{V_2}{10^4} + \frac{V_1}{10^6/s} = 0 \end{cases}$$

solving the equations

$$V_o(s^2+100s+10000) = 10000V_{in}$$

$$H(s) = \frac{V_o}{V_{in}} = \frac{10000}{s^2+100s+10000}$$

13.36 $V_{in}(s) = \mathcal{L}\{u(t) - u(t-1)\} = 1 - e^{-s}$

$$V_o = H(s)\cdot V_{in}(s) = \frac{10000}{s^2+100s+10000}(1-e^{-s})$$

$$= \frac{10000}{s^2+100s+10000} - e^{-s}\frac{10000}{s^2+100s+10000}$$

$$= \frac{10000}{\sqrt{7500}}\,\frac{\sqrt{7500}}{(s+50)^2+7500} - e^{-s}\frac{10000}{\sqrt{7500}}\,\frac{\sqrt{7500}}{(s+50)^2+7500}$$

$$= \frac{200}{\sqrt{3}}e^{-50t}\sin[\sqrt{7500}\,t] - \frac{200}{\sqrt{3}}e^{-50(t-1)}\sin[\sqrt{7500}\,(t-1)]$$

Chapter 14
Frequency response

14.1 The frequency response function

14.1 For the transfer function

$$H(s) = \frac{1}{s^3 + 2s^2 + 2s + 1}$$

find the gain and phase shift at frequency $\omega = 10\text{rad/sec}$.

14.2 For the transfer function

$$H(s) = \frac{s}{s^2 - 2s + 5}$$

find the maximum gain, its corresponding phase shift and the frequency ω where the maximum gain occurs.

14.3 For the circuit shown, find the gain and phase shift of the voltage transfer ratio $\frac{V_3}{V_1}$ as a function of frequency.

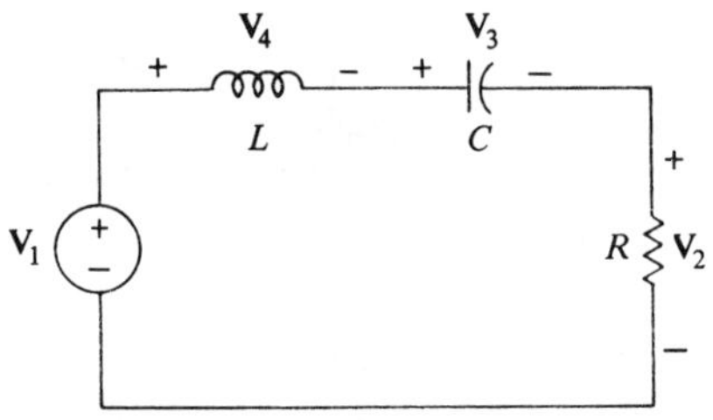

Problem 14.3

14.4 Find $H(s) = \frac{V_2(s)}{V_1(s)}$ and sketch the amplitude and phase responses. Show that the peak amplitue and zero phase occur at $\omega = 10\text{rad/sec}$.

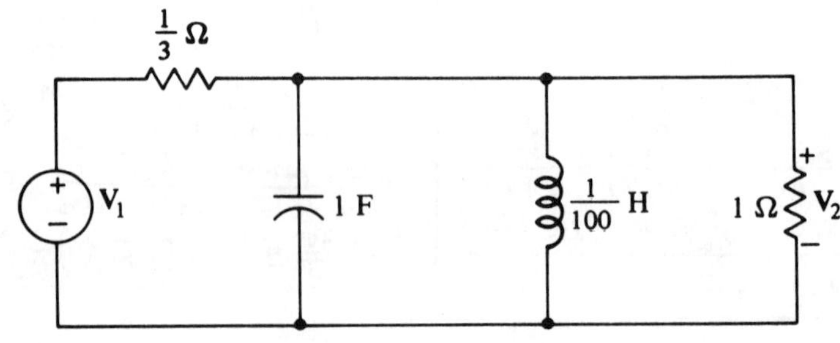

Problem 14.4

14.5 For the circuit shown, find the network function, $H(s) = \frac{V_2(s)}{V_1(s)}$. Show that the peak amplitude and zero phase occur at $\omega = 0$.

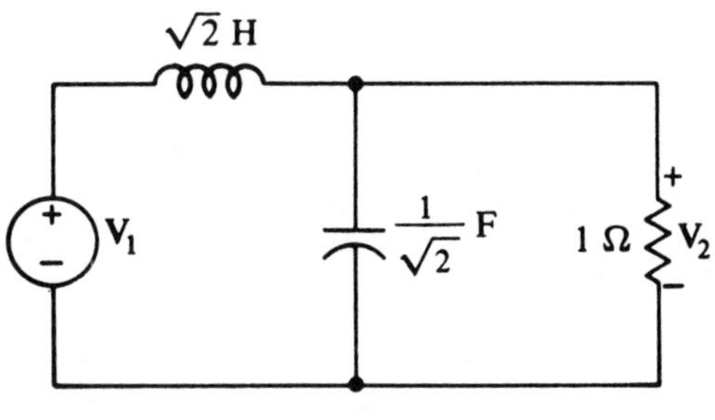

Problem 14.5

14.6 For the circuit shown, $R = 10\ \Omega$, $C_1 = 0.9$ F and, $C_2 = 0.1$ F, if the input and output are ν_1 and ν_2 respectively, find the network function and sketch the amplitude and phase responses.

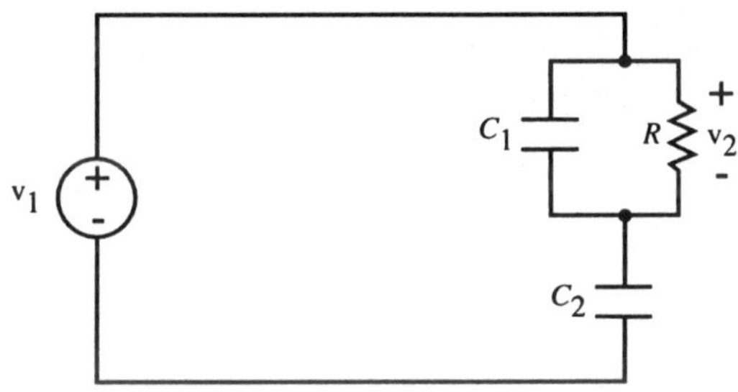

Problem 14.6

14.7 For the circuit shown, $R_1 = R_2 = 0.1\Omega$ and $R_3 = 0.005\Omega$. If the input and output are V_1 and V_2 respectively, find the network function and sketch the amplitude and phase responses.

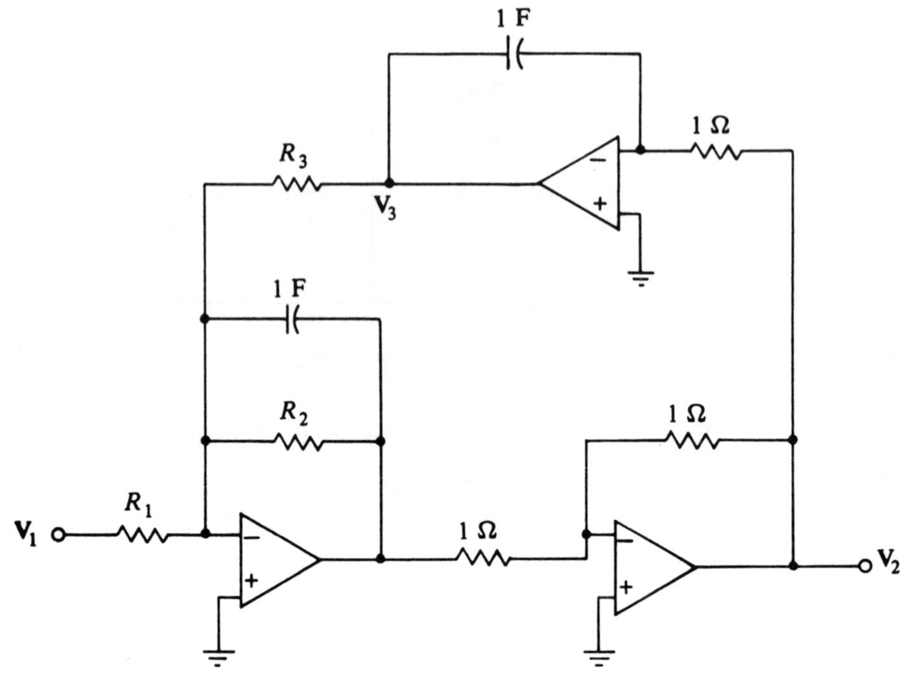

Problem 14.7

14.2 The decibel scale

14.8 Given the transfer function

$$H(s) = \frac{s}{s^2 + s + 4}$$

find the gains in decibel at $\omega = 0.02$, 1.56, 2, 2.56, 20, and 200rad/sec.

14.9 A 4-stage amplifier has the gains 10dB, 6dB, -4dB, and 0 dB for stage 1, 2, 3 and 4 respectively. If the input of the amplifier is 5V, what is the output voltage?

14.10 For the Prob.14.4, find the bandwidth that the gain is greater than -3dB.

14.3 Bode gain plots

14.11 For each component factor in the transfer function

$$H(s) = \frac{-7(s+10)}{s(s+4)}$$

specify the initial slope, final slope, initial value and location of break frequency. Sketch the uncorrected and corrected Bode gain plots of H(s).

14.12 Sketch the uncorrected and corrected Bode gain plots of the transfer function

$$H(s) = \frac{s}{s^2 + 6s + 9}$$

14.13 Sketch the uncorrected and corrected Bode gain plots of the transfer function

$$H(s) = \frac{s^2 - 16}{2s^2 + 201s + 100}$$

14.14 Determine the natural frequency ω_n and damping ratio ζ of the circuit with transfer function

$$H(s) = \frac{1}{(s^2 + 2s + 400)^2}$$

sketch the uncorrected and corrected Bode gain plots.

14.4 Resonance

14.15 Find (a) the resonant frequency ω_r (b) the bandwidth B and (c) quality factor Q for the series RLC circuit described by R = 5Ω, L = 2H, and C = 2F.

14.16 Repeat the Prob.14.15 for the parallel RLC circuit.

14.17 Find the resonance frequency and the frequency at which the network function Y(jω) is real.

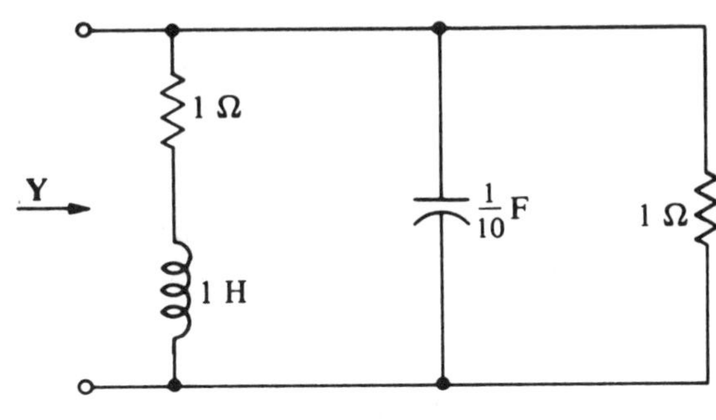

Problem 14.17

14.18 Design a series or parallel RLC resonant circuit whose admittance goes through a minimum at frequency f = 40KHz and which has a quality factor Q of 100. Use a 1pF capacitor.

14.19 A complex circuit is observed to have the transfer function

$$H(s) = \frac{s}{[(s+0.001)^2+10^2][(s+0.008)^2+237^2]+[(s+0.0012)^2+777^2]^4}.$$

Estimate the resonant frequency by its natural frequency.

14.5 Frequency response of op amps

14.20 An op amp with 200MHz gain bandwidth product is used as an inverting amplifier. Find the bandwidth of the circuit in Hz if the gain $\frac{Rf}{RA}$ is set to -20dB, 0dB, 3dB, 20dB, 40dB, and 60 dB.

14.21 Suppose two copies of the inverting amplifier of Prob.14.20 are put in cascade creating a circuit with gain $(\frac{Rf}{RA})^3$. What is the bandwidth of this new circuit for gains $(\frac{Rf}{RA})^3 = 1000$. Note that for correct bandwidth, one has to consider the frequency ω at which corrected gain of +1dB occur.

14.22 In chapter 3, it is shown that for noninverting amplifier

$$H(j\omega) = \frac{V_{out}}{V_{in}} = \frac{A(j\omega)(R_A+R_f)}{[1+A(j\omega)]R_A+R_f}$$

Show that the voltage follower, which is a special case of noninverting amplifier, has both the bandwidth ω_c and gain-bandwidth product of $A_0\omega_0$. A_0, ω_0 are the open loop gain and open loop bandwidth of an op amp respectively.

14.23 For the difference amplifier shown, determine its bandwidth ω_c and gain-bandwidth product. [Hint: Using superposition principle and results of Prob.14.22 and section 14.5]

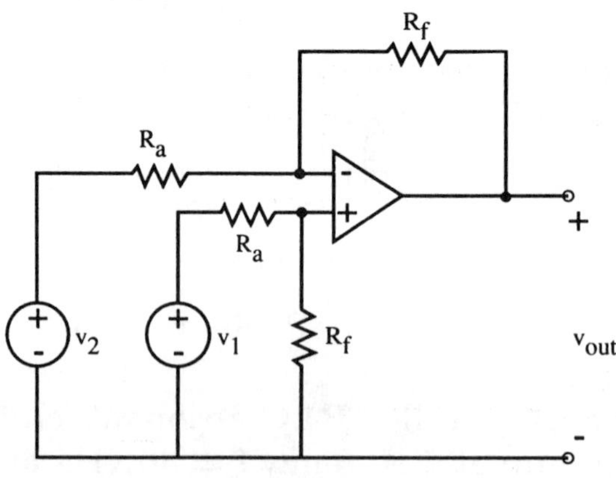

Problem 14.23

14.6 Filters

14.24 Show that

$$H(s) = \frac{6s}{s^2 + s + 4}$$

is the network function of a bandpass filter, and find ω_o, ω_{c1}, ω_{c2}, and B.

14.25 Show that

$$H(s) = \frac{6s^2}{s^2 + s + 4}$$

is the network function of a high-pass filter, and find $| H(j\omega) |_{max}$ and ω_c.

14.26 Show that

$$H(s) = \frac{6(s^2 + 4)}{s^2 + s + 4}$$

is the network function of a band-reject filter, and find $| H(j\omega) |_{max}$, ω_o, ω_{c1}, and ω_{c2}.

14.27 Show that the given circuit is a low-pass filter with $\omega_c = 1$rad/sec by finding $H(s) = \frac{V_2(s)}{V_1(s)}$ and the amplitude response.

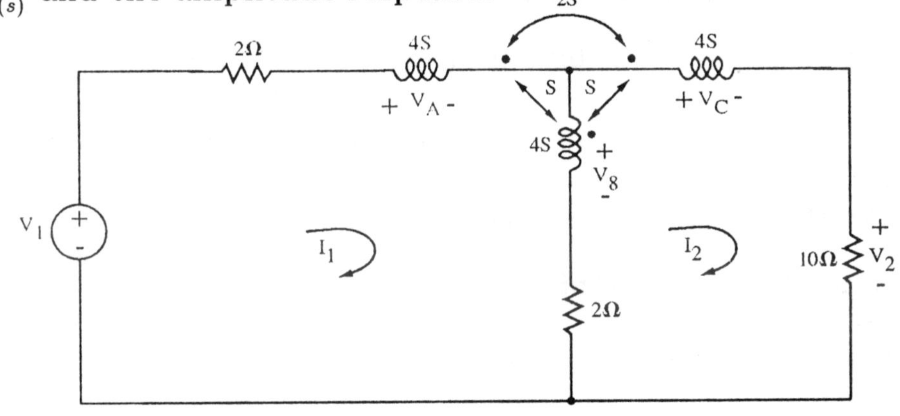

Problem 14.27

14.28 Show by finding $\frac{V_2}{V_1}$ that the circuit is a band pass filter, and find the gain, the bandwidth, and center frequency.

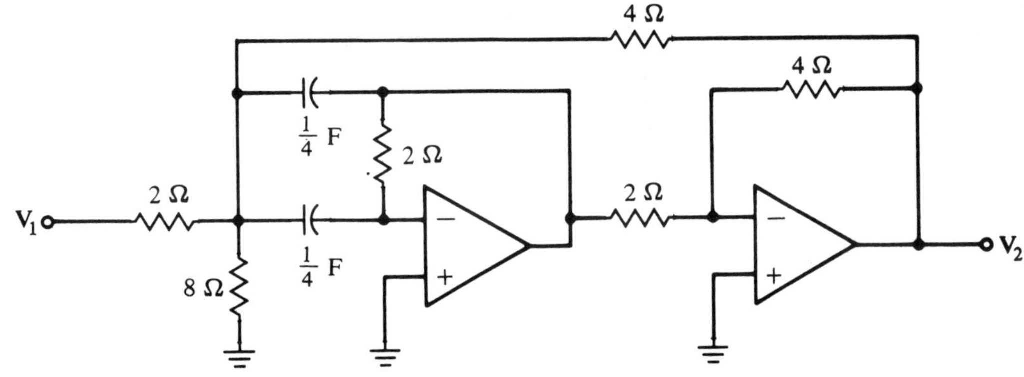

Problem 14.28

14.7 Active filter design

14.29 Design a circuit with bilinear transfer function

$$H(s) = \frac{5s - 6}{s + 2}$$

Make the resistances between 10KΩ and 100KΩ.

14.30 Design a circuit with bilinear transfer function

$$H(s) = \frac{\frac{1}{4}s}{3s + 1}$$

Make the resistances between 10KΩ and 100KΩ.

14.31 Design a circuit with bilinear transfer function

$$H(s) = \frac{2s + 1}{s + 9}$$

Make the resistances between 10KΩ and 100KΩ.

14.32 Design a circuit with biquadratic transfer function

$$H(s) = \frac{1}{s^2 + 2s + 1}$$

Make the resistances between 10KΩ and 100KΩ.

14.33 Design a circuit with biquadratic transfer function

$$H(s) = \frac{(s^2 + s + 1)}{(3s^2 + s + 2)(s + 1)}$$

Make the resistances between 10KΩ and 100KΩ.

14.8 Scaling

14.34 Frequency- and impedance-scale the circuit to obtain $\omega_c = 400\pi$rad/sec, using capacitors of 2μF and 1μF.

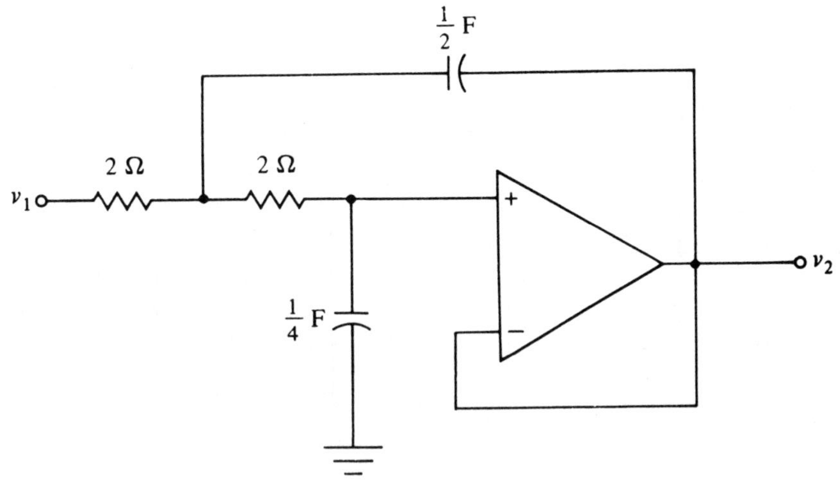

Problem 14.34

14.35 Determine R_1 and R_2 so that the circuit is a first-order low-pass filter ($H = \frac{V_2}{V_1}$) with ω_c=10^5rad/sec and H(0) = -2.0. Use a 1nF capacitor.

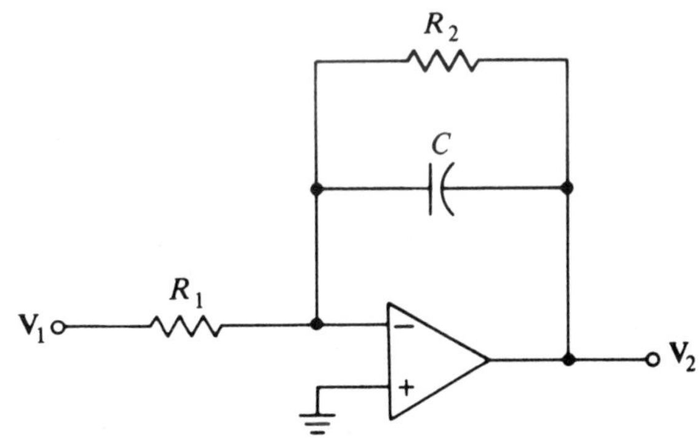

Problem 14.35

14.36 The circuit is a third-order low-pass Butterworth filter with $\omega_c = 1$rad/sec and a gain of 1. Scale the circuit so that the capacitors are 1μF each and $\omega_c = 2 \times 10^4$rad/sec.

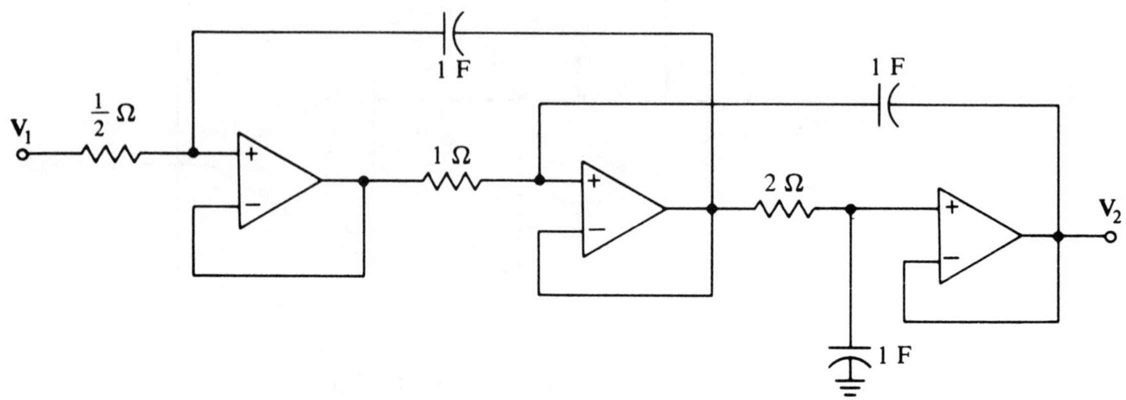

Problem 14.36

14.37 Scale the network so that $\omega_o = 10^6$rad/sec, Q = 5, K = 0.5, and the capacitance is 10nF.

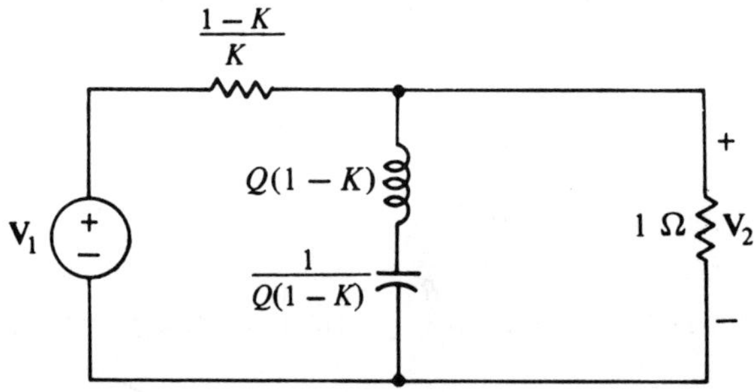

Problem 14.37

14.38 The circuit is a band-reject filter with a gain of 1, Q = 1, and a center frequency ω_o = 1rad/sec. Scale the network to obtain a center frequency f_o = 1000Hz using capacitances of 5 and 10nF.

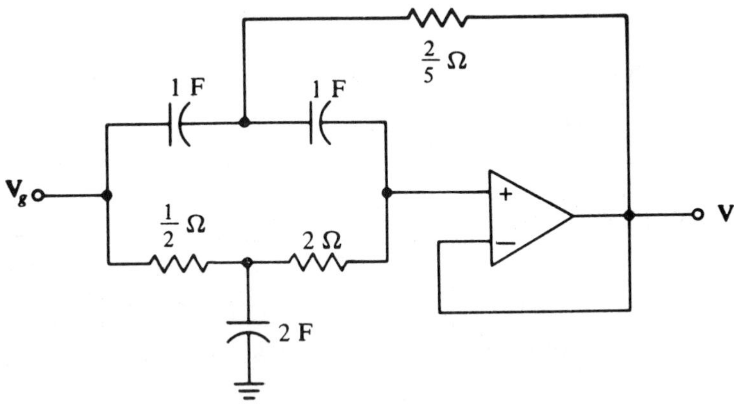

Problem 14.38

14.9 SPICE and frequency response

14.39 Use SPICE to create plots of the gain and phase shift curves for the circuit of Prob.14.6 for $0 < \omega < 10$, using log frequency and output scales with a sample every 0.1rad/sec.

14.40 Use the op amp model of Fig.3.9 and SPICE to plot the gain in dB vs. log frequency for the circuit of Prob.14.28.

14.10 SPICE and active filter design

14.41 Use the op amp model of Fig.3.9 and SPICE to plot the gain in dB vs. log frequency for the circuit of Prob.14.29. Repeat the simulation by changing all feedback resisitor values of ±10%. Compare the three cases of their Bode gain plots.

14.42 Consider the buffered bandpass filter as shown, using SPICE to produce its Bode gain plot with the op amp model of Fig.3.9. Remove the voltage follower, and repeat the SPICE simulation.

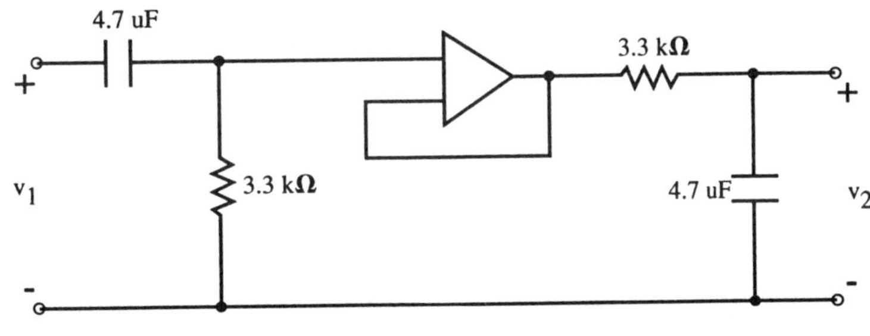

Problem 14.42

14.1 $H(j10) = \dfrac{1}{(j10)^3 + 2(j10)^2 + 2(j10) + 1}$

$= \dfrac{1}{-j1000 - 400 + j20 + 1} = \dfrac{1\angle 0^\circ}{1058\angle 247.8^\circ}$

$= 0.000945\angle -247.8^\circ$

gain $|H(j\omega)| = 0.000945$

phase shift $\angle H(j\omega) = -247.8^\circ = 112.2^\circ$

14.2 $H(j\omega) = \dfrac{j\omega}{(j\omega)^2 - 2(j\omega) + 5} = \dfrac{j\omega}{(5-\omega)^2 - j2\omega}$

$|H(j\omega)|^2 = \dfrac{\omega^2}{(5-\omega^2)^2 + (2\omega)^2} = \dfrac{1}{\left(\frac{5-\omega^2}{\omega}\right)^2 + 4}$

at $\omega^2 - 5 = 0 \Rightarrow \omega = \sqrt{5}$ rad/sec

The maximum gain occurs with the value $G = \sqrt{\frac{1}{4}} = \frac{1}{2}$

The phase shift corresponding with the maximum gain is always 0°

14.3 $\dfrac{V_2}{V_1}(s) = \dfrac{1/sC}{sL + \frac{1}{sC} + R} = \dfrac{1}{s^2LC + 1 + sRC}$

$\dfrac{V_2}{V_1}(j\omega) = \dfrac{1}{(1-\omega^2LC) + j\omega RC}$

$\left|\dfrac{V_2}{V_1}(j\omega)\right| = \sqrt{\dfrac{1}{(1-\omega^2LC)^2 + j\omega RC}} = \dfrac{1}{\sqrt{(1-\omega^2LC)^2 + (\omega RC)^2}}$

$\angle \dfrac{V_2}{V_1}(j\omega) = -\tan^{-1}\dfrac{\omega RC}{1-\omega^2LC}$

14.4 KCL yields

$V_2\left(1 + 3 + \frac{1}{s} + \frac{s}{100}\right) = V_1 3$

$H(s) = \dfrac{V_2}{V_1} = \dfrac{s300}{s^2 + s400 + 100}$

$|H(j\omega)|^2 = \dfrac{9\times10^4\omega^2}{(100-\omega^2)^2 + 16\times10^4\omega^2}$

$= \dfrac{9\times10^4}{\left(\frac{100-\omega^2}{\omega}\right)^2 + 16\times10^4}$

$|H(j\omega)|_{max}$ occurs when

$100 - \omega^2 = 0 \Rightarrow \omega_0 = 10$ rad/s

$H(j10) = \dfrac{j3000}{j4000} = \dfrac{3}{4} \Rightarrow \phi(\omega) = 0$

14.4 Cont.

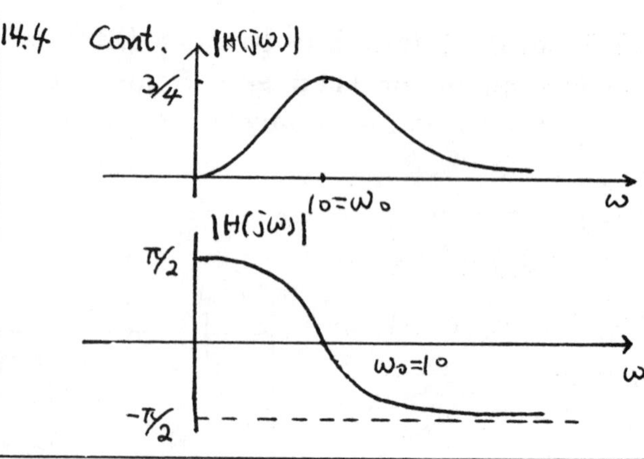

14.5 KCL yields

$V_2\left(1 + \frac{s}{\sqrt{2}} + \frac{1}{s\sqrt{2}}\right) = V_1/s\sqrt{2}$

$H(s) = \dfrac{V_2}{V_1} = \dfrac{1}{s^2 + s\sqrt{2} + 1}$

$|H(j\omega)| = \dfrac{1}{\sqrt{(1-\omega^2)^2 + 2\omega^2}} = \dfrac{1}{\sqrt{1+\omega^4}}$

$|H(j\omega)|_{max}$ occurs when $\omega = 0$

$H(0) = \frac{1}{1} \Rightarrow \phi(0) = 0^\circ$

14.6 $\dfrac{V_2}{V_1} = \dfrac{R/sC_1}{\dfrac{R/sC_1}{\frac{1}{sC_1} + R} + \dfrac{1}{sC_2}} = \dfrac{R/1+sC_1R}{\dfrac{R}{1+sC_1R} + \dfrac{1}{sC_2}}$

$= \dfrac{sC_2R}{sC_2R + 1 + sC_1R} = \dfrac{sC_2R}{1 + sR(C_1 + C_2)}$

$|H(j\omega)| = \sqrt{\dfrac{\omega^2}{1 + (10\omega)^2}} = \sqrt{\dfrac{\omega^2}{1 + 100\omega^2}} = \dfrac{1}{\sqrt{100 + \frac{1}{\omega^2}}}$

$\angle H(j\omega) = -\tan^{-1} 10\omega$

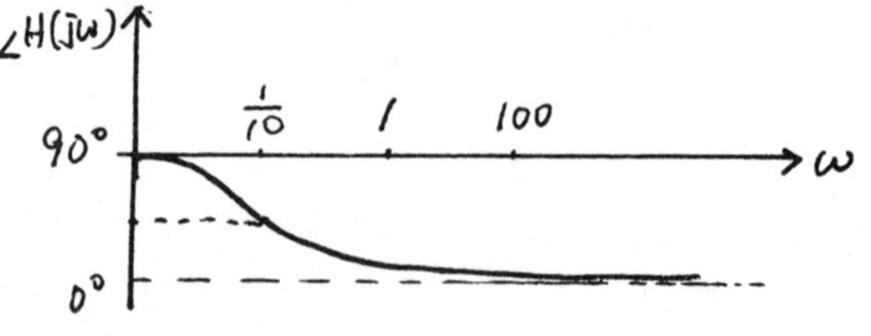

$\angle H(j\omega)$

90°

1/10 1 100

ω

0°

14.7 V_4 = output of lossy integrator

KCL at the inverting terminals gives:

$V_2 + V_4 = 0 \Rightarrow V_4 = -V_2$;

$V_2 + sV_3 = 0 \Rightarrow V_3 = -\frac{1}{s} V_2$

$\frac{1}{R_1} V_1 + (s + \frac{1}{R_2}) V_4 + \frac{1}{R_3} V_3 = 0$ or

$[(s + \frac{1}{R_2})(-1) + \frac{1}{R_3}(-\frac{1}{s})] V_2 = -\frac{1}{R_1} V_1$

$\therefore H(s) = \frac{V_2}{V_1} = \frac{1/R_1 \, s}{s^2 + 1/R_2 \, s + 1/R_3} = \frac{10s}{s^2 + 10s + 200}$

$\omega_0 = 10\sqrt{2}$ rad/s, $H(j10\sqrt{2}) = 1$

Amplitude and phase responses are similarly to those in Prob. 14.4 with $\omega_0 = 10\sqrt{2}$ and $|H(j\omega)|_{max} = 1$

14.8 $H(j\omega) = \frac{j\omega}{(j\omega)^2 + (j\omega) + 4} = \frac{j\omega}{(4-\omega)^2 + j\omega}$

$|H(j\omega)| = \sqrt{\frac{\omega^2}{(4-\omega^2)^2 + \omega^2}} = \frac{1}{\sqrt{(\frac{4-\omega^2}{\omega})^2 + 1}}$

$|H(j0.02)| = 20 \log\left(\frac{1}{\sqrt{(\frac{4-\omega^2}{\omega})^2 + 1}}\right) = -10 \log_{10}\left\{\left(\frac{4-\omega^2}{\omega}\right)^2 + 1\right\}$

$= \underline{-46 \text{ dB}}$

$|H(j1.56)| = \underline{-3 \text{ dB}}$ $\quad |H(j2.56)| = \underline{-3 \text{ dB}}$

$|H(j2)| = \underline{0 \text{ dB}}$ $\quad |H(j20)| = \underline{-26 \text{ dB}}$

$|H(j200)| = \underline{-46 \text{ dB}}$

14.9 gain $= 10 + 6 - 4 + 0 = 12$ dB

$12 = 20 \log\left(\frac{V_{out}}{5}\right)$

$10^{\frac{12}{20}} = \frac{V_{out}}{5}$

$V_{out} = 5(10)^{\frac{12}{20}} = 5(3.98) = \underline{19.9 \text{ V}}$

14.10 From the result of Prob. 14.4

$|H(j\omega)| = \sqrt{\frac{9 \times 10^4 \omega^2}{(100-\omega^2)^2 + 16 \times 10^4 \omega^2}}$

$-3 \text{dB} = \frac{1}{\sqrt{2}} = \sqrt{\frac{9 \times 10^4 \omega^2}{(100-\omega^2)^2 + 16 \times 10^4 \omega^2}}$

$\frac{1}{2} = \frac{90000\omega^2}{\omega^4 - 200\omega^2 + 10000 + 160000\omega^2}$

or $\omega^4 - 19800\omega^2 + 10000 = 0$

$\omega = 1$ or 140.7

The bandwidth of -3dB is

$\bar{\omega} = 140.7 - 1 = \underline{139.7 \text{ rad/s}}$

14.11 $20 \log_{10} |-7| = 16.9$ dB

Initial value, final slope, initial value, break frequency.

-7: 0dB/dec, 0dB/dec, 16.9dB, no break frequency

$s+10$: 0dB/dec, 20dB/dec, 20dB, 10 rad/s

$\frac{1}{s}$: -20dB/dec, -20dB/dec, no initial value, no break frequency.

$\frac{1}{s+4}$: 0dB/dec, -20dB/dec, -12dB, 4 rad/s.

$16.9 + 20 - 12 = 24.9$ dB.

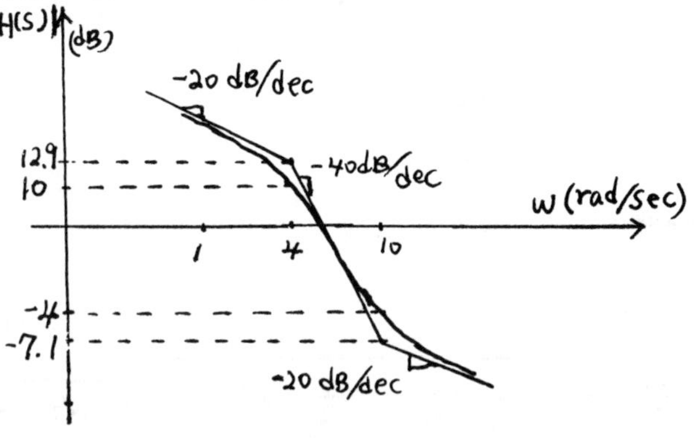

14.12

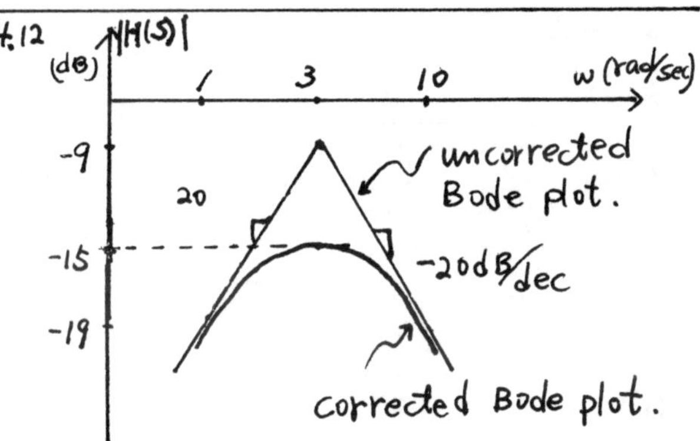

14.13 $H(s) = \frac{(s-4)(s+4)}{(2s+1)(s+100)}$

$20 \log_{10}|-4| + 20 \log_{10}|4| - 20 \log_{10}|1| - 20 \log_{10}|100|$

$= 12 + 12 - 0 - 40 = -26$ dB

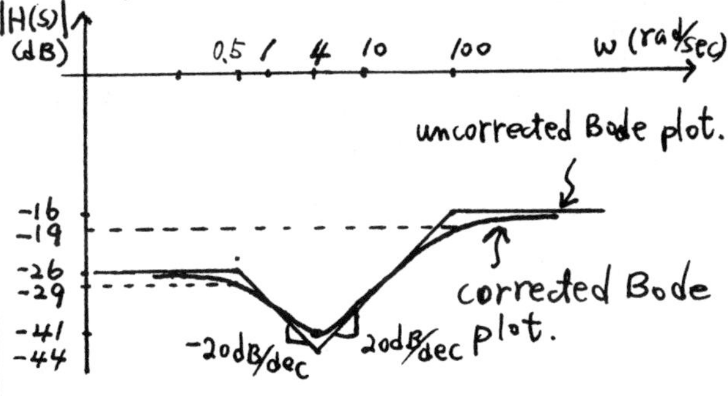

14.14 $H(s) = \frac{1}{(s^2+2s+400)^2}$

$\omega_n^2 = 400 \Rightarrow \omega_n = \underline{20 \text{ rad/sec}}$

$2\zeta\omega_n = 2 \Rightarrow \zeta = \underline{0.05}$

Initial value: $-20 \cdot 2 \log_{10} 400 = -104\text{dB}$

slope: -40dB/dec

the correction $= (2)(20 \log_{10} 2|\zeta|)$

$= 2 \cdot 20 \log_{10} 0.1 = -40\text{dB}$

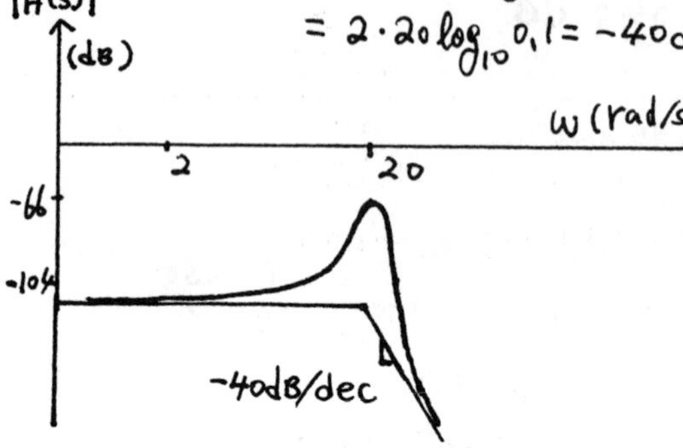

14.15 (a) $\omega_r = \frac{1}{\sqrt{LC}} = \frac{1}{\sqrt{2 \times 2}} = \underline{\frac{1}{2} \text{ rad/s}}$

(b) $B = \frac{R}{L} = \frac{5}{2} = \underline{2.5 \text{ rad/s}}$

(c) $Q = \frac{1/2}{2.5} = \underline{0.2}$

14.16 (a) $\omega_r = \frac{1}{\sqrt{LC}} = \underline{\frac{1}{2} \text{ rad/s}}$

(b) $B = \frac{1}{RC} = \frac{1}{5 \times 2} = \underline{\frac{1}{10} \text{ rad/s}}$

(c) $Q = \frac{1/2}{1/10} = \underline{5}$

14.17 $Y(j\omega) = \frac{1}{1+j\omega} + \frac{j\omega}{10} + 1$

$= \frac{(20-\omega^2) + j11\omega}{10(1+j\omega)}$

$|Y(j\omega)|^2 = \frac{(20-\omega^2)^2 + 121\omega^2}{100(1+\omega^2)}$

maximum occurs when

$\frac{d}{d\omega^2}|Y(j\omega)|^2 = 0 = \frac{d}{d\omega^2}(100)|Y(j\omega)|^2$

$(100)\frac{d}{d\omega^2}|Y(j\omega)|^2 = \frac{(2\omega^2+81)(\omega^2+1) - (\omega^4+81\omega^2+400)(1)}{(1+\omega^2)^2}$

$= \frac{\omega^4 + 2\omega^2 - 319}{(1+\omega^2)^2}$

$\therefore \omega^4 + 2\omega^2 - 319 = 0$ and

$\omega_0 = \underline{4.11 \text{ rds/sec}}$

14.17 Cont.

To find real part

$Y(j\omega) = \frac{[(20-\omega^2)+j11\omega](1-j\omega)}{10(1+j\omega)(1-j\omega)}$

$= \frac{20+10\omega^2 + j(\omega^3 - 9\omega)}{10(1+\omega^2)}$

real when $\omega^3 - 9\omega = 0$ or

when $\omega = \underline{3 \text{ rad/s}}$

14.18 Use a parallel RLC circuit.

$\omega_r = (2\pi)(40 \times 10^3) = \frac{1}{\sqrt{LC}} = \frac{1}{\sqrt{L \times 10^{-9}}}$

$L = 0.016 \text{ H} = \underline{16 \text{ mH}}$

$Q = R\sqrt{\frac{C}{L}} \Rightarrow 100 = R\sqrt{\frac{10^{-9}}{0.016}} \Rightarrow R = \underline{400 \text{k}\Omega}$

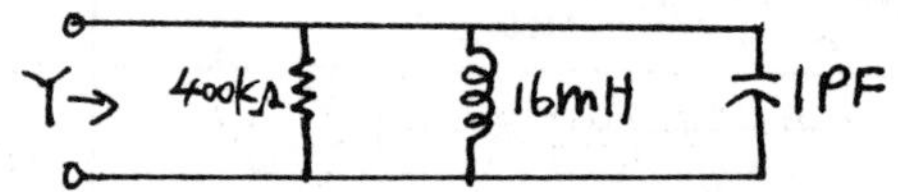

14.19 The natural frequency of the system are the poles of the system.

$H(s) = \frac{s}{[(s+0.001)^2+10^2][(s+0.008)^2+237^2][(s+0.0012)^2+777^2]}$

Poles

$s = -0.001 \pm j10$

$s = -0.008 \pm j237$

$s = -0.0012 \pm j777$

$\because$ It is a slightly damped system, the natural frequencies are good approximation of its resonance frequency. It has three peaks in its Bode gain plot. Maximum occurs at $\omega = 777$ because it is where the quatripple poles occur.

$\omega_r = \underline{777 \text{ rad/s}}$

14.20 $-20\text{dB} = 0.1$, $0\text{dB} = 1$, $3\text{dB} = \sqrt{2}$

$20\text{dB} = 10$, $40\text{dB} = 100$, $60\text{dB} = 1000$

The bandwidth, ω_c, of inverting amplifier $\omega_c(-20\text{dB}) = \frac{A_0\omega_0}{1+R_f/R_A} = \underline{181.8 \text{MHz}}$

$\omega_c(0\text{dB}) = \underline{100 \text{ MHz}}$

$\omega_c(3\text{dB}) = \underline{82.84 \text{ MHz}}$, $\omega_c(20\text{dB}) = \underline{18.18 \text{MHz}}$

$\omega_c(40\text{dB}) = \underline{1.98 \text{ MHz}}$, $\omega_c(60\text{dB}) = \underline{199.8 \text{kHz}}$

14.21 Since the 1 dB correction occurs at frequency ω

$$1\ dB = 20 \log \sqrt{\left(\frac{\omega}{\omega_c}\right)^2 + 1}$$

$$\omega = 0.509\,\omega_c$$

The bandwidth of this cascaded circuit is $\omega = 0.509 \times 18.18\,MHz$

$$= \underline{9.25\,MHz}$$

14.22 $H(j\omega) = \dfrac{A(j\omega)(R_A + R_f)}{[1 + A(j\omega)](R_A + R_f)} = \left(\dfrac{R_A + R_f}{R_A}\right)\dfrac{A(j\omega)}{A(j\omega) + (1 + \frac{R_f}{R_A})}$

$$= \left(1 + \frac{R_f}{R_A}\right)\frac{A(j\omega)}{A(j\omega) + (1 + \frac{R_f}{R_A})}$$

$$A(j\omega) = \frac{A_o\omega_o}{j\omega + \omega_o}$$

$$H(j\omega) = \left(1 + \frac{R_f}{R_A}\right)\left[\frac{1}{1 + \left(\frac{j\omega + \omega_o}{A_o\omega_o}\right)\left(1 + \frac{R_f}{R_A}\right)}\right]$$

$$= \left(1 + \frac{R_f}{R_A}\right)\left[\frac{1}{j\omega\left(\frac{1 + R_f/R_A}{A_o\omega_o}\right) + \left(1 + \frac{1 + R_f/R_A}{A_o}\right)}\right]$$

For voltage follower, $R_f = 0$, $R_A = \infty$

$$\Rightarrow H(j\omega) = \left(1 + \frac{0}{\infty}\right)\left\{\frac{1}{j\omega\left(\frac{1 + 0/\infty}{A_o\omega_o}\right) + \left[1 + \frac{1 + 0/\infty}{A_o}\right]}\right\}$$

$$= \frac{1}{j\omega\left(\frac{1}{A_o\omega_o}\right) + 1}$$

bandwidth $\omega_c = A_o\omega_o$

gain-bandwidth product $= 1 \cdot \omega_c = \underline{A_o\omega_o}$

14.23 From section 14.5 of the text book, for inverting amplifier

$$V_{out} = -\frac{R_f}{R_A}\left[\frac{1}{j\omega\left(\frac{1 + R_f/R_A}{A_o\omega_o}\right) + 1}\right]V_2$$

from the result of Prob. 14.23

$$V_{out} = \left(1 + \frac{R_f}{R_A}\right)\left[\frac{1}{j\omega\left(\frac{1 + R_f/R_A}{A_o\omega_o}\right) + 1}\right]V_1$$

and by superposition principle

$$V_{out} = \left(1 + \frac{R_f}{R_A}\right)\left(\frac{R_f}{R_f + R_A}\right)\left[\frac{1}{j\omega\left(\frac{1 + R_f/R_A}{A_o\omega_o}\right) + 1}\right]V_1$$

14.23 Cont.

$$V_{out} = \frac{R_f}{R_A}\left[\frac{1}{j\omega\left(\frac{1 + R_f/R_A}{A_o\omega_o}\right) + 1}\right](V_1 - V_2)$$

Hence the difference amplifier has $\omega_c = \dfrac{A_o\omega_o}{1 + R_f/R_A}$, and gain-bandwidth product $GBP = A_o\omega_o\left(\dfrac{R_f/R_A}{1 + \frac{R_f}{R_A}}\right)$

14.24 $|H(j\omega)|^2 = \dfrac{36\omega^2}{(4 - \omega^2)^2 + \omega^2} = \dfrac{36}{1 + \left(\frac{4 - \omega^2}{\omega}\right)^2}$

$|H(j0)| = |H(j\infty)| = 0$; $\Rightarrow$ bandpass

$|H(j\omega)|_{max} = |H(j2)| = 6$ $\therefore \omega_o = \underline{2\ rad/s}$

To find ω_{c1} and ω_{c2} set

$$\frac{36}{2} = \frac{36}{1 + \left(\frac{4 - \omega^2}{\omega}\right)^2} \quad \therefore \frac{4 - \omega^2}{\omega} = \pm 1 :$$

$\omega^2 \pm \omega - 4 = 0 \Rightarrow \omega_{c1} = \underline{1.56}\ rad/s$;

$\omega_{c2} = \underline{2.56\ rad/s}$; $B = \omega_{c2} - \omega_{c1} = \underline{1\ rad/s}$

14.25 $|H(j\omega)|^2 = \dfrac{36\omega^4}{(4 - \omega^2) + \omega^2} = \dfrac{36}{1 + \left(\frac{16 - 7\omega^2}{\omega^4}\right)}$

$|H(j0)| = 0$; $|H(j\infty)| = 6$; $\Rightarrow$ high pass

$|H(j\omega)|_{max} = |H(j\omega)| = 6$

$$|H(j\omega_c)| = \left(\frac{6}{\sqrt{2}}\right)^2 = \frac{36}{1 + \frac{16 - 7\omega^2}{\omega^4}}$$

$$\therefore 1 = \frac{16 - 7\omega^2}{\omega^4} : \omega^4 + 7\omega^2 - 16 = 0$$

$\omega_c = \underline{1.347\ rad/s}$

14.26 $|H(j\omega)|^2 = \dfrac{36(4 - \omega^2)^2}{(4 - \omega^2)^2 + \omega^2} = \dfrac{36}{1 + \left(\frac{\omega}{4 - \omega^2}\right)^2}$

$|H(j2)| = 0$; $\Rightarrow \omega_o = 2\ rad/s$

$|H(j0)| = |H(j\infty)| = 6 \Rightarrow$ band-reject

ω_{c1} and ω_{c2} satisfy

$$\frac{36}{2} = \frac{36}{1 + \left(\frac{\omega}{4 - \omega^2}\right)^2} : \pm 1 = \frac{\omega}{4 - \omega^2}$$

$\therefore \omega^2 \pm \omega - 4 = 0$: $\omega_{c1} = \underline{1.56\ rad/s}$

$\omega_{c2} = \underline{2.56\ rad/s}$

14.27 V_3 = left side node voltage
KCL gives

$V_2(1+s+\frac{1}{2s}) - V_3(\frac{1}{2s}) = 0$ or

$V_3 = V_2(2s^2+2s+1)$

$V_3(s+\frac{1}{2s}) - V_2(\frac{1}{2s}) = V_1$ or

$V_2\left[\frac{(2s^2+2s+1)(2s^2+2s+1)}{2s} - \frac{1}{2s}\right] = V_1$

$$H(s) = \frac{V_2}{V_1} = \frac{2s}{(2s^2+2s+1)^2-1} = \frac{1}{2s^3+4s^2+s+2}$$

$$|H(j\omega)|^2 = \frac{1}{(2-4\omega^2)^2+(4\omega-2\omega^3)^2} = \frac{1/4}{1+\omega^6}$$

$|H(j0)| = \frac{1}{2}$; $H(j\infty) = 0$; $\Rightarrow$ low pass

$|H(j\omega)|_{max} = H(j0) = \frac{1}{2}$

$|H(j\omega_c)| = \frac{1/4}{2} = \frac{1/4}{1+\omega^6}$;

$\therefore \omega^6 = 1 \Rightarrow \omega_c = \underline{1 \text{ rad/s}}$

14.28 V_3 = output voltage of first op amp
V_4 = voltage across 8Ω resistor

At inverting input of second op amp KCL gives

$\frac{V_2}{4} + \frac{V_3}{2} = 0 \Rightarrow V_3 = -\frac{1}{2}V_2$

At inverting input of first op amp KCL gives

$\frac{V_3}{2} + \frac{s}{4}V_4 = 0 \Rightarrow V_4 = -\frac{2}{s}V_3 = \frac{1}{s}V_2$

KCL at V_4:

$V_4(\frac{1}{2}+\frac{1}{8}+\frac{s}{4}+\frac{s}{4}+\frac{1}{4}) - \frac{s}{4}V_3 - \frac{1}{4}V_2 = \frac{1}{2}V_1$

$\left[\frac{4s+7}{8s}(\frac{1}{s}) - \frac{s}{4}(-\frac{1}{2}) - \frac{1}{4}\right]V_2 = \frac{1}{2}V_1$

$$\therefore H(s) = \frac{V_2}{V_1} = \frac{4s}{s^2+2s+7}$$

$k=4$, $B=\underline{2}$, $\omega_0^2=\underline{7}$,

$(a=2)$, $G = \frac{k}{a} = \frac{4}{2} = \underline{2}$

14.29 One needs an input summer with coefficients +1 and −1, and an output summer with coefficients −5 and −6.

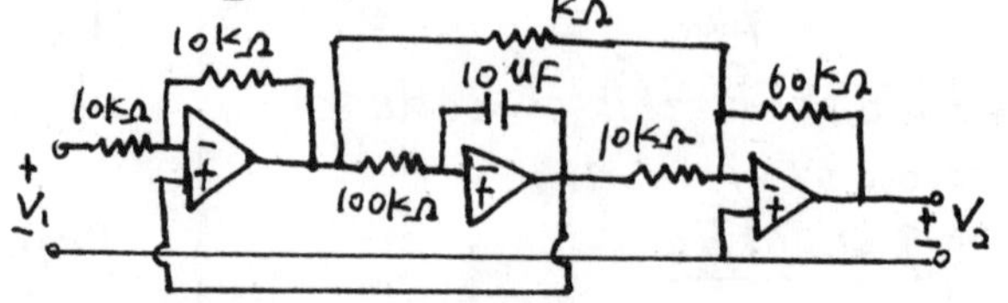

14.30 One needs an input summer with coefficients +1 and −3, and an output stage of gain −¼.

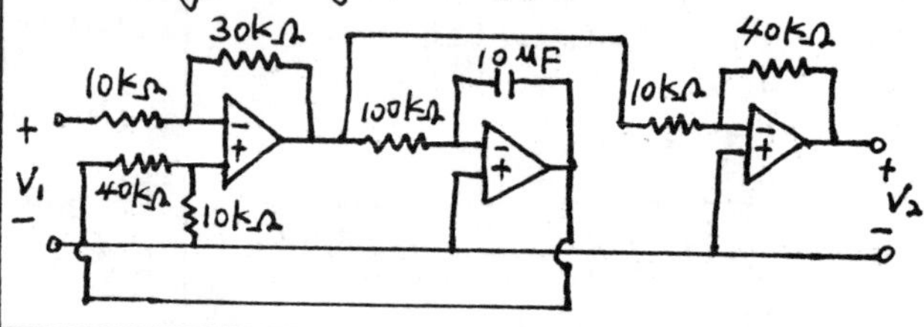

14.31 We need an input summer with coefficients +9 and −1, and an output summer with coefficients 7 and −2.

$1+\frac{R_f}{R_a/\!/R_b} = 9 \Rightarrow \frac{R_f}{R_a/\!/R_b} = 8 \Rightarrow R_a/\!/R_b = \frac{1}{8}R_f$

Let $R_a = R_f = 100k\Omega \Rightarrow \frac{100k \cdot R_b}{100k+R_b} = \frac{100k}{8}$

$7R_b = 100k \Rightarrow R_b = \frac{100}{7}k = 14.3k\Omega$

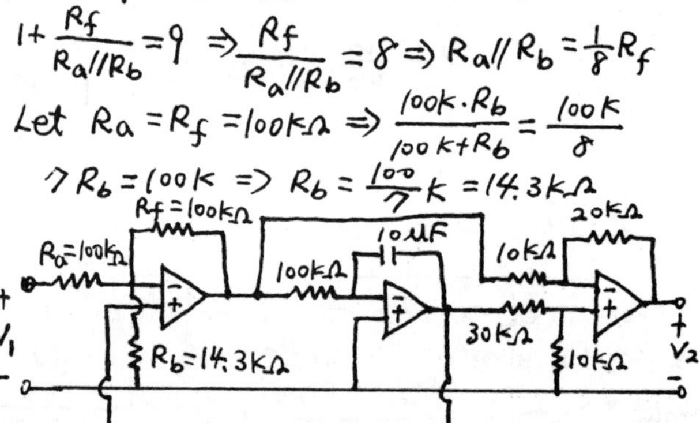

14.32 A biquadratic transfer function can be realized by a single op-amp section.

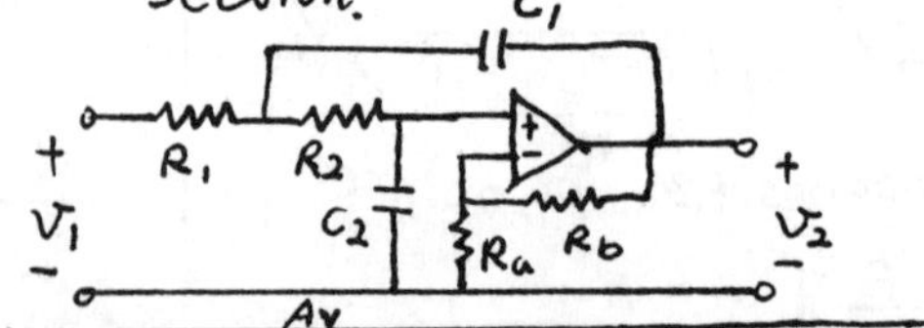

$$H(s) = \frac{A_V}{R_1R_2C_1C_2s^2+s[C_2(R_1+R_2)+R_1C_1(1-A_V)]+1}$$

Let $R_b = \underline{0}$, $R_a = \underline{100\,k\Omega} \Rightarrow A_V = 1$

$R_1R_2C_1C_2 = 1 \Rightarrow$ Let $R_1 = R_2 = \underline{50k\Omega}$, $C_1 = C_2 = \underline{20\,\mu F}$

14.33 $H(s)=\dfrac{s^2+s+1}{(3s^2+s+2)(s+1)}$ $\quad \left(\dfrac{s^2+s+1}{3s^2+s+2}\right)\left(\dfrac{1}{s+1}\right)$

The solution is a cascade of two circuits.

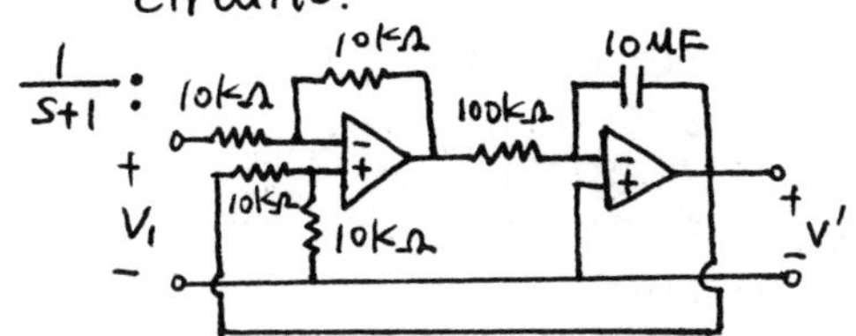

14.34 $W_c=\Omega_c k_f$; $k_f=\dfrac{W_c}{\Omega_c}$

$$k_f=\frac{400\pi}{\sqrt{2}}=200\pi\sqrt{2}=888.6$$

$$C=\frac{C'}{k_i k_f} \Rightarrow k_i=\frac{C'}{C k_f}=\frac{\frac{1}{2}\,F}{(2\mu F)(888.6)}=281.3$$

$$R=k_i R'=(281.3)(2)=\underline{562.7\,\Omega}$$

14.35 KCL at inverting input gives

$$\frac{V_1}{R_1}+V_2\left(\frac{1}{R_2}+sC\right)=0$$

$$H(s)=\frac{V_2}{V_1}=\frac{-1/R_1}{1/R_2+sC}=\frac{-R_2/R_1}{1+sR_2C}$$

$H(0)=-R_2/R_1=-2$, Since

$$\frac{4}{2}=\frac{4}{1+(W_c R_2 C)^2}\Rightarrow W_c R_2 C=1$$

$$R_2=1/W_c C=\frac{1}{(10^5)(10^{-9})}=\underline{10\,k\Omega}$$

14.36 $k_f=\dfrac{W_c}{\Omega_c}=2\times10^4$

$$k_i=\frac{C'}{C k_f}=\frac{1}{(10^{-6})(2\times10^4)}=50$$

$R_1'=\frac{1}{2}\,\Omega$, $R_2'=1\,\Omega$ and $R_3'=2\,\Omega$

then $R_1=k_i R_1'=\underline{25\,\Omega}$

$R_2=k_i R_2'=\underline{50\,\Omega}$, $R_3=k_i R_3'=\underline{100\,\Omega}$

14.37 KCL yields

$$V_2\left[1+\frac{1}{Q(1-K)(s+1/s)}+\frac{K}{1-K}\right]=\frac{K}{1-K}V_1$$

$$\therefore H(s)=\frac{V_2}{V_1}=\frac{K(s^2+1)}{s^2+(1/Q)s+1}\ ;\ W_o=1\text{ rad/s}$$

$$k_f=10^6\ ;\ k_i=\frac{C'}{C k_f}=\frac{1/Q(1-K)}{(10^6)(10^{-8})}=\frac{1}{5(0.5)(10^6)(10^{-8})}=40$$

$$R'=\frac{1-K}{K}=1\,\Omega\rightarrow\underline{40\,\Omega}$$

$$L'=Q(1-K)=2.5\,H\rightarrow 2.5\frac{k_i}{k_f}=\underline{100\,\mu H}$$

14.38 $k_f=2\pi(10^3)$; $k_i=\dfrac{C'}{C k_f}=\dfrac{2}{2\pi(10^3)(10^{-8})}=10^5/\pi$

$\frac{1}{2}\,\Omega\rightarrow \underline{50/\pi\ k\Omega}$; $2\,\Omega\rightarrow\underline{\frac{200}{\pi}\,k\Omega}$;

$\frac{2}{5}\,\Omega\rightarrow\underline{40/\pi\ k\Omega}$

14.39

```
*GAIN AND PHASE PLOT
*OF A RC CIRCUIT
C_C1        1 2 0.9
C_C2        0 1 0.1
R_R1        1 2 10
V_V1        2 0 AC 1
*Control statement
.AC LIN 10 0.0001 10
*GAIN=VM(2,1)/VM(2)
*PHASE SHIFT=VP(2,1)-VP(2)
.END
**************************
*THE GAIN AND PHASE PLOT
*VS. LOG FREQUENCY ARE
*SHOWN BELOW:
```

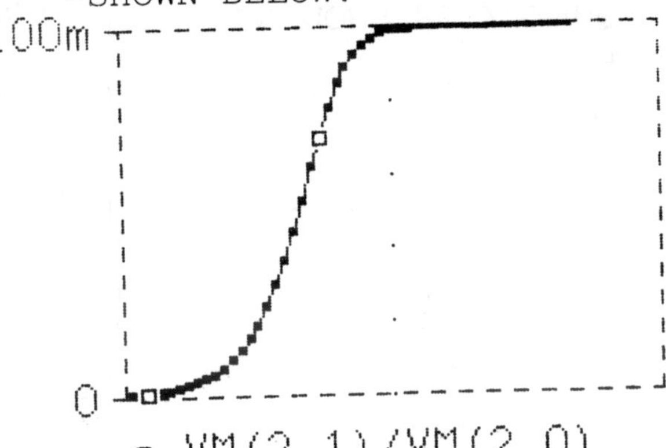

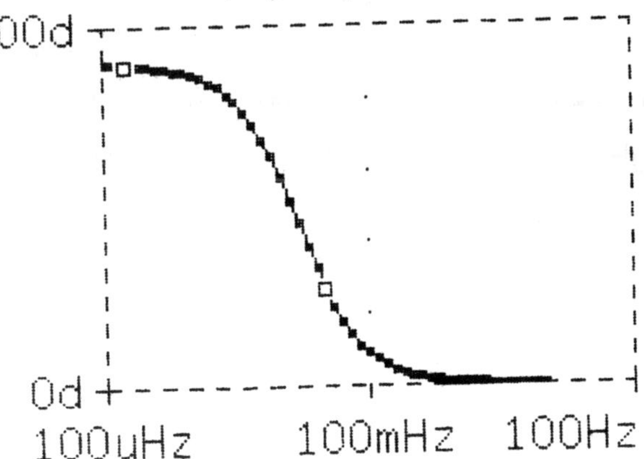

14.40

```
*GAIN VS. LOG FREQUENCY
*OF A BANDPASS FILTER
X_U1     0 1 2 OPAMP
X_U2     0 3 4 OPAMP
R_R1     2 3 2
R_R2     3 4 4
R_R3     5 4 4
R_R4     6 5 2
R_R5     0 5 8
R_R6     1 2 2
C_C1     5 1 0.25
C_C2     5 2 0.25
V_V1     6 0 DC 0 AC 1
*Control statement
.INC C:\PS\OPAMP.LIB
.AC DEC 10 0.0001 1000
*Output GAIN=VM(4)/VM(6)
.END
*************************
*PLOT OF GAIN VS. LOG
*FREQUENCY IS SHOW AS
*FOLLOWING:
*************************
```

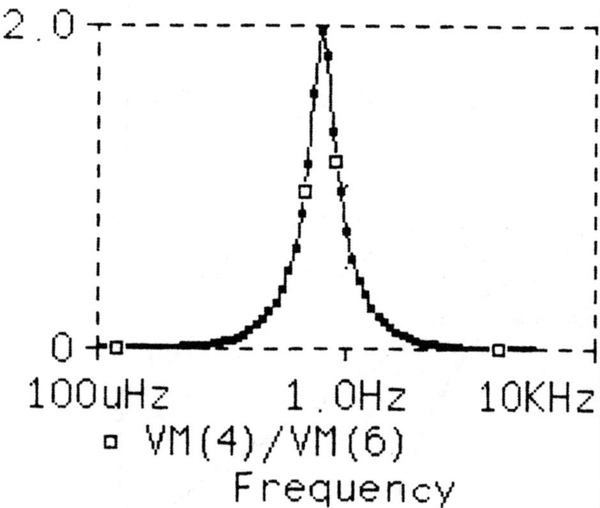

14.41

```
*GAIN VS LOG FREQUENCY
*OF AN HIGH-PASS FILTER
*USING FEEDBACK OP AMP
*CIRCUIT
V_V1      1 0 DC 0 AC 1
X_U1      7 2 3 OPAMP
X_U2      0 4 7 OPAMP
X_U3      0 5 6 OPAMP
C_C1      4 7 10U
R_R1      1 2 10K
R_R2      2 3 RMOD 10K
R_R3      3 5 RMOD 50K
R_R4      3 4 100K
R_R5      7 5 10K
R_R6      5 6 RMOD 60K
*Control statement
.INC C:\PS\OPAMP.LIB
.MODEL RMOD RES(R=1.0)
.AC DEC 10 10 100000000
*Output GAIN=V(6)/1
.END
****************************
*THREE GAIN PLOTS OF
*FEEDBACK RESISTANCE
*SCALING FACTOR: 1,.9,1.1
*ARE SHOWN AS THE FOLLOWING
****************************
```

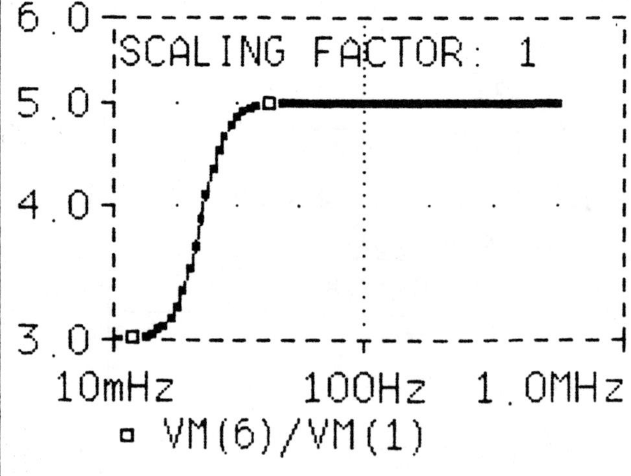

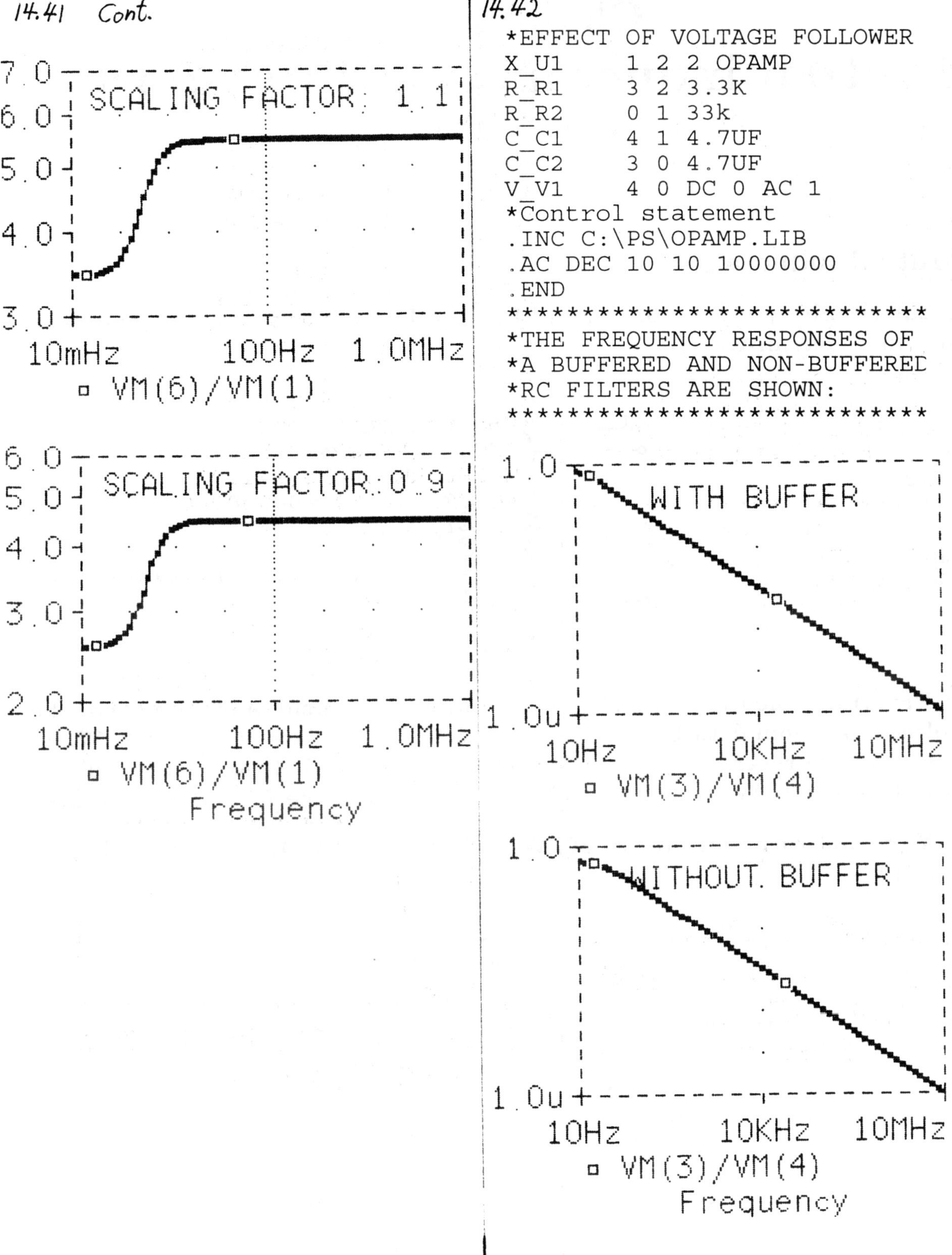

14.41 Cont.

14.42

```
*EFFECT OF VOLTAGE FOLLOWER
X_U1     1 2 2 OPAMP
R_R1     3 2 3.3K
R_R2     0 1 33k
C_C1     4 1 4.7UF
C_C2     3 0 4.7UF
V_V1     4 0 DC 0 AC 1
*Control statement
.INC C:\PS\OPAMP.LIB
.AC DEC 10 10 10000000
.END
****************************
*THE FREQUENCY RESPONSES OF
*A BUFFERED AND NON-BUFFERED
*RC FILTERS ARE SHOWN:
****************************
```

Chapter 15
Mutual inductance and two-port circuits

15.1 Mutual inductance

15.1 **In the circuit $L_1 = L_2 = 0.1$H and M = 10mH. Find ν_1 and ν_2 if $i_1 = 0$, and $i_2 = 10\sin(100t)$mA.**

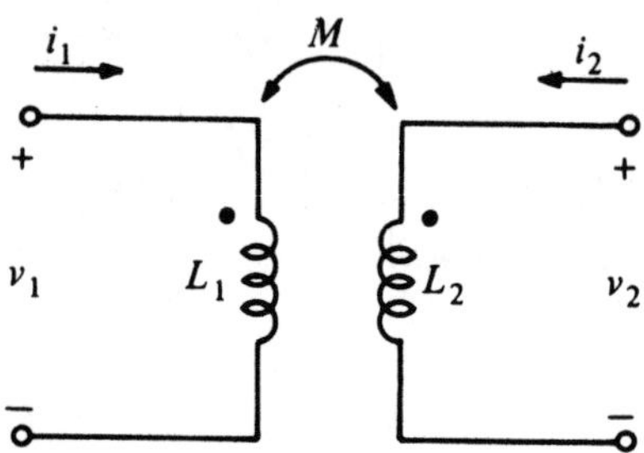

Problem 15.1

15.2 **In Prob.15.1 when $i_1 = 4\sin(6t + 30^o)$A and coil 2 is held open circuited, the voltages are found to be $\nu_1 = 2\cos(6t + 30^o)$V, $\nu_2 = 3\cos(6t - 150^o)$V. When the roles are reversed $i_2 = 4\sin(6t + 30^o)$A and coil 1 is held open circuited, $\nu_2 = 8\cos(6t + 30^o)$V. Find L_1, L_2, M and the sign of the mutual inductance.**

15.3 **If $N_2 = 500$ turns and $\phi_{21} = 30\mu$Wb when $i_1 = 4$A. Determine ν_2 if $i_1 = 5\cos(10t)$A.**

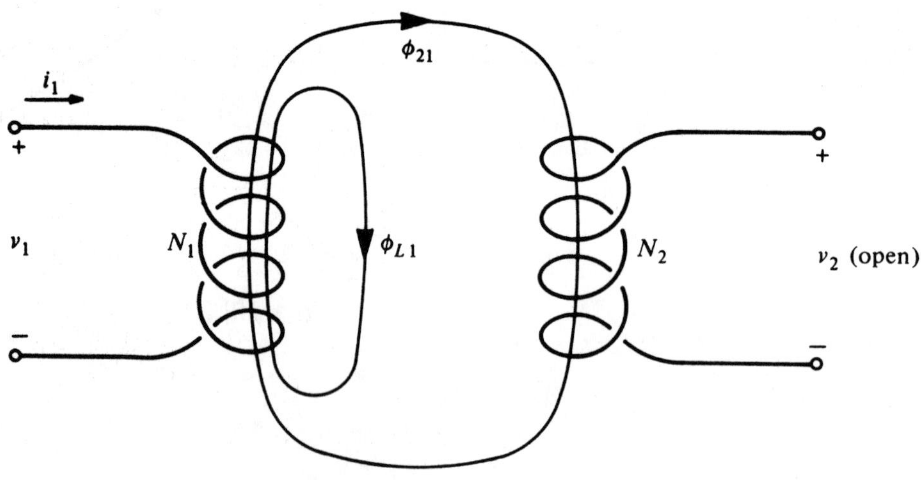

Problem 15.3

15.4 Determine the coefficient if coupling for Prob.15.2.

15.5 If the inductance measured between terminals a and d is 4H when terminals b and c are connected, and the inductance measured between terminals a and c is 10H when terminals b and d are connected, find the mutual inductance M between the two coils, and the position of the dots.

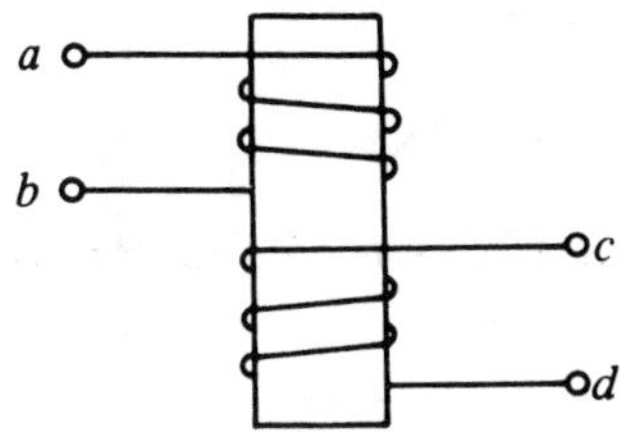

Problem 15.5

15.6 Find ν_1 and ν_2 if $L_1 = 6$H, $L_2 = 4$H, M $= 2$H, $i_1 = \sin(t)$**A**, and $i_2 = -2\cos(2t)$**A**.

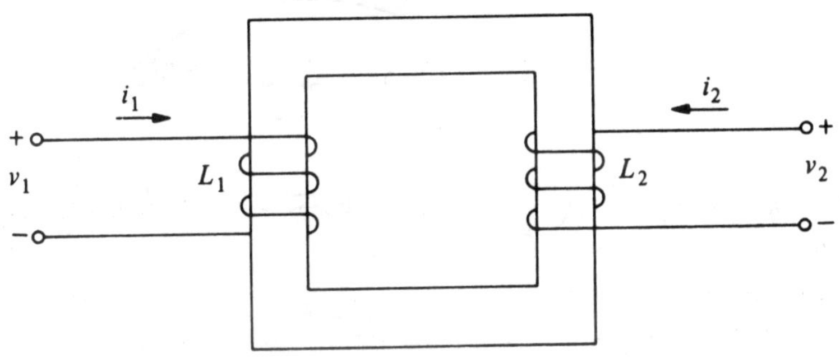

Problem 15.6

15.7 (a)Find the energy stored in the transformer at a time when $i_1 = 3$A and $i_2 = 2$A if $L_1 = 1$H, $L_2 = 8$H, and M $= 3$H.
(b)Repeat part(a) if one of the dots is moved to another terminal.

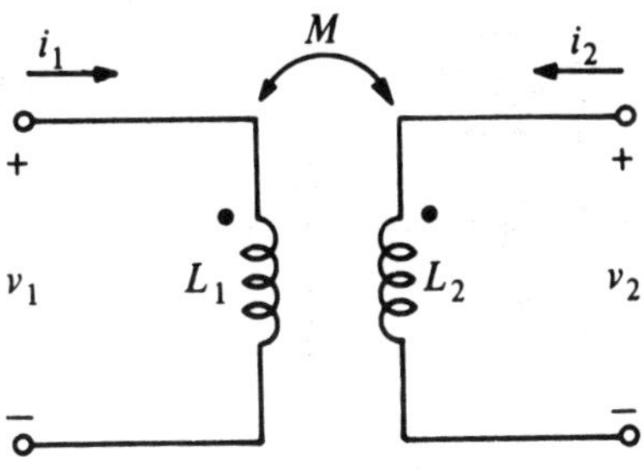

Problem 15.7

15.2 Circuits with mutual inductance

15.8 **Determine i_1 for t > 0 in the network, given $M = \frac{1}{2}$H. Assume the circuit is in steady state at $t = 0^-$.**

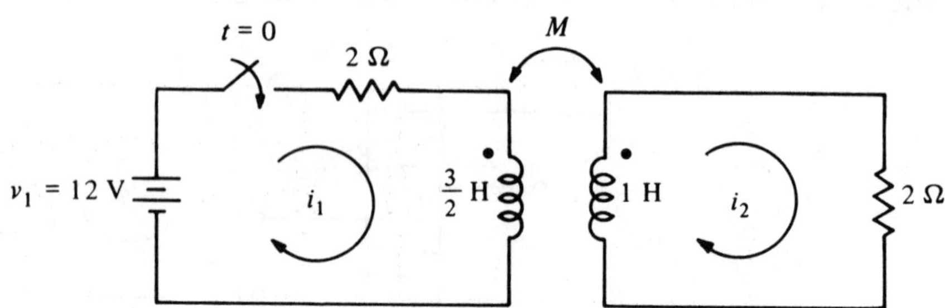

Problem 15.8

15.9 **Find the forced response v_2 if $v_1 = 3e^{-t}\cos(3t)$V.**

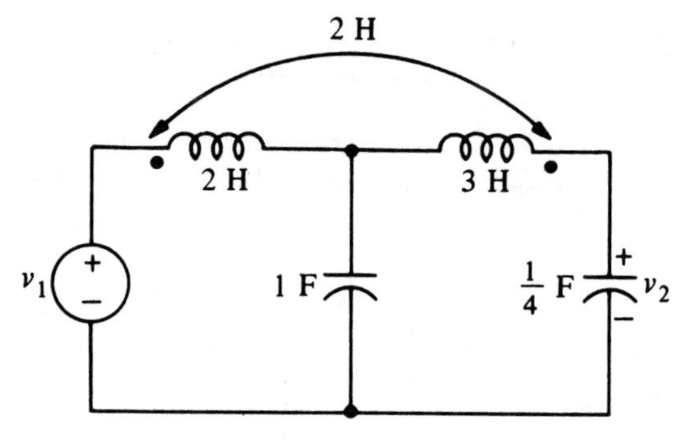

Problem 15.9

15.10 **Find the network function $\frac{V_2}{V_1}$ for the phasor circuit shown.**

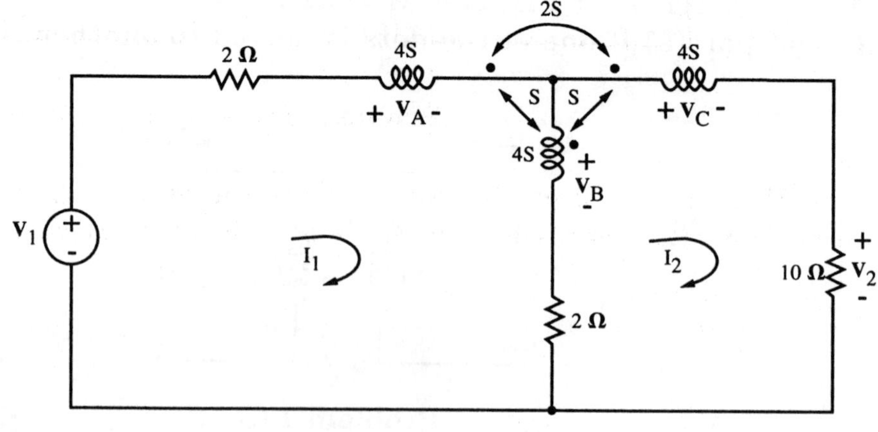

Problem 15.10

15.3 Mutual inductance and transformers

15.11 Given $V_g = 20\angle 0°$V, $Z_g = 20\Omega$, $L_1 = 1$H, $L_2 = 0.5$H, M $= 0.5$H, and $\omega =$ 10rad/sec. If $Z_2 = -(\frac{j10}{\omega})\Omega$, find (a)$Z_{in}$, (b)$I_1$, (c)$I_2$, (d)$V_1$, and (e)$V_2$.

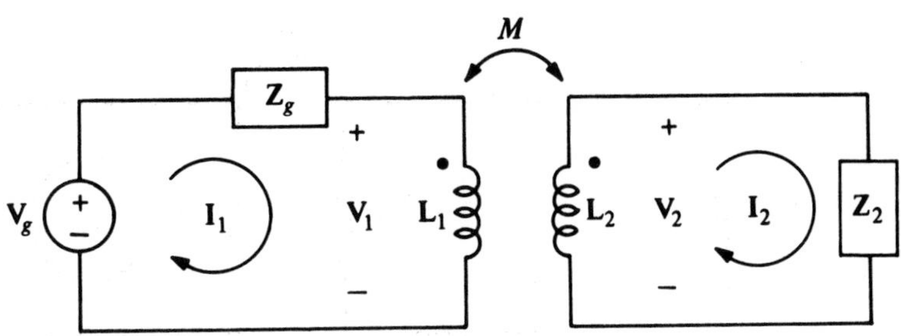

Problem 15.11

15.12 Repeat Prob.15.10 if the polarity dot is on the lower terminal of the secondary.

15.13 If $Z_2 = (10 - \frac{j10}{\omega})\Omega$ in Prob.15.10, find the frequency for which the reflected impedance is real.

15.14 For the circuit shown, find the steady-state current i_1 using reflected impedance.

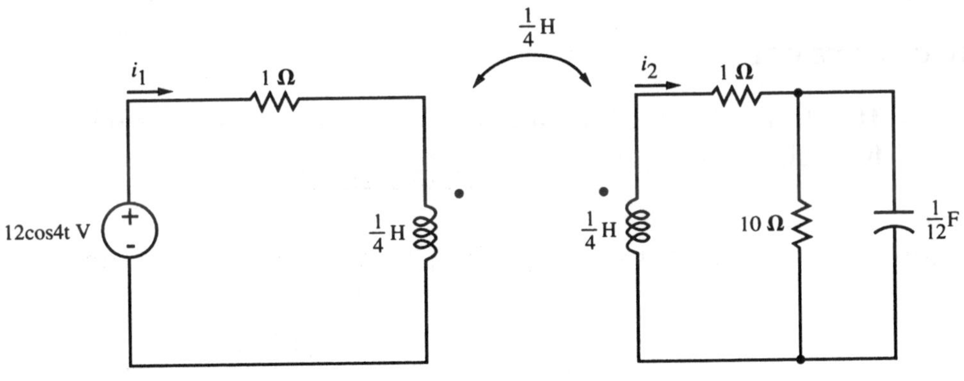

Problem 15.14

15.15 For the Prob.15.14, find the steady-state current i_2 by replacing the subnetwork from the secondary to the left with its Thevenin equivalent.

15.16 Find the input impedance seen by the independent voltage source V_1 for the circuit shown.

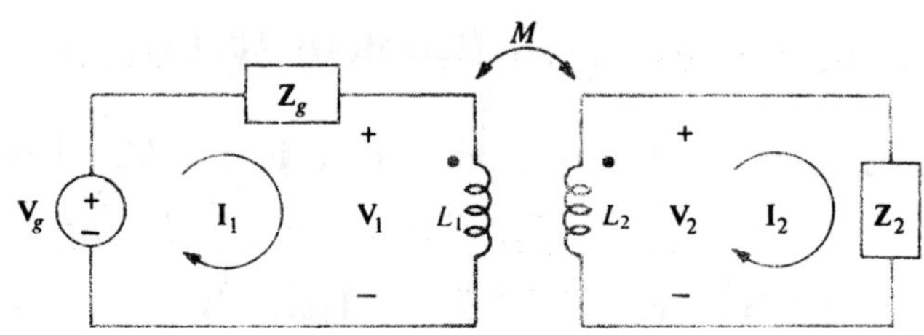

Problem 15.16

15.4 Ideal transforms

15.17 $V_g = 100\angle 0^o$V, $Z_g = 100\Omega$, and $Z_2 = 90\text{K}\Omega$. Find n such that $Z_1 = Z_g$, and then find the power delivered to Z_2.

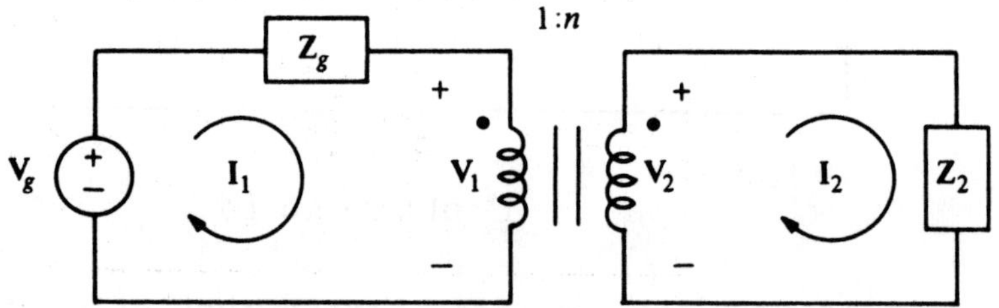

Problem 15.17

15.18 Find the average power delivered to the 8Ω resistor.

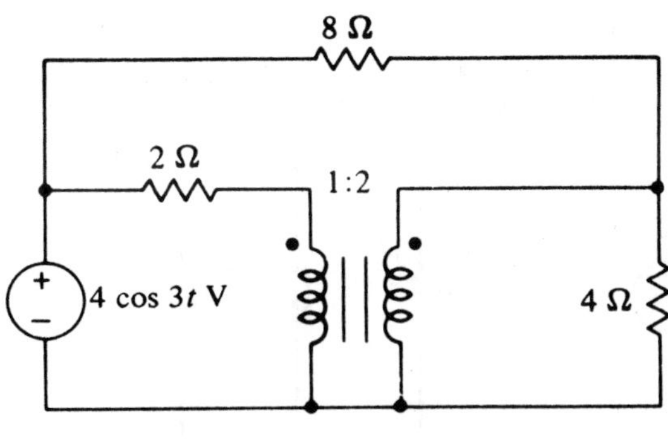

Problem 15.18

15.19 Find V_1, V_2, I_1 and I_2 using the method of reflected impedance.

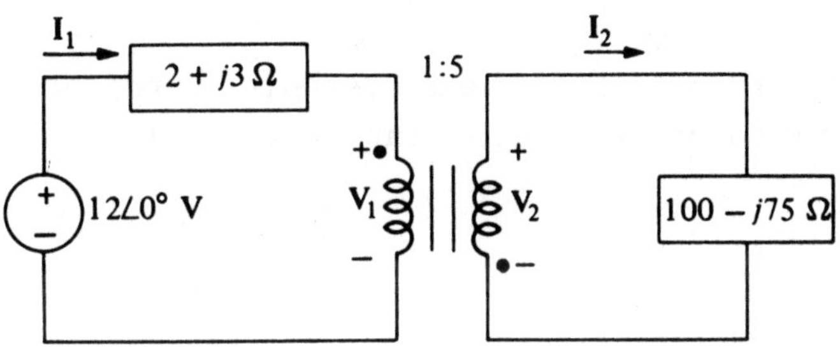

Problem 15.19

15.20 Find the turn ratio n so that the maximum power is delivered to the 10KΩ resistor.

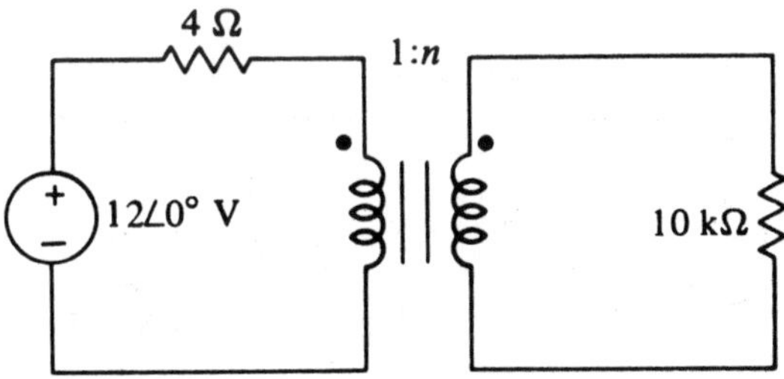

Problem 15.20

15.21 For the circuit shown, find v_0.

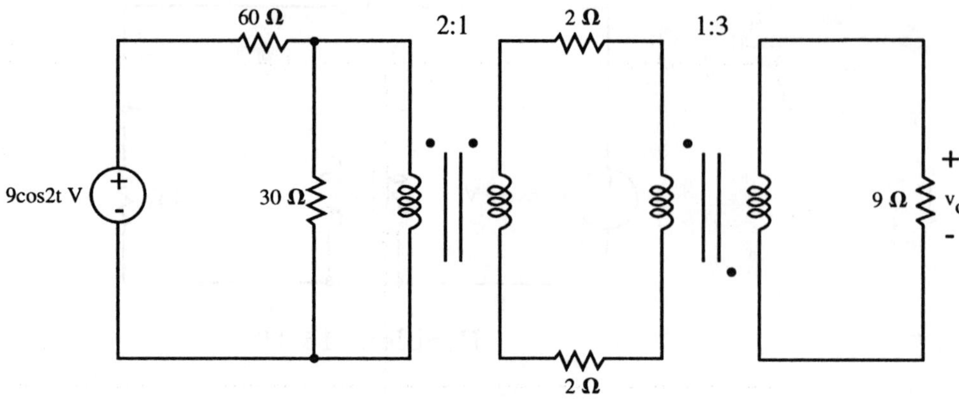

Problem 15.21

15.22 For the circuit shown, if the 2Ω resisitor is replaced by a load with resistance R, find the value of R such that maximum power is delivered to the load.

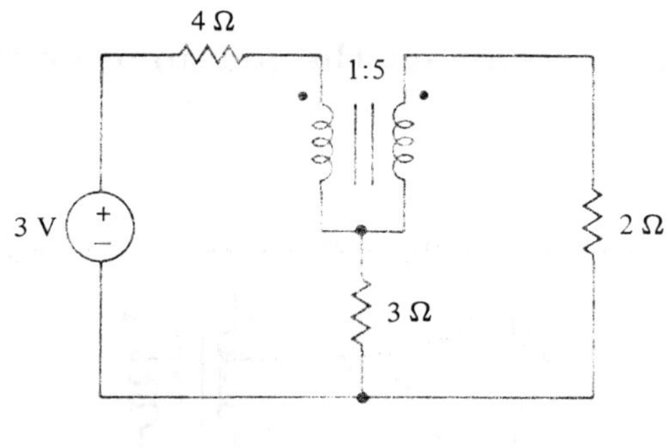

Problem 15.22

15.5 Two-port circuits

15.23 Find the Z-parameters for the two-port circuit shown.

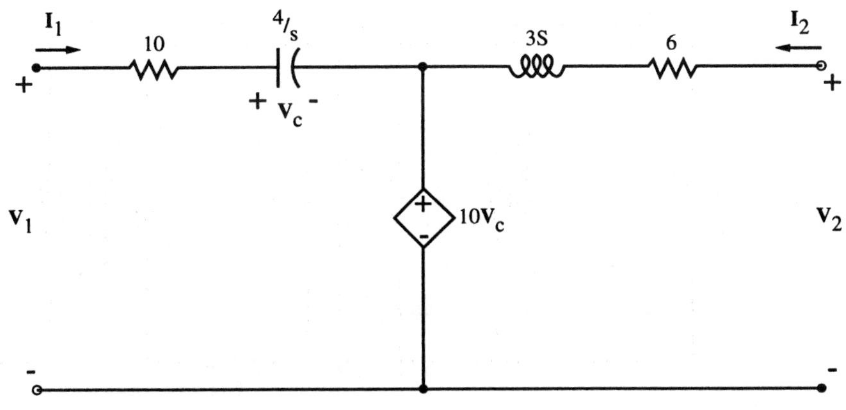

Problem 15.23

15.24 Find the Z-parameters for the two-port circuit shown.

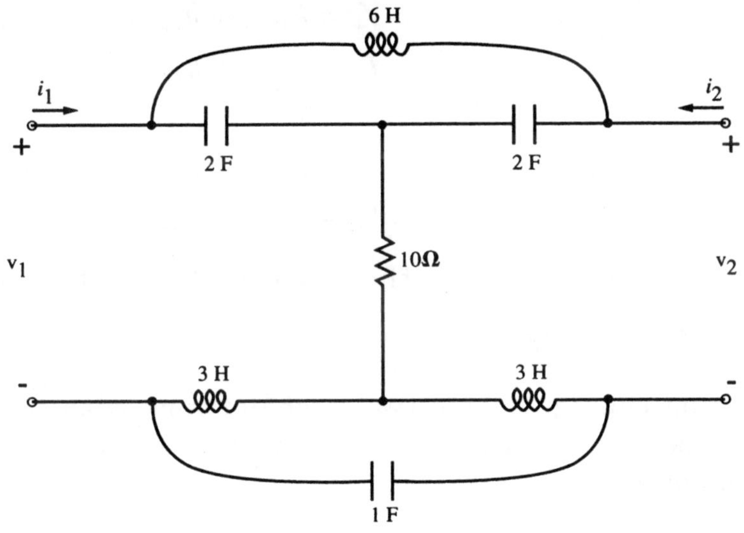

Problem 15.24

15.6 Two-port parameters

15.25 Find the Z-parameters and Y-parameters of the two-port circuit shown.

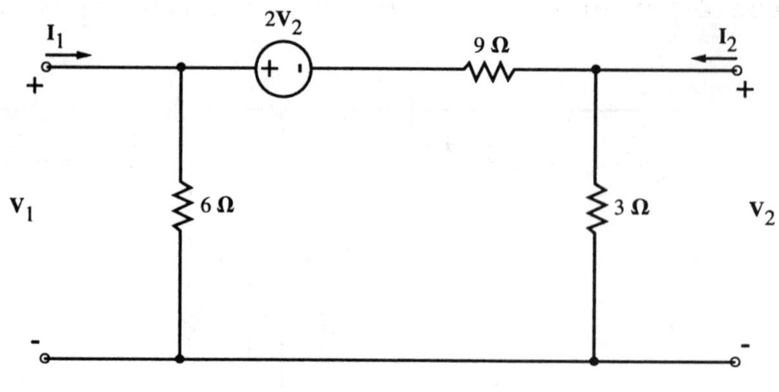

Problem 15.25

15.26 Draw an equivalent circuit for a two-port network with parameters $Z_{11} = 8\Omega$, $Z_{12} = 2\Omega$, and $Z_{22} = 4\Omega$.

15.27 Two sets of hybrid parameters h_{11}, h_{12}, h_{21}, h_{22}, and g_{11}, g_{12}, g_{21}, g_{22} are defined by

$$V_1 = h_{11}I_1 + h_{12}V_2$$
$$I_2 = h_{21}I_1 + h_{22}V_2$$

and

$$I_1 = g_{11}V_1 + g_{12}I_2$$
$$V_2 = g_{21}V_1 + g_{22}I_2$$

Find the h-parameters and the g-parameters of the network of Prob.15.26.

15.7 Two-port models

15.28 Considering the figure as a two-port network with terminals from the right of the voltage source to the left of the 1Ω resistor, find the Z-parameters as functions of s.

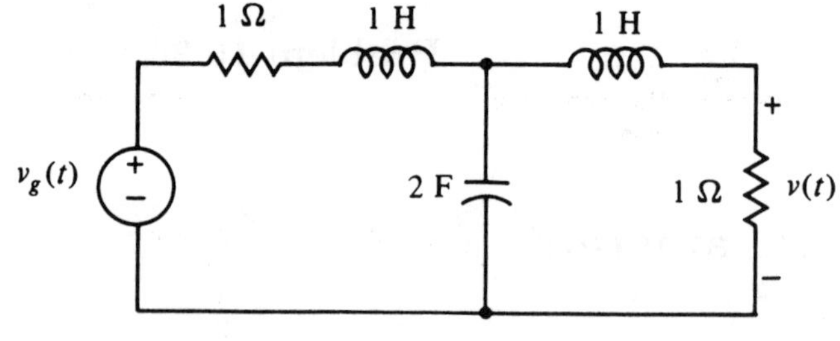

Problem 15.28

15.29 Find the h-parameters of the two-port networks of Prob.15.28.

15.30 Find the transmission parameters of the two-port network of the Prob.15.28.

15.31 Find the Z-parameters for the two-port circuit shown.

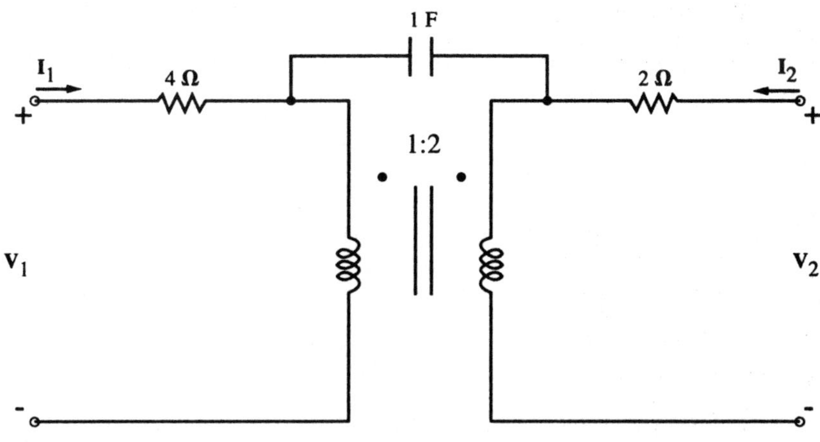

Problem 15.31

15.32 Considering the figure as a two-port network with terminals as shown, find the Z- and Y-parameters as function of s.

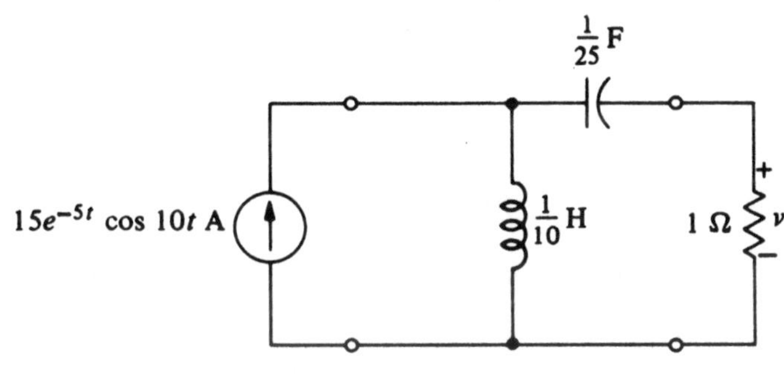

Problem 15.32

15.33 Let $h_{11} = 3\text{K}\Omega$, $h_{12} = 10^{-3}$, $h_{21} = 200$, and $h_{22} = 10^{-3}$s, and find the network function $\frac{V_2}{V_1}$ if port 2 is open circuited.

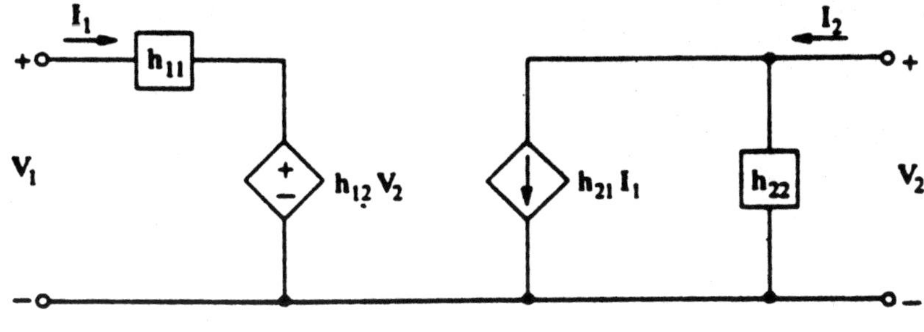

Problem 15.33

15.9 SPICE, transformers and two-ports

15.34 Verify the impedance found of Prob.15.16 at $\omega = 4\text{rad/sec}$ and $\omega = 7\text{rad/sec}$ using SPICE.

15.35 Find the transient and steady-state current of Prob.15.14 using SPICE.

15.1 $v_1 = L_1 \frac{di_1}{dt} + M\frac{di_2}{dt}$

$= (0.1)(0) + (10^{-2})(10^3 \cos 100t)\,mA$

$= \underline{10 \cos 100t \; mV}$

$v_2 = M\frac{di_1}{dt} + L_2\frac{di_2}{dt}$

$= 0 + (0.1)(10^3 \cos 100t \; mA)$

$= \underline{100 \cos 100t \; mV}$

15.2 $v_1 = L_1\frac{di_1}{dt} + M\frac{di_2}{dt}$

$v_2 = L_2\frac{di_2}{dt} + M\frac{di_1}{dt}$

$i_1 = 4\sin(6t+30°) = 2\sqrt{3}\sin 6t + 2\cos 6t$

$\frac{di_1}{dt} = 12\sqrt{3}\cos 6t - 12\sin 6t$

$v_1 = L_1(12\sqrt{3}\cos 6t - 12\sin 6t) + M(0)$

$= \sqrt{3}\cos 6t - \sin 6t \Rightarrow L_1 = \underline{\frac{1}{12}H}$

$v_2 = 0 + M\frac{di_1}{dt} = M(12\sqrt{3}\cos 6t - 12\sin 6t)$

$= -\frac{3\sqrt{3}}{2}\cos 6t + \frac{3}{2}\sin 6t$

$M = \underline{-\frac{1}{8}H}$

$v_2 = L_2(12\sqrt{3}\cos 6t - 12\sin 6t)$

$= 4\sqrt{3}\cos 6t - 4\sin 6t$

$L_2 = \underline{\frac{1}{3}H}$

15.3 $M_{21} = \frac{N_2 \Phi_{21}}{i_1} = \frac{(500)(30\times10^{-6})}{4}$

$= 3.75 \; mH$

$v_2 = M_{21}\frac{di_1}{dt} = (3.75\times10^{-3})\frac{d}{dt}(5\cos 10t)$

$= \underline{-0.1875 \sin 10t \; V}$

15.4 $K = \frac{M}{\sqrt{L_1L_2}} = \frac{1/8}{\sqrt{\frac{1}{12}\times\frac{1}{3}}} = \underline{\frac{6}{8}}$

15.5 With b and c connected

$v_{ad} = (L_{ab} + L_{cd} - 2M)\frac{di}{dt} = L_{ad}\frac{di}{dt}$

With b and d connected

$v_{ac} = (L_{ab} + L_{cd} + 2M)\frac{di}{dt} = L_{ac}\frac{di}{dt}$

$\therefore L_{ab} + L_{cd} - 2M = L_{ad} = 4H$

$L_{ab} + L_{cd} + 2M = L_{ac} = 10H$

subtracting equations

$4M = 10 - 4 \Rightarrow M = \underline{3/2 \; H}$

15.6 $v_1 = L_1\frac{di_1}{dt} - M\frac{di_2}{dt}$

$= 6\cos t - 2(4\sin 2t)$

$= \underline{6\cos t - 8\sin 2t \; V}$

$v_2 = -M\frac{di_1}{dt} + L_2\frac{di_2}{dt} = (-2)\cos t + 4(4\sin 2t)$

$= \underline{-2\cos t + 16\sin 2t \; V}$

15.7 (a) $W = \frac{1}{2}L_1i_1^2 + Mi_1i_2 + \frac{1}{2}L_2i_2^2$

$= \frac{1}{2}(1)(3)^2 + (3)(3)(2) + \frac{1}{2}(8)(2)^2$

$= \underline{38.5 \; J}$

(b) $W = \frac{1}{2}L_1i_1^2 - Mi_1i_2 + \frac{1}{2}L_2i_2^2$

$= \underline{2.5 \; J}$

15.8 $V_1(s) = (3/2\,s + 2)I_1 + s/2\,I_2$

$0 = -\frac{s}{2}I_1 + (s+2)I_2$ or

$I_2 = \frac{s}{2(s+2)}I_1$, substitution

$H(s) = \frac{I_1}{V_1} = \frac{4(s+2)}{5s^2+20s+16}$

$= \frac{4/5\,(s+2)}{(s+1.11)(s+2.89)}$

$i_{1f} = H(0)V_1 = \frac{8}{16}(12) = 6$

$i_1 = 6 + A_1e^{-1.11t} + A_2e^{-2.89t}$

$i_1(0^+) = i_1(0^-) = 0 = 6 + A_1 + A_2$

From KVL

$\frac{3}{2}\frac{di_1(0^+)}{dt} - \frac{1}{2}\frac{di_2(0^+)}{dt} + 2i_1(0^+) = 12$

$-\frac{1}{2}\frac{di_1}{dt}(0^+) + \frac{di_2(0^+)}{dt} + 2i_2(0^+) = 0$

Since $i_1(0^+) = i_2(0^+) = 0$ We obtain

$\frac{di_1(0^+)}{dt} = \frac{2di_2(0^+)}{dt} = \frac{4}{5}(12) = 9.6$

$\frac{di_1(0^+)}{dt} = 9.6 = -1.11A_1 - 2.89A_2$

$\therefore A_1 = -4.34$ & $A_2 = -1.66$

$i_1 = \underline{6 - 4.34e^{-1.11t} - 1.66e^{-2.89t} \; A}$

15.9 Let i_1 and i_2 be clockwise mesh currents then

$V_1 = (2s + 1/s)I_1 - (2s + \frac{1}{s})I_2$

$0 = -(2s + 1/s)I_1 + (3s + \frac{1}{s} + \frac{4}{s})I_2$

$V_2 = \frac{4}{s}I_2$. $\therefore I_2 = \frac{sV_1}{s^2+4}$, $V_2 = \frac{4V_1}{s^2+4}$,

$s = -1+j3$, $V_1 = 3$.

15.9 Cont.

$$V_2 = \frac{4(3)}{(-1+j3)^2+4} = 1.66\angle 123.7^\circ$$

$$v_{2f} = \underline{1.66e^{-t}\cos(2t+123.7^\circ)\,V}$$

15.10 $V_A = (4s-s)I_1 - (2s-s)I_2 = 3sI_1 - sI_2$

$V_B = (4s-s)I_1 - (4s+s)I_2 = 3sI_1 - 5sI_2$

$V_C = (-2s+s)I_1 + (4s-s)I_2 = -sI_1 + 3sI_2$

Applying KVL in loop 1 and loop 2 gives

$V_1 = 4I_1 + V_A + V_B - 2I_2$

$0 = -2I_1 - V_B + V_C + 12I_2$

$V_1 = 4I_1 + (3sI_1 - sI_2) + (3sI_1 - 5sI_2) - 2I_2$

$= I_1(6s+4) - I_2(6s+2)$

$0 = -2I_1 - (3sI_1 - 5sI_2) + (-sI_1 + 3sI_2)$

$+12I_2 = I_1(-4s-2) + I_2(12+8s)$

$I_2(6s+2) - V_1 = I_1(6s+4)$

$I_2(12+8s) = I_1(4s+2)$

$$I_2 = \frac{\begin{vmatrix} 6s+4 & -1 \\ 4s+2 & 0 \end{vmatrix}}{\begin{vmatrix} 6s+2 & -1 \\ 8s+12 & 0 \end{vmatrix}} = \frac{4s+2}{8s+12}$$

$$V_1 = \frac{\begin{vmatrix} 6s+2 & 6s+4 \\ 8s+12 & 4s+2 \end{vmatrix}}{\begin{vmatrix} 6s+2 & -1 \\ 8s+12 & 0 \end{vmatrix}} = \frac{(6s+2)(4s+2)-(8s+12)(6s+4)}{8s+12}$$

$$H(s) = \frac{V_2}{V_1} = \frac{10I_2}{V_1} = \frac{10(4s+2)}{(6s+2)(4s+2)-(8s+12)(6s+4)}$$

$$= \frac{20(2s+1)}{24s^2+20s+4-48s^2-104s+48}$$

$$= \underline{\frac{-5(2s+1)}{6s^2+21s+11}}$$

15.11 $Z_1 = j\omega L_1 + \frac{\omega^2 M^2}{Z_2 + j\omega L_2}$

$= j(10 - \frac{25}{4}) = j3.75\,\Omega$

(a) $Z_{in} = Z_g + Z_1 = \underline{20 + j3.75\,\Omega}$

(b) $I_1 = V_g/Z_{in} = \frac{20}{20+j3.75} = 0.97 - j0.18 = \underline{0.983\angle -10.62^\circ A}$

(c) $I_2 = \frac{j\omega M}{Z_2 + j\omega L_2} = \frac{j5(0.97-j0.18)}{-j+j5}$

$= 1.21 - j0.23 = \underline{1.23\angle -10.6^\circ A}$

(d) $V_1 = Z_1 I_1 = (j3.75)(0.97 - j0.18) = \underline{3.686\angle 79.38^\circ V}$

15.11 Cont.

(e) $V_2 = Z_2 I_2 = (-j)(1.23\angle -10.62^\circ)$

$= \underline{0.983\angle -100.62^\circ V}$

15.12 Z_1 is the same as in Prob.16.19

(a) $Z_{in} = \underline{20 + j3.75\,\Omega}$

(b) $I_1 = \underline{0.983\angle -10.62^\circ A}$

(c) $I_2 = \frac{(-j\omega M)}{Z_2 + j\omega L_2} = \underline{1.23\angle 169.4^\circ A}$

(d) $V_1 = \underline{3.686\angle 79.38^\circ V}$

(e) $V_2 = Z_2 I_2 = \underline{0.983\angle 79.38^\circ V}$

15.13 $Z_R = \frac{\omega^2 M^2}{Z_2 + j\omega L_2}$

$Z_2 + j\omega L_2 = 10 - j10/\omega + j\,\omega/2$

Z_R is real when $Z_2 + j\omega L_2$ is real.

$\therefore \frac{\omega}{2} - \frac{10}{\omega} = 0 \Rightarrow \omega^2 = 20$

$\omega = \underline{\sqrt{20}}$ rad/s

15.14

$$Z_2 = 1 + \frac{10(-j3)}{10+(-j3)} = 1 + \frac{-j30}{10-j3}$$

Reflected impedance

$$Z_R = \frac{\omega^2 M^2}{Z_2 + j\omega L_2} = \frac{(4)^2(\frac{1}{4})^2}{1 - \frac{j30}{10-j3} + j(4)(\frac{1}{4})} = \frac{1}{1 - \frac{j30}{10-j3} + 1}$$

$$= \frac{10-j3}{(1+j)(10-j3)-j30} = \frac{10-j3}{10+3+j7-j30} = \frac{10-j3}{13-j23}$$

$$I_1 = \frac{V_1}{Z_1} = \frac{12\angle 0^\circ}{1 + \frac{10-j3}{13-j23}} = \frac{13-j37}{23-j26} = \frac{39.22\angle -70.6^\circ}{34.7\angle -48.5^\circ}$$

$= 1.13\angle -22.14^\circ$

Hence $i_1 = \underline{1.13\cos(4t - 22.14^\circ)\,A}$

15.15 Open circuit secondary

$12 = (1+j)I_1 \Rightarrow I_1 = \frac{12}{1+j}$

$V_{oc} = jI_1 = \frac{12(1-j)j}{2} = 6(1+j)\,V$

15.15 Cont.

short circuit secondary

$$\begin{cases} 12 = (1+j)I_1 - jI_{sc} \\ 0 = -jI_1 + jI_{sc} \end{cases}$$

$$I_{sc} = \frac{\begin{vmatrix} 12 & 1+j \\ 0 & -j \end{vmatrix}}{\begin{vmatrix} -j & 1+j \\ j & -j \end{vmatrix}} = \frac{j12}{j} = 12\angle 0^\circ$$

$$Z_{th} = \frac{V_{oc}}{I_{sc}} = \frac{6(1+j)}{12} = \frac{1+j}{2}$$

$$I_2 = \frac{6(1+j)}{\frac{1+j}{2} + \left(1 - \frac{j30}{10-j3}\right)} = \frac{6(1+j)}{\frac{(1+j)(10-j3)+2(10-j3)-j60}{2(10-j3)}}$$

$$= \frac{12(1+j)(10-j3)}{13+j7+20-j3-j60} = \frac{12(13-j7)}{33-j56}$$

$$= \frac{12(13.9\angle -28.3^\circ)}{65\angle -59.5^\circ} = 2.6\angle 31.2^\circ$$

$$i_2(t) = \underline{2.6\cos(4t-31.2^\circ)}$$

15.16 KVL yields

$$\begin{cases} V_1 = (1+j\omega)I_1 - j\frac{\omega}{2}I_2 - j\frac{\omega}{2}I_3 & (1) \\ 0 = -j\frac{\omega}{2}I_1 + (1+j\omega)I_2 + j\frac{\omega}{2}I_3 & (2) \\ 0 = -j\frac{\omega}{2}I_1 + j\frac{\omega}{2}I_2 + (1+j\omega)I_3 & (3) \end{cases}$$

(1)+(2) $V_1 = (1+j\frac{\omega}{2})I_1 + (1+j\frac{\omega}{2})I_2$

(1)+(3) $V_1 = (1+j\frac{\omega}{2})I_1 + (1+j\frac{\omega}{2})I_3$

(1)+(2)+(3) $V_1 = I_1 + (1+j\omega)I_2 + (1+j\omega)I_3$

$$I_1 = \frac{\begin{vmatrix} V_1 & 1+j\frac{\omega}{2} & 0 \\ V_1 & 0 & 1+j\frac{\omega}{2} \\ V_1 & 1+j\omega & 1+j\omega \end{vmatrix}}{\begin{vmatrix} 1+j\frac{\omega}{2} & 1+j\frac{\omega}{2} & 0 \\ 1+j\frac{\omega}{2} & 0 & 1+j\frac{\omega}{2} \\ 1 & 1+j\omega & 1+j\omega \end{vmatrix}}$$

$$= \frac{V_1\left[(1+j\frac{\omega}{2})(1+j\frac{\omega}{2}) - 2(1+j\omega)(1+j\frac{\omega}{2})\right]}{(1+j\frac{\omega}{2})(1+j\frac{\omega}{2}) - (1+j\omega)(1+j\frac{\omega}{2})(1+j\frac{\omega}{2})}$$

$$= \frac{V_1(1+j\frac{\omega}{2})(1+j\frac{\omega}{2}-2-j2\omega)}{(1+j\frac{\omega}{2})(1+j\frac{\omega}{2}-2+\omega^2-j3\omega)}$$

$$= \frac{V_1(-1-j\frac{3}{2}\omega)}{-1+\omega^2-j\frac{5}{2}\omega}$$

$$Z_{in} = \frac{V_1}{I_1} = \frac{1+j\frac{3}{2}\omega}{1-\omega^2+j\frac{5}{2}\omega} = \underline{\frac{2+j3\omega}{2-2\omega^2+j5\omega}\ \Omega}$$

15.17

$$Z_1 = \frac{Z_2}{n^2} = \frac{90\times10^3}{n^2};$$

$$Z_g = Z_1 \Rightarrow 100 = \frac{90\times10^3}{n^2} \Rightarrow n = \underline{30}$$

$$I_1 = \frac{V_g}{Z_g+Z_1} = \frac{V_g}{2Z_g} = \frac{100}{2(100)} = 0.5\,A$$

$$I_2 = -\frac{I_1}{n} = \frac{-0.5}{30} = -\frac{1}{60}A$$

$$P_{Load} = |I_2|^2(90\times10^3) = \underline{25\,W}$$

15.18

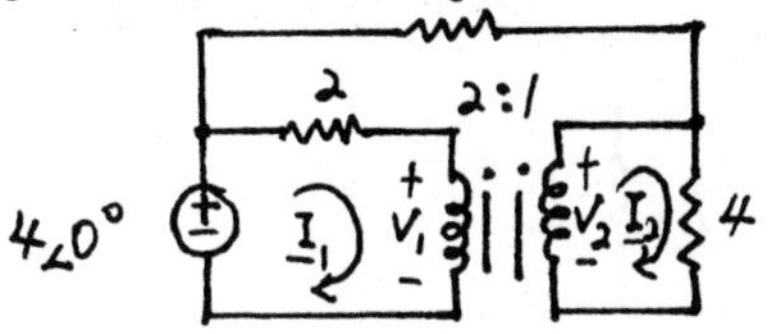

KVL: $4 = 2I_1 + V_1 = 2(2I_2) + \frac{V_2}{2}$ or

$V_2 + 8I_2 = 8$

KCL: $\frac{V_2-4}{8} - I_2 + \frac{V_2}{4} = 0$ or

$3V_2 - 8I_2 = 4$

Adding, $4V_2 = 12 \Rightarrow V_2 = 3V$

$$P_{8\Omega} = \frac{1}{2}\left|\frac{V_2-4}{8}\right|^2(8) = \underline{\frac{1}{16}\,W}$$

15.19

$$\underline{Z}_1 = \frac{Z_2}{n^2} = \frac{100-j75}{(5)^2} = 4-j3\,\Omega$$

$$\underline{I}_1 = \frac{V_g}{Z_g+Z_1} = \frac{12}{2+j3+4-j3} = \underline{2\angle 0^\circ A}$$

$$\underline{I}_2 = -\frac{I_1}{n} = \underline{-\tfrac{2}{5}\angle 0^\circ A}$$

$$\underline{V}_1 = \underline{Z}_1\underline{I}_1 = (4-j3)2 = \underline{10\angle -36.87^\circ V}$$

$$\underline{V}_2 = -nV_1 = \underline{50\angle 143.13^\circ V}$$

15.20

$$Z_g = \frac{Z_2}{n^2} \Rightarrow 4 = \frac{10000}{n^2}$$

$$n^2 = \frac{10^4}{4} \Rightarrow n = \underline{50}$$

15.21

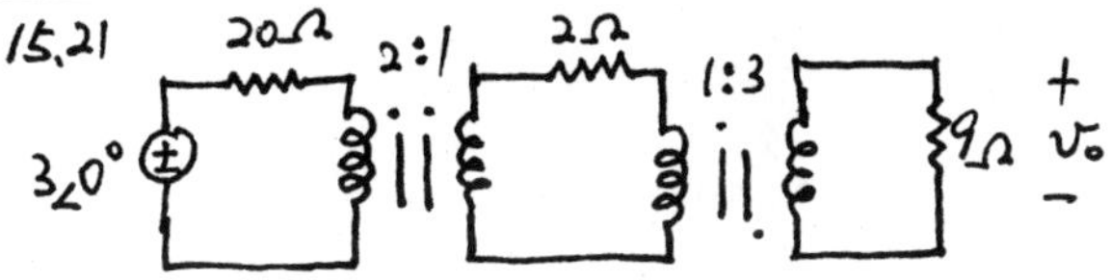

15.21 Cont.

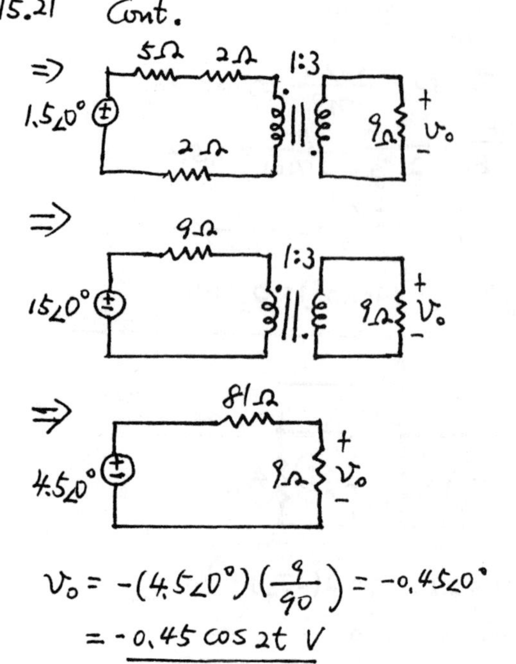

$$v_o = -(4.5\angle 0^\circ)\left(\frac{9}{90}\right) = -0.45\angle 0^\circ$$

$$= -0.45 \cos 2t \text{ V}$$

15.22 By the Thevenin equivalent

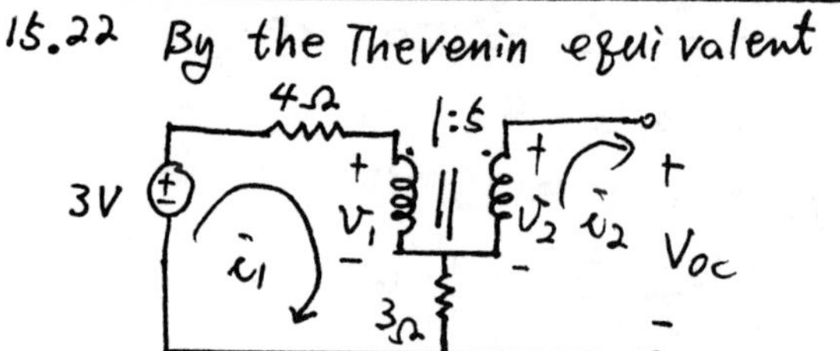

At DC steady state, inductors are short circuited.

$$V_{oc} = 3 \times \frac{3}{4+3} = \frac{9}{7} \text{ V}$$

By the Norton equivalent

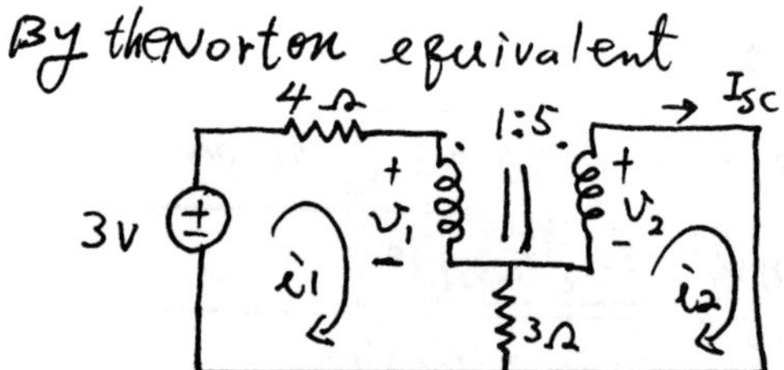

Inductors shorted.

$$I_{sc} = \frac{3}{4} \text{ A}$$

$$R_{th} = \frac{V_{oc}}{I_{sc}} = \frac{\frac{9}{7}}{\frac{3}{4}} = \frac{12}{7}\,\Omega$$

Hence to have maximum power delivered, the load is $\frac{12}{7}\,\Omega$

15.23

$$V_1 = \left(10+\frac{4}{s}\right)I_1 + 10V_c = \left(10+\frac{4}{s}\right)I_1 + 10\frac{4}{s}I_1 = \left(10+\frac{44}{s}\right)I_1$$

$$V_2 = 10V_c + (6+3s)I_2 = \frac{40}{s}I_1 + (6+3s)I_2$$

$$\Rightarrow z_{11} = 10 + \frac{44}{s}$$

$$z_{12} = 0$$

$$z_{21} = \frac{40}{s}$$

$$z_{22} = 6+3s$$

15.24 S-domain circuit

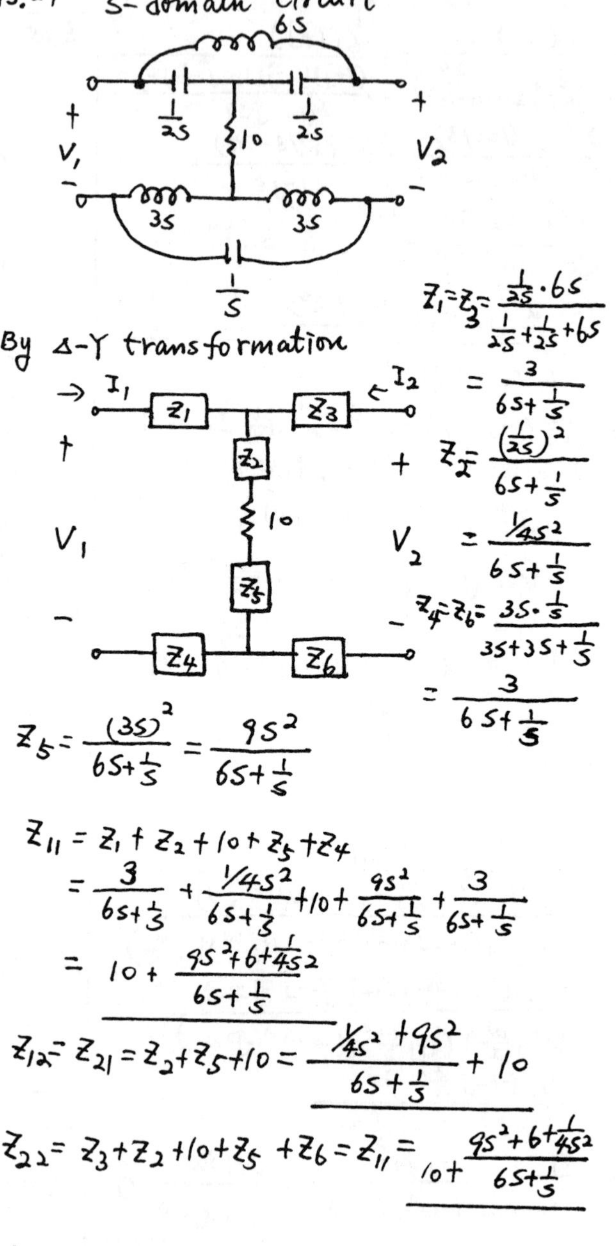

By Δ-Y transformation

$$Z_1 = Z_3 = \frac{\frac{1}{2s}\cdot 6s}{\frac{1}{2s}+\frac{1}{2s}+6s} = \frac{3}{6s+\frac{1}{s}}$$

$$Z_2 = \frac{\left(\frac{1}{2s}\right)^2}{6s+\frac{1}{s}} = \frac{\frac{1}{4s^2}}{6s+\frac{1}{s}}$$

$$Z_4 = Z_6 = \frac{3s\cdot\frac{1}{s}}{3s+3s+\frac{1}{s}} = \frac{3}{6s+\frac{1}{s}}$$

$$Z_5 = \frac{(3s)^2}{6s+\frac{1}{s}} = \frac{9s^2}{6s+\frac{1}{s}}$$

$$Z_{11} = Z_1 + Z_2 + 10 + Z_5 + Z_4$$

$$= \frac{3}{6s+\frac{1}{s}} + \frac{\frac{1}{4s^2}}{6s+\frac{1}{s}} + 10 + \frac{9s^2}{6s+\frac{1}{s}} + \frac{3}{6s+\frac{1}{s}}$$

$$= 10 + \frac{9s^2+6+\frac{1}{4s^2}}{6s+\frac{1}{s}}$$

$$Z_{12} = Z_{21} = Z_2 + Z_5 + 10 = \frac{\frac{1}{4s^2}+9s^2}{6s+\frac{1}{s}} + 10$$

$$Z_{22} = Z_3 + Z_2 + 10 + Z_5 + Z_6 = Z_{11} = 10 + \frac{9s^2+6+\frac{1}{4s^2}}{6s+\frac{1}{s}}$$

15.25 Find the y-parameters

$I_1 = (1)\left(\frac{1}{6}+\frac{1}{9}\right) = \frac{5}{18} = y_{11}$

$I_2 = \left(-\frac{6}{15}\right)\left(\frac{5}{18}\right) = -\frac{1}{3} = y_{21}$

$I_2 = \frac{1-(-2)}{9} + \frac{1}{3} = \frac{2}{3} = y_{22}$

$I_1 = \frac{-2-1}{9} = -\frac{1}{3} = y_{12}$

y-parameter matrix

$$\begin{bmatrix} y_{11} & y_{12} \\ y_{21} & y_{22} \end{bmatrix} = \begin{bmatrix} \frac{5}{18} & -\frac{1}{3} \\ -\frac{1}{3} & \frac{2}{3} \end{bmatrix}$$

$$\begin{bmatrix} V_1 \\ V_2 \end{bmatrix} = \begin{bmatrix} z_{11} & z_{12} \\ z_{21} & z_{22} \end{bmatrix}\begin{bmatrix} I_1 \\ I_2 \end{bmatrix} \Rightarrow \begin{bmatrix} z_{11} & z_{12} \\ z_{21} & z_{22} \end{bmatrix} = \begin{bmatrix} y_{11} & y_{12} \\ y_{21} & y_{22} \end{bmatrix}^{-1}$$

$$\begin{bmatrix} z_{11} & z_{12} \\ z_{21} & z_{22} \end{bmatrix} = \begin{bmatrix} \frac{5}{18} & -\frac{1}{3} \\ -\frac{1}{3} & \frac{2}{3} \end{bmatrix}^{-1} = \frac{\begin{bmatrix} \frac{2}{3} & \frac{1}{3} \\ \frac{1}{3} & \frac{5}{18} \end{bmatrix}}{\frac{5}{18}\times\frac{2}{3} - \frac{1}{3}\frac{1}{3}}$$

$$= \frac{\begin{bmatrix} \frac{2}{3} & \frac{1}{3} \\ \frac{1}{3} & \frac{5}{18} \end{bmatrix}}{\frac{2}{27}} = \frac{9}{2}\begin{bmatrix} 2 & 1 \\ 1 & \frac{5}{6} \end{bmatrix} = \begin{bmatrix} 9 & \frac{9}{2} \\ \frac{9}{2} & \frac{15}{4} \end{bmatrix}$$

15.26 $z_{11} - z_{12} = 6\,\Omega$, $z_{22} - z_{12} = 2\,\Omega$

$z_{12} = 2\,\Omega$

15.27 $I_2 = -\frac{z_{21}}{z_{22}} I_1 + \frac{1}{z_{22}} V_2$,

$V_1 = z_{11} I_1 + z_{12}\left(-\frac{z_{21}}{z_{22}} I_1 + \frac{1}{z_{22}} V_2\right)$

$= \frac{\Delta}{z_{22}} I_1 + \frac{z_{12}}{z_{22}} V_2$ where

$\Delta = z_{11} z_{22} - z_{12} z_{21}$. Comparing these with the relations for the h-parameters gives

$h_{11} = \frac{\Delta}{z_{22}} = \frac{28}{4} = 7\,\Omega$, $h_{12} = \frac{z_{12}}{z_{22}} = \frac{2}{4} = \frac{1}{2}$

15.27 Cont.

$h_{21} = -\frac{z_{21}}{z_{22}} = -\frac{1}{2}$, $h_{22} = \frac{1}{z_{22}} = \frac{1}{4}\,\mho$

$I_1 = \frac{1}{z_{11}} V_1 - \frac{z_{12}}{z_{11}} I_2$,

$V_2 = z_{21}\left(\frac{1}{z_{11}} V_1 - \frac{z_{12}}{z_{11}} I_2\right) + z_{22} I_2$

$= \frac{z_{21}}{z_{11}} V_1 + \frac{\Delta}{z_{11}} I_2$, By compare

$g_{11} = \frac{1}{z_{11}} = \frac{1}{8}\,\mho$, $g_{12} = \frac{-z_{12}}{z_{11}} = -\frac{2}{8} = -\frac{1}{4}$

$g_{21} = \frac{z_{21}}{z_{11}} = \frac{1}{4}$, $g_{22} = \frac{\Delta}{z_{11}} = \frac{28}{8} = \frac{7}{2}\,\Omega$

15.28 $z_{12} = \frac{1}{2s}$

$z_{11} = (1+s) + z_{12} = \frac{2s^2+2s+1}{2s}$

$z_{22} = s + z_{12} = \frac{2s^2+1}{2s}$

15.29 $\Delta = z_{11}z_{22} - z_{12}z_{21}$

$= \frac{(2s^2+2s+1)}{2s}\frac{(2s^2+1)}{2s} - \left(\frac{1}{2s}\right)\left(\frac{1}{2s}\right)$

$= \frac{2s^3+2s^2+2s+1}{2s}$

$h_{11} = \frac{\Delta}{z_{22}} = \frac{2s^3+2s^2+2s+1}{2s^2+1}\,\Omega$

$h_{12} = \frac{z_{12}}{z_{22}} = \frac{1}{2s^2+1}$, $h_{21} = -\frac{z_{21}}{z_{22}} = -\frac{1}{2s^2+1}$

$h_{22} = \frac{1}{z_{22}} = \frac{2s}{2s^2+1}\,\mho$

15.30 $I_1 = \frac{1}{z_{21}} V_2 - \frac{z_{22}}{z_{21}} I_2 = CV_2 - DI_2$

$V_1 = z_{11}\left(\frac{1}{z_{21}} V_2 - \frac{z_{22}}{z_{21}} I_2\right) + z_{12} I_2$

$= \frac{z_{11}}{z_{21}} V_2 - \frac{\Delta}{z_{21}} I_2 = AV_2 - BI_2$

$A = \frac{z_{11}}{z_{21}} = 2s^2+2s+1$

$B = \frac{\Delta}{z_{21}} = 2s^3+2s^2+2s+1\,\Omega$

$C = \frac{1}{z_{21}} = 2s\,\mho$

$D = \frac{z_{22}}{z_{21}} = 2s^2+1$

15.31 s-domain circuit

$$\begin{cases} V_3 = V_1 - 4I_1 \\ V_4 = V_2 - 2I_2 \end{cases}$$

$$\begin{cases} I_3 = I_1 - \frac{V_3 - V_4}{1/s} \Rightarrow I_3 - sV_3 = I_1 \\ I_4 = \frac{V_4 - V_3}{1/s} - I_2 \Rightarrow I_3 - 2sV_3 = -2I_2 \end{cases}$$

and $I_3 = 2I_4$

$V_4 = 2V_3$

$$I_3 = \frac{\begin{vmatrix} I_1 & -s \\ -2I_2 & -2s \end{vmatrix}}{\begin{vmatrix} 1 & -s \\ 1 & -2s \end{vmatrix}} = \frac{-2sI_1 - 2sI_2}{-2s+s} = \frac{-s(2I_1+2I_2)}{-s}$$

$$= 2I_1 + 2I_2$$

$$V_3 = \frac{\begin{vmatrix} 1 & I_1 \\ 1 & -2I_2 \end{vmatrix}}{-s} = \frac{-2I_2 - I_1}{-s} = \frac{2I_2 + I_1}{s}$$

$$\Rightarrow \begin{cases} V_1 - 4I_1 = \frac{I_1 + 2I_2}{s} \\ V_2 - 2I_2 = \frac{2I_1 + 4I_2}{s} \end{cases} \Rightarrow \begin{matrix} V_1 = (4+\frac{1}{s})I_1 + \frac{2}{s} I_2 \\ V_2 = \frac{2}{s} I_1 + (2+\frac{4}{s})I_2 \end{matrix}$$

$$Z = \begin{bmatrix} Z_{11} & Z_{12} \\ Z_{21} & Z_{22} \end{bmatrix} = \begin{bmatrix} 4+\frac{1}{s} & \frac{2}{s} \\ \frac{2}{s} & 2+\frac{4}{s} \end{bmatrix}$$

15.32

The loop equations are

$$V_1 = \frac{s}{10} I_1 + \frac{s}{10} I_2$$

$$V_2 = \frac{s}{10} I_1 + \left(\frac{s}{10} + \frac{25}{s}\right) I_2$$

$$\therefore Z_{11} = \frac{s}{10}, \quad Z_{12} = Z_{21} = \frac{s}{10}, \quad Z_{22} = \frac{s}{10} + \frac{25}{s}$$

The nodal equations are

$$I_1 = \left(\frac{10}{s} + \frac{s}{25}\right) V_1 - \frac{s}{25} V_2$$

$$I_2 = -\frac{s}{25} V_1 + \frac{s}{25} V_2$$

$$\therefore y_{11} = \frac{10}{s} + \frac{s}{25}, \quad y_{12} = y_{21} = -\frac{s}{25}, \quad y_{22} = \frac{s}{25}$$

15.33 $I_2 = 0$ then from Prob. 15.27

$$I_2 = 0 = h_{21} I_1 + h_{22} V_2 \Rightarrow I_1 = -\frac{h_{22}}{h_{21}} V_2$$

$$V_1 = h_{11}\left(-\frac{h_{22}}{h_{21}} V_2\right) + h_{22} V_2$$

$$\frac{V_2}{V_1} = \frac{h_{21}}{h_{12}h_{21} - h_{11}h_{22}} = \frac{200}{(200)10^{-3} - 3(10^3)(10^3)}$$

$$= -71.43$$

15.34

```
*INPUT IMPEDANCE SEEN
*BY THE VOLTAGE SOURCE
L_PRI     2 0 1
L_SEC1    3 0 1
L_SEC2    4 0 1
K_TX1     L_PRI L_SEC1 .5
K_TX2     L_PRI L_SEC2 .5
K_TX3     L_SEC1 L_SEC2 .5
R_R1      2 1 1
R_R2      3 0 1
R_R3      4 0 1
V_V1      1 0 DC 0 AC 1
*Control statement
.AC       LIN 2 0.6369 1.1146
.PRINT    AC IM(R_R1) IP(R_R1)
.END
****AC ANALYSIS
FREQ        IM(R_R1)   IP(R_R1)
6.369E-01 3.373E-01 1.142E+02
1.115E+00 2.064E-01 1.046E+02

*******************************
*INPUT IMPEDANCE IS DETERMINED
*******************************
Z=1/(IM(R_R1)*COS(IP(R_R1))+
           J*SIN(IP(R_R1)))
AT W=4 RAD/SEC (F=0.6369 HZ)
I=(.337*COS(114.2)+J*.337*SIN(114.2
Z=V/I= -1.21-J2.71

AT W=7 RAD/SEC (F=1.115HZ)
I=(.206*COS(104.6)+J*.206*SIN(1.046
Z=V/I= -1.22-J4.70
```

15.35

```
*PROB.15_35
*TRANSIENT ANS STEADY
*STATE CURRENTS OF
*TRANSFORMER CIRCUIT
L_PRI   1 0 0.25
L_SEC   2 0 0.25
K_TX1   L_PRI L_SEC 0.25
R_R1    0 3 10
R_R2    4 1 1
V_V1    4 0 DC 0 AC 12
+SIN (0 12 4 0 0 90)
C_C1    0 3 0.08333
R_R3    2 3 1
*Control statement
.TRAN   0.01 10
.END
***************************
*THE PLOTS OF TRANSIENT AND
*STEADY STATE CURRENT I(R_R2)
*ARE SHOW AS FOLLOWING:
***************************
```

Chapter 16
Fourier Series and Transform

16.1 Periodic functions

16.1 **Show that if $f_1(t)$ is periodic of period T_1 and f_2 is periodic of period T_2, then $f_1(t) + f_2(t)$ is also periodic of period T if positive integers K and L exist such that $T = KT_1 = LT_2$.**

16.2 **Use the result of Prob.16.1 to show (a) given $f(t) = \cos(t) + \sin(wt)$ with $\omega = \frac{6}{7}$ is a periodic function with periodic 14π, (b) $f(t)$ is not a periodic function with $\omega = \pi$.**

16.3 **Find the even and odd parts of the function $f(t) = \sin(7t + 60°)$**

16.2 The trigonometric Fourier series

16.4 **Find the trigonometric Fourier series for the function**

$$\begin{aligned} f(t) &= 4, \qquad -1 < t < 1 \\ f(t) &= -1, \qquad 1 < t < 3 \\ f(t+4) &= f(t) \end{aligned}$$

16.5 **Find the trigonometric Fourier series for the function**

$$\begin{aligned} f(t) &= -1, \qquad -1 < t < 0 \\ f(t) &= 2, \qquad 0 < t < 2 \\ f(t) &= -1, \qquad 2 < t < 3 \\ f(t+4) &= f(t) \end{aligned}$$

16.6 **Find the trigonometric Fourier series for the function**

$$\begin{aligned} f(t) &= 5, \qquad 0 < t < 2 \\ f(t) &= 0, \qquad 2 < t < 4 \\ f(t+4) &= f(t) \end{aligned}$$

By first translating the axes to move the origin to the point (1, 1) and using the result of Prob.16.4.

16.7 **Find the trigonometric Fourier series for the function with a period of 4, where**

$$\begin{aligned} f(t) &= 2, \qquad -2 < t < 0 \\ &= t, \qquad 0 < t < 2 \end{aligned}$$

16.8 **Find the trigonometric Fourier series for the function f(x) with a period of 2π, where $f(t) = t^2$, $-\pi < t < \pi$.**

16.9 **Find the trigonometric Fourier series for the function**

$$\begin{aligned} f(t) &= \cos(t^2), \qquad -\pi < t < \pi \\ f(t+2\pi) &= f(t). \end{aligned}$$

16.10 **Find the trigonometric Fourier series for the function $f(t) = t(2-t)$,$0 < t < 2$, where f(t) is an even function with period T=4.**

16.3 The exponential Fourier series

16.11 **Find the exponential Fourier series for the function f(t) where**

$$\begin{aligned} f(t) &= e^{-t} - 1, \qquad -1 < t < 0 \\ f(t) &= e^{t} - 1, \qquad 0 < t < 1 \\ f(t+2) &= f(t). \end{aligned}$$

16.12 **Find the exponential Fourier series for the function f(t) where**

$$\begin{aligned} f(t) &= 2\cos(2\pi t), \qquad 0 < t < 1 \\ f(t+1) &= f(t). \end{aligned}$$

16.13 **Find the exponential Fourier series for the function f(t) as shown.**

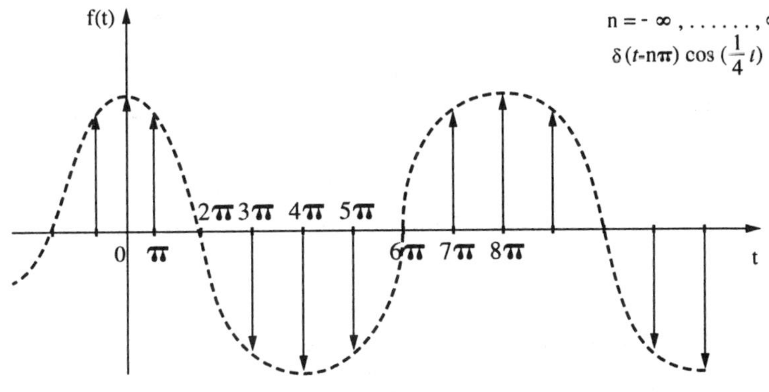

Problem 16.13

16.14 Find the exponential Fourier series for the function

$$\begin{aligned} f(t) &= 4, \qquad -1 < t < 1 \\ f(t) &= 0, \qquad 1 < |t| < 2 \\ f(t+4) &= f(t). \end{aligned}$$

16.15 The function $sinc(x) = \frac{\sin \pi x}{\pi x}$ is called the sinc function. Show that if one defines $c_o = \lim_{n \to 0} C_n$, then the result of Prob16.14 may be given by $f(t) = 2\sum_{n=-\infty}^{\infty} sinc(\frac{n}{2})e^{jn\pi t/2}$.

16.16 Find the exponential Fourier series for the function f(t) of Prob.16.9.

16.4 Response to periodic inputs

16.17 Find the forced component of the voltage across the inductor if $v(t) = f(t)$V from Prob.16.8.

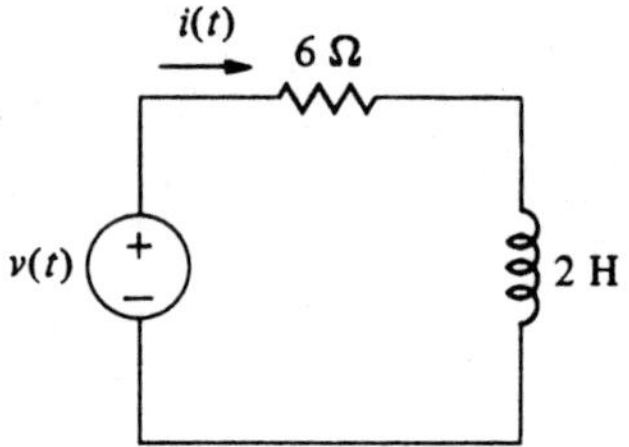

Problem 16.17

16.18 Find the forced component of the voltage across the capacitor if $v(t) = f(t)$V from Prob.16.10.

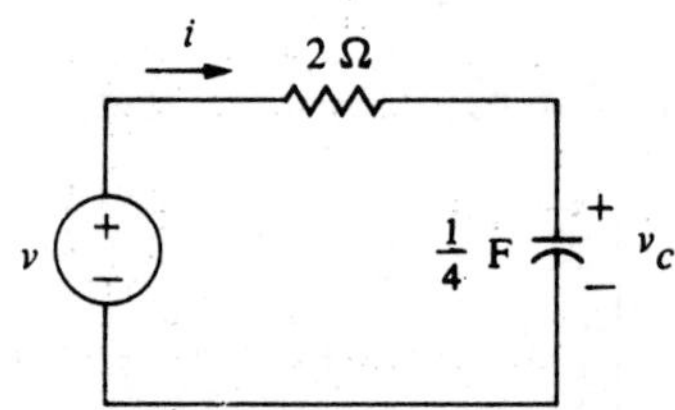

Problem 16.18

16.19 The periodic signal f(t) of period T=2 which equals $sinh(t)$ for $-1 < t < 1$ is passed through an ideal bandpass filter

$$\begin{aligned} H(j\omega) &= 1, \qquad 2 < |j\omega| < 4 \\ &= 0, \qquad elsewhere \end{aligned}$$

Find the forced response $g(t)$.

16.5 Discrete spectra and phase plots

16.20 Plot the discrete amplitude and phase spectra for Prob.16.8.

16.21 Suppose the waveform f(t) of Prob.16.16 is passed through a high passed filter with frequency response $H(j\omega) = \frac{j\omega}{1+j\omega^2}$. Find the discrete line spectrum and phase spectrum of the output g(t).

16.6 The Fourier transforms

16.22 Find the Fourier transform of the function

$$\begin{aligned} f(t) &= e^t, & 0 < t < 10 \\ &= 0, & elsewhere. \end{aligned}$$

16.23 Find the Fourier transform of the rectangular pulse

$$\begin{aligned} f(t) &= K, & -T < t < T \\ &= 0, & |t| \geq T \end{aligned}$$

Sketch the Fourier transform in ω domain.

16.24 Find the Fourier transform of the function

$$\begin{aligned} f(t) &= K \cos \frac{\pi}{T} t, & -T < t < T \\ &= 0, & elsewhere. \end{aligned}$$

16.25 Find the Fourier transform of the triangular pulse

$$\begin{aligned} f(t) &= K(\frac{t}{T} + 1), & -T < t < 0 \\ f(t) &= K(-\frac{t}{T} + 1), & 0 < t < T \\ &= 0, & elsewhere. \end{aligned}$$

16.26 Find the Fourier transform of the function

$$\begin{aligned} f(t) &= 2\cosh(t), & -2 < t < 2 \\ &= 0, & elsewhere. \end{aligned}$$

16.27 Find the Fourier transform of the function $f(t) = \cos(\pi t)u(t+5) - \cos(\pi t)u(t-5)$.

16.28 Using the properties of even and odd functions, find the Fourier transform of the function

$$\begin{aligned} f(t) &= |t|, \qquad -3 < t < 3 \\ &= 0, \qquad \textit{elsewhere.} \end{aligned}$$

16.29 Find the Fourier transform of the function (a)$\delta(t)$, (b)$\delta'(t)$.

16.30 Find the inverse Fourier transform of $F(\omega) = \frac{\sin\omega}{\omega} e^{-j\omega/2}$.

16.31 Find the inverse Fourier transform of $F(\omega) = -\frac{\omega^2}{1+j\omega}$.

16.7 Fourier transforms properities

16.32 If x(t) is the input and y(t) is the output, find the network function using Fourier transforms of $4y'' + 6y' + y = 2x$.

16.33 If $x(t) = e^{-(t-2)}u(t-2)$ in Prob.16.32, find $Y(j\omega)$.

16.34 If x(t) is the input and y(t) is the output, find the network function using Fourier transforms of $y'' + 3y' = 4x$.

16.35 If $x(t) = 1$, $-1/2 < t < 1/2$ in Prob.16.34, find $Y(j\omega)$.

16.8 SPICE and Fourier analysis

16.36 Using SPICE, find the Fourier coefficients for the waveform shown.

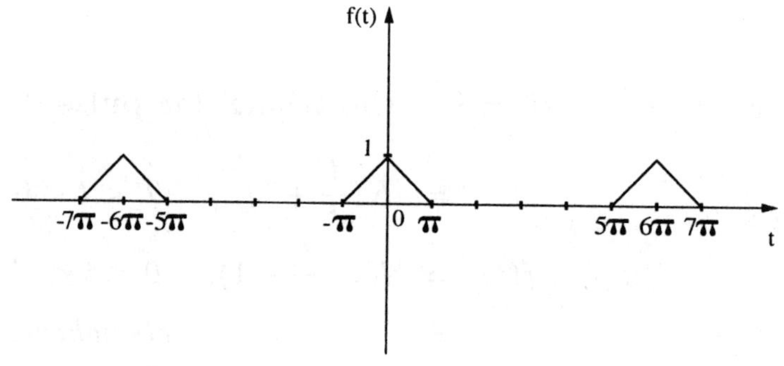

Problem 16.36

16.37 Using SPICE, repeat Prob.16.9 for the first 10 harmonics with $T = 2\pi$.

16.1 By definition $f_1(t+T_1)=f_1(t)$

$f_2(t+T_2)=f_2(t)$

$g(t)=f_1(t)+f_2(t)=f_1(t+T_1)+f_2(t+T_2)$

$=f_1(t+kT_1)+f_2(t+LT_2)=f_1(t+T)+f_2(t+T)$

$=g(t+T)$

Hence $f_1(t)+f_2(t)$ is periodic of period T if k and L are positive integers.

16.2 (a) $\cos t$ is periodic with period $T_1=\frac{2\pi}{\omega}=\frac{2\pi}{1}=2\pi$. $\sin\omega t$ is periodic with period $T_2=\frac{2\pi}{\omega}=\frac{2\pi}{6/7}=\frac{7}{3}\pi$.

Hence $f(t)=\cos t+\sin\frac{6}{7}t$ is periodic with $T=14\pi$

(b) $\sin\pi t$ is periodic with period $T_2=\frac{2\pi}{\pi}=2$

$\because$ There exists no positive integers k and L such that $2k=(2\pi)L$

$f(t)=\cos t+\sin\pi t$ is not periodic

16.3 $f_e(t)=\frac{f(t)+f(-t)}{2}=\frac{\sin(7t+60°)+\sin(\quad+60°)}{2}$

$=\frac{1}{2}[\cos 7t\sin 60°+\cos 7t\sin 60°]$

$=\frac{\sqrt{3}}{2}\cos 7t$

$f_o(t)=\frac{f(t)-f(-t)}{2}=\frac{2\sin 7t\cos 60°}{2}=\frac{1}{2}\sin 7t$

16.4 $T=4,\ \omega_0=2\pi/T=\pi/2$

$a_0=\frac{2}{4}\int_{-1}^{1}4dt+\frac{2}{4}\int_{1}^{3}(-1)dt=\frac{2}{4}(8-2)=3$

$a_n=\frac{2}{4}\int_{-1}^{1}4\cos\left(\frac{n\pi t}{2}\right)dt+\frac{2}{4}\int_{1}^{3}(-1)\cos\left(\frac{n\pi t}{2}\right)dt$

$=\frac{2}{4}\left\{4\left(\frac{2}{n\pi}\right)\sin\left(\frac{n\pi t}{2}\right)\Big|_{-1}^{1}+(-1)\left(\frac{2}{n\pi}\right)\sin\left(\frac{n\pi}{2}t\right)\Big|_{1}^{3}\right\}$

$=\frac{2}{4}\left\{\frac{16}{n\pi}\sin\frac{n\pi}{2}-\frac{2}{n\pi}\left[\sin\frac{n\pi}{2}\cos n\pi-\sin\frac{n\pi}{2}\right]\right\}$

It is evident that there is no even homonics.

For n is odd $\Rightarrow a_n=\frac{40}{}\sin\frac{n\pi}{2}=\frac{10}{n\pi}\sin\frac{n\pi}{2}$

$b_n=0$ since $f(t)$ is an even function.

$f(t)=\frac{3}{2}+\frac{10}{\pi}\left(\cos\frac{\pi t}{2}-\frac{1}{3}\cos\frac{3\pi t}{2}+\frac{1}{5}\cos\frac{5\pi t}{2}-\cdots\right)$

$=\frac{3}{2}+\frac{10}{\pi}\sum_{n=1}^{\infty}\frac{(-1)^{2n+1}\cos\{[(2n-1)\pi t]/2\}}{2n-1}$

16.5 $T=4,\ \omega_0=\frac{2\pi}{4}=\pi/2$

$a_0=2\left(\frac{2}{4}\right)=1\Rightarrow a_0/2=\frac{1}{2}$

$\because f(t)$ is an even function (if x-axis shifted by $1/2$) $\Rightarrow a_n=0$.

$b_n=\frac{2}{4}\int_{-1}^{3}f(t)\sin\frac{n\pi}{2}t\,dt$

$=\frac{2}{4}\left[\int_{-1}^{0}(-1)\sin\frac{n\pi t}{2}dt+\int_{0}^{2}(2)\sin\frac{n\pi}{2}t\,dt\right.$

$\left.+\int_{2}^{3}(-1)\cos\frac{n\pi}{2}t\,dt\right]$

$=\frac{2}{4}\frac{2}{n\pi}\left\{\cos\frac{n\pi t}{2}\Big|_{-1}^{0}-(2)\cos\frac{n\pi t}{2}\Big|_{0}^{2}+\cos\frac{n\pi t}{2}\Big|_{2}^{3}\right\}$

$=\frac{1}{n\pi}\left\{\left(1-\cos\frac{n\pi}{2}\right)-(2)(\cos n\pi-1)+\left(\cos\frac{3n\pi}{2}-\cos n\pi\right)\right\}$

$=\frac{1}{n\pi}\left\{3-\cos\frac{n\pi}{2}-3\cos n\pi+\cos\frac{3n\pi}{2}\right\}$

$=\frac{1}{n\pi}\{3-3\cos n\pi\}=\frac{3}{n\pi}(1-\cos n\pi)$

Hence $f(t)=\frac{1}{2}+\frac{3}{\pi}\sum_{n=1}^{\infty}\frac{1}{n}\left[(1-\cos n\pi)\sin\frac{n\pi t}{2}\right]$

16.6 From the result of Prob. 16.4

$f(t')=\frac{3}{2}+\frac{10}{\pi}\left(\cos\frac{\pi t'}{2}-\frac{1}{3}\cos\frac{3\pi t'}{2}+\frac{1}{5}\cos\frac{5\pi t'}{2}-\cdots\right)$

since the origin is shift from $(0,0)\to(1,1)$

$f(t)=\left(\frac{3}{2}+1\right)+\frac{10}{\pi}\left[\cos\frac{\pi}{2}(t-1)-\frac{1}{3}\cos\frac{3\pi}{2}(t-1)+\cdots\right]$

$=\frac{5}{2}+\frac{10}{\pi}\left[\sin\frac{\pi t}{2}+\frac{1}{3}\sin\frac{\pi t}{2}+\frac{1}{5}\sin\frac{5\pi t}{2}+\cdots\right]$

16.7 $T=4,\ \omega_0=\frac{2\pi}{T}=\frac{\pi}{2}$

$a_0=\frac{2}{4}\left(\int_{-2}^{0}2dt+\int_{0}^{2}t\,dt\right)$

$=\frac{1}{2}\left[2t\Big|_{-2}^{0}+\frac{t^2}{2}\Big|_{0}^{2}\right]=\frac{1}{2}(4+2)=3.$

$a_n=\frac{2}{T}\int_{t_0}^{t_0+T}f(t)\cos n\omega_0t\,dt$

$=\frac{1}{2}\left[\int_{-2}^{0}2\cos n\omega_0t\,dt+\int_{0}^{2}t\cos\omega_0t\,dt\right]$

$=\frac{1}{2}\left[\frac{\cos n\pi-1}{(n\omega_0)^2}\right]=\begin{cases}-\frac{4}{n^2\pi^2}, & n\text{ odd}\\ 0 & n\text{ even}\end{cases}$

$b_n=\frac{2}{T}\int_{t_0}^{t_0+T}f(t)\sin n\omega_0t\,dt$

$=\frac{1}{2}\left[\int_{-2}^{0}2\sin n\omega_0t\,dt+\int_{0}^{2}t\sin n\omega_0t\,dt\right]$

$=\frac{1}{2}\left[-\frac{2}{n\omega_0}(1-\cos n\pi)-\frac{2}{n\omega_0}\cos n\pi\right]$

$=-\frac{2}{n\pi}$

$f(t)=\frac{3}{2}+\frac{2}{\pi}\sum_{n=1}^{\infty}\left(\frac{(-1)^n-1}{n^2\pi^2}\cos\frac{n\pi}{2}t-\frac{1}{n}\sin\frac{n\pi t}{2}\right)$

16.8 $T=2\pi,\ \omega_0=\frac{2\pi}{T}=1$

$a_0=\frac{2}{2\pi}\left[\int_{-\pi}^{\pi}t^2dt\right]=\frac{2}{3}\pi^2$

$a_n=\frac{2}{2\pi}\left[\int_{-\pi}^{\pi}t^2\cos nt\,dt\right]$

$=\frac{1}{\pi}\left[\frac{2\pi\cos n\pi}{n^2}+\frac{2\pi}{n^2}\cos n\pi\right]=\frac{4}{n^2}(-1)^n$

$b_n=\frac{1}{\pi}\int_{-\pi}^{\pi}t^2\sin nt=0$

$f(t)=\frac{\pi^2}{3}+4\sum_{n=1}^{\infty}\frac{(-1)^n}{n^2}\cos nt$

16.9 $T=2\pi,\ \omega_0=\frac{2\pi}{T}=1$

$\because f(t)=f(-t)$. $f(t)$ is an even ft.

$b_n=0$.

$a_n=\frac{4}{T}\int_0^{\pi}f(t)\cos n\omega_0 t\,dt \quad ,n=0,1,2,\cdots$

$=\frac{4}{2\pi}\int_0^{\pi}\cos t^2\cos nt\,dt$

$=\frac{4}{2\pi}\int_0^{\pi}\frac{1}{2}[\cos(t^2+nt)+\cos(t^2-nt)]dt$

$=\frac{1}{\pi}\left\{\frac{1}{2\pi+n}\sin(t^2+nt)\Big|_0^{\pi}+\frac{1}{2\pi-n}\sin(t^2-nt)\Big|_0^{\pi}\right\}$

$=\frac{1}{\pi}\left\{\frac{1}{2\pi+n}\sin(\pi^2+n\pi)+\frac{1}{2\pi-n}\sin(\pi^2-n\pi)\right\}$

$=\frac{1}{\pi}\left\{\frac{1}{2\pi+n}\sin\pi^2\cos n\pi+\frac{1}{2\pi-n}\sin\pi^2\cos n\pi\right\}$

$=\frac{1}{\pi}\left\{\frac{(-1)^n}{2\pi+n}\sin\pi^2+\frac{(-1)^n}{2\pi-n}\sin\pi^2\right\}$

$=\frac{1}{\pi}\left[\frac{4\pi}{(2\pi+n)(2\pi-n)}\right](-1)^n\sin\pi^2$

$=\frac{4(-1)^n}{4\pi^2-n^2}\sin\pi^2$

$f(t)=\sum_{n=0}^{\infty}\frac{4(-1)^n}{4\pi^2-n^2}\sin\pi^2\cos nt \quad n=0,1,2,\cdots$

16.10 $T=4,\ \omega_0=\frac{2\pi}{4}=\frac{\pi}{2}$,

Since $f(t)$ is even, $b_n=0$.

$a_0=\frac{4}{4}\int_0^2 t(2-t)\,dt=4/3$.

$a_n=\int_0^2 t(2-t)\cos\frac{n\pi}{2}t\,dt$

$=\frac{8}{n^2\pi^2}\cos n\pi-1)-\frac{16}{n^2\pi^2}\cos n\pi$

$=\begin{cases}0 & n \text{ odd}\\ \frac{-16}{n^2\pi^2}, & n \text{ even.}\end{cases}$

$f(t)=\frac{2}{3}-\frac{16}{\pi^2}\sum_{n=1}^{\infty}\frac{\cos(2n)\pi/2\,t}{(2n)^2}$

$=\frac{2}{3}-\frac{4}{\pi^2}\sum_{n=1}^{\infty}\frac{\cos n\pi t}{n^2}$

16.11 $T=2,\ \omega_0=2\pi/2=\pi$

$C_n=\frac{1}{2}\left[\int_{-1}^{0}(e^{-t}-1)e^{-jn\pi t}dt+\int_0^1(e^{t}-1)e^{-jn\pi t}dt\right]$

$=\frac{1}{2}\left[\frac{-2+e(e^{jn\pi}+e^{-jn\pi})}{1+n^2\pi^2}\right]$

$=\frac{-1+e\cos n\pi}{1+n^2\pi^2},\quad n\neq 0.$

$C_0=\frac{1}{2}\left[\int_{-1}^{0}(e^{-t}-1)dt+\int_0^1(e^{t}-1)dt\right]=e-2$

$f(t)=e-2+\sum_{\substack{n=-\infty\\ n\neq 0}}^{\infty}\frac{e\cos n\pi-1}{1+n^2\pi^2}e^{-jn\pi t}$

16.12 $T=1,\ \omega_0=2\pi$

$C_n=\int_0^1 2\cos 2\pi t\, e^{-jn2\pi t}dt$

$=\int_0^1(e^{j2\pi t}+e^{-j2\pi t})e^{-jn2\pi t}dt$

$=\frac{e^{j2\pi(1-n)}-1}{j2\pi(1-n)}+\frac{1-e^{-j2\pi(n+1)}}{j2\pi(n+1)}$

$=0$ when $n\neq 1,-1$.

By L'Hosiptol rule for $n=1$,

$\lim_{n\to 1}\left[\frac{d/dn\,(e^{j2\pi(1-n)})}{d/dn\,(j2\pi(1-n))}\right]=1$

$\lim_{n\to -1}\left[\frac{d/dn\,(e^{j2\pi(1+n)})}{d/dn\,(j2\pi(1+n))}\right]=1$

$f(t)=e^{-j2\pi t}+e^{j2\pi t}=2\cos 2\pi t$

16.13 $T=\frac{2\pi}{\omega_0}=\frac{2\pi}{1/4}=8\pi$

$C_m=\frac{1}{8\pi}\int_0^{8\pi}\left[\sum_{n=\infty}^{7}\delta(t-n\pi)\cos\frac{t}{4}\right]e^{-jm\omega_0 t}dt,\ m=0,1,2\cdots$

$=\frac{1}{8\pi}\int_0^{8\pi}\sum_{n=0}^{7}\delta(t-n\pi)\cos(\frac{t}{4})e^{-j\frac{mt}{4}}dt,\ m=0,1,2,\cdots$

$=\frac{1}{8\pi}\sum_{n=0}^{7}\cos\frac{n\pi}{4}e^{-j\frac{nm\pi}{4}}$

$=\sum_{m=-\infty}^{\infty}C_m e^{jmt/4}$

$=\frac{1}{8\pi}\sum_{m=-\infty}^{\infty}\sum_{n=0}^{7}\cos\frac{n\pi}{4}e^{\frac{jm}{4}(t-n\pi)}$

16.14 $T=4,\ \omega_0=2\pi/T=\pi/2$

$C_n=\frac{1}{4}\int_{-1}^{3}f(t)e^{-jn\omega_0 t}dt$

$=\frac{1}{4}\left[\int_{-1}^{1}4e^{-j\frac{n\pi t}{2}}dt+\int_1^3 0\,dt\right]$

$=-\frac{2}{jn\pi}e^{-j\frac{n\pi t}{2}}\Big|_{-1}^{1}$

$=\frac{2}{jn\pi}(e^{-j\frac{n\pi}{2}}-e^{j\frac{n\pi}{2}})$

$=-\frac{2}{jn\pi}\left[-j(2\sin\frac{n\pi}{2})\right]=\frac{4}{\pi}\frac{\sin\frac{n\pi}{2}}{n}$

16.14 Cont.

at $n=0$, $C_0 = 4\dfrac{\sin\frac{n\pi}{2}}{n\pi}\Big|_{n=0} = 2\lim_{n\to 0}\text{Sinc}\left(\frac{n}{2}\right) = 2$

$$f(t) = 2 + \frac{4}{\pi}\sum_{\substack{n=-\infty \\ n\neq 0}}^{\infty} \frac{\sin(n\pi/2)}{n} e^{jn\pi t/2}$$

16.15 From the result of Prob. 16.14

$$f(t) = 2 + \frac{4}{\pi}\sum_{\substack{n=-\infty \\ n\neq 0}}^{\infty} \frac{\sin(n\pi/2)}{n} e^{jn\pi t/2}$$

$$= 2 + 2\sum_{\substack{n=-\infty \\ n\neq 0}}^{\infty} \frac{\sin(n\pi/2)}{\pi n/2} e^{jn\pi t/2}$$

$$= 2 + 2\sum_{\substack{n=-\infty \\ n\neq 0}}^{\infty} \text{Sinc}\left(\frac{n}{2}\right) e^{jn\pi t/2}$$

$C_0 = \lim_{n\to 0} C_n = \lim_{n\to 0} \text{Sinc}\left(\frac{n}{2}\right) = 1$

$\Rightarrow 2 = 2C_0$

$$\Rightarrow f(t) = 2C_0 + 2\sum_{\substack{n=-\infty \\ n\neq 0}}^{\infty} \text{Sinc}\left(\frac{n}{2}\right) e^{jn\pi t/2}$$

$$= 2\sum_{n=-\infty}^{\infty} \text{Sinc}\left(\frac{n}{2}\right) e^{jn\pi t/2}$$

16.16 $f(t) = \cos(t^2) \qquad -\pi < t < \pi$

$f(t+2\pi) = f(t)$

$T = 2\pi \Rightarrow \omega_0 = \frac{2\pi}{T} = 1$

$$C_n = \frac{1}{2\pi}\int_{-\pi}^{\pi} \cos(t^2) e^{-jnt} dt, \quad n = 0, 1, 2, \ldots$$

$$= \frac{1}{2\pi}\int_{-\pi}^{\pi} \frac{e^{jt^2} + e^{-jt^2}}{2} e^{-jnt} dt$$

$$= \frac{1}{4\pi}\int_{-\pi}^{\pi}\left[e^{j(t^2 - nt)} + e^{-j(t^2+nt)}\right] dt$$

$$= \frac{1}{4\pi}\left[\frac{1}{j(2t-n)} e^{j(t^2-nt)}\Big|_{-\pi}^{\pi} - \frac{1}{j(2t+n)} e^{-j(t^2+n\pi)}\Big|_{-\pi}^{\pi}\right]$$

$$= \frac{1}{j4\pi}\left\{\frac{1}{2\pi-n} e^{j(\pi^2-n\pi)} - \frac{1}{-2\pi-n} e^{j(\pi^2+n\pi)} - \left[\frac{1}{2\pi+n} e^{-j(\pi^2+n\pi)} - \frac{1}{-2\pi+n} e^{-j(\pi^2-n\pi)}\right]\right\}$$

$$= \frac{1}{j4\pi}\left\{\frac{1}{2\pi-n}\left[e^{j(\pi^2-n\pi)} - e^{-j(\pi^2-n\pi)}\right] + \frac{1}{2\pi+n}\left[e^{j(\pi^2+n\pi)} - e^{-j(\pi^2+n\pi)}\right]\right\}$$

$$= \frac{1}{j(4\pi)}\left\{\frac{1}{2\pi-n}\left[j(2)\sin(\pi^2-n\pi)\right] + \frac{1}{2\pi+n}\left[j(2)\sin(\pi^2+n\pi)\right]\right\}$$

$$= \frac{2}{4\pi}\left[\frac{1}{2\pi-n}\sin\pi^2\cos n\pi + \frac{1}{2\pi+n}\sin\pi^2\cos n\pi\right]$$

$$= \frac{2(-1)^n}{4\pi^2 - n^2}\sin\pi^2$$

$$f(t) = \sum_{n=-\infty}^{\infty} \frac{2(-1)^n}{4\pi^2 - n^2}\sin\pi^2 \cos nt.$$

16.17 $v(t) = \pi^2/3 + \sum_{n=1}^{\infty} \frac{4(-1)^n}{n^2}\cos nt$

$v_n = \frac{4(-1)^n}{n^2}\cos nt \Rightarrow \underline{V}_n = \frac{4(-1)^n}{n^2}\angle 0°$,

$\omega_n = n$

$$\underline{V}_2 = \frac{j2n}{6+j2n}(\underline{V}_n) = \frac{4(-1)^n(n+j3)}{n^2(9+n^2)}$$

$$= \frac{4(-1)^n}{n(n^2+9)^{1/2}}\angle\tan^{-1}(3/n)$$

$$v_2 = \sum_{n=-\infty}^{\infty} v_2(n)$$

$$= \sum_{n=1}^{\infty} \frac{4(-1)^n}{n(n^2+9)^{1/2}}\cos\left[nt + \tan^{-1}\left(\frac{3}{n}\right)\right] \text{ V}$$

16.18 $v(t) = 2/3 - 4/\pi^2 \sum_{n=1}^{\infty} \frac{\cos n\pi t}{n^2}$

$v_0 = 2/3$ (capacitor open circuit)

$v_n = -\frac{4}{\pi^2}\frac{\cos n\pi t}{n^2} \Rightarrow \underline{V}_n = -\frac{4}{\pi^2 n^2}\angle 0$

$\omega_n = n\pi$

$$\underline{V}_c = \frac{-j\,4/n\pi}{2 - j\,4/n\pi}(\underline{V}_n) = \frac{8(-2+jn\pi)}{n\pi(n^2\pi^2+4)}$$

$$= \frac{8}{n\pi(4+n^2\pi^2)^{1/2}}\angle\tan^{-1}\left(\frac{n\pi}{-2}\right)$$

$$v_c(t) = \frac{2}{3} + \sum_{n=1}^{\infty} \frac{8\cos\left[n\pi t + \tan\left(-\frac{n\pi}{2}\right)\right]}{n\pi(\qquad + 4)^{1/2}} \text{ V}$$

16.19 $T = 2$, $\omega_0 = 2\pi/T = 2\pi/2 = \pi$

$\sin ht = \frac{e^t - e^{-t}}{2}$

$$C_n = \frac{1}{2}\int_{-1}^{1} \frac{e^t - e^{-t}}{2} e^{jn\pi t} dt$$

$$= \frac{1}{4}\left[\int_{-1}^{1} e^{t(1-jn\pi)} dt + \int_{-1}^{1} e^{t(-1-jn\pi)} dt\right]$$

$$= \frac{1}{4}\left\{\frac{1}{1-jn\pi} e^{t(1-jn\pi)}\Big|_{-1}^{1} - \frac{1}{-1-jn\pi} e^{t(-1-jn\pi)}\Big|_{-1}^{1}\right\}$$

$$= \frac{1}{4}\left\{\frac{1}{1-jn\pi}\left[e^{(1-jn\pi)} - e^{-1+jn\pi}\right] + \frac{1}{1+jn\pi}\left[e^{-1-jn\pi} - e^{1+jn\pi}\right]\right\}$$

∵ The ideal banpass filter has a passband of $2 < |j\omega| < 4$ only the frequency component of $n = \pm 1$ will pass the filter without attenuation while others will be blocked.

$$\Rightarrow g(t) = \frac{1}{4}\left\{\frac{1}{1-j\pi}\left[e^{1-j\pi} - e^{-1+j\pi}\right] + \frac{1}{1+j\pi}\left[e^{-1-j\pi} - e^{1+j\pi}\right]\right\} e^{j\pi t}$$

$$+ \frac{1}{4}\left\{\frac{1}{1+j\pi}\left[e^{1+j\pi} - e^{-1-j\pi}\right] + \frac{1}{1-j\pi}\left[e^{-1+j\pi} - e^{1-j\pi}\right]\right\} e^{-j\pi t}$$

$$= \frac{\pi(e - e^{-1})}{1+\pi^2}\sin\pi t$$

16.20 $C_n = \frac{a_n - jb_n}{2} = \frac{4(-1)^n}{n^2/2} = \frac{2(-1)^n}{n^2}\angle 0^\circ$

$C_0 = \frac{\pi^2}{3}$

n	$\lvert C_n \rvert$	ϕ_n
0	3.29	0°
−1,1	2.0	180°
−2,2	0.5	0°
−3,3	0.22	180°
−4,4	0.125	0°

$\lvert C_n \rvert$

−4 −3 −2 −1 0 1 2 3 4 n

ϕ_n : 180°

−4 −3 −2 −1 0 1 2 3 4 n

16.21 From the result of Prob. 16.16

$f(t) = \sum_{n=-\infty}^{\infty} \frac{2(-1)^n}{4\pi^2 - n^2} \sin\pi^2 \cos nt$

$g(t) = \sum_{n=-\infty}^{\infty} \frac{jn}{1+jn^2} \frac{2(-1)^n}{4\pi^2 - n^2} \sin\pi^2 \cos nt$

$= \sum_{n=-\infty}^{\infty} \frac{n\angle 90^\circ}{\sqrt{1+n^2}\angle \tan^{-1} n^2} \frac{2(-1)^n}{4\pi^2 - n^2} \sin\pi^2 \cos nt$

$\Rightarrow \lvert C_n \rvert = \frac{n}{\sqrt{1+n^2}} \frac{2}{4\pi^2 - n^2} \sin\pi^2$

$= \frac{2n}{\sqrt{1+n^2}(4\pi^2 - n^2)} \sin\pi^2$

$\angle C_n = 90^\circ - \angle \tan^{-1} n^2$

16.22 $F(j\omega) = \int_0^{10} e^t e^{-j\omega t}\, dt$

$= \frac{e^{(1-j\omega)10}}{1-j\omega} - \frac{1}{1-j\omega}$

$= \frac{e^{10-j10\omega} - 1}{1-j\omega}$

16.23 $F(j\omega) = \int_{-\infty}^{\infty} f(t) e^{-j\omega t}\, dt$

$= k\int_{-T}^{T} e^{-j\omega t}\, dt$

$= k\left(-\frac{1}{j\omega} e^{-j\omega t}\Big|_{-T}^{T}\right)$

$= k\left(\frac{1}{j\omega}\right)\left(-e^{-j\omega T} + e^{j\omega T}\right)$

$= k\left(\frac{1}{j\omega}\right) j2\sin\omega T$

$= 2kT\left(\frac{\sin\omega T}{\omega T}\right)$

16.23 Cont.

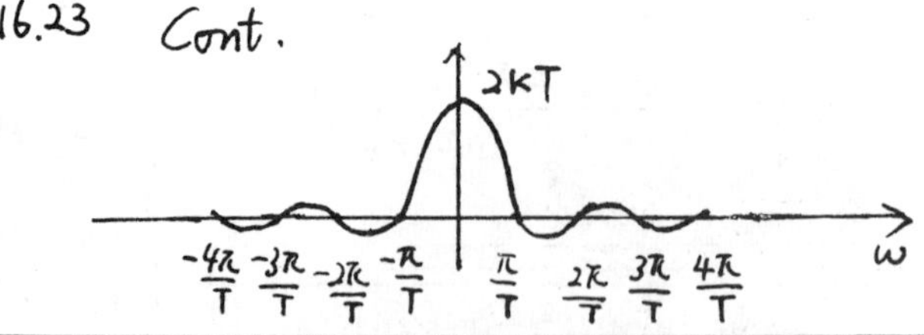

16.24 $F(j\omega) = k\int_{-\infty}^{\infty} \frac{e^{j\frac{\pi}{T}t} - e^{-j\frac{\pi}{T}t}}{2j} e^{-j\omega t}\, dt$

$= \frac{k}{2j}\left\{\int_{-T}^{T} e^{jt(\frac{\pi}{T}-\omega)}\, dt - \int_{-T}^{T} e^{-jt(\frac{\pi}{T}+\omega)}\, dt\right\}$

$= \frac{k}{2j}\left\{\frac{1}{j(\frac{\pi}{T}-\omega)} e^{jt(\frac{\pi}{T}-\omega)}\Big|_{-T}^{T} + \frac{1}{j(\frac{\pi}{T}+\omega)} e^{-jt(\frac{\pi}{T}+\omega)}\Big|_{-T}^{T}\right\}$

$= \frac{k}{2j}\left[\frac{1}{j(\frac{\pi}{T}-\omega)} e^{j(\pi-\omega T)} - \frac{1}{j(\frac{\pi}{T}-\omega)} e^{j(-\pi+\omega T)} + \frac{1}{j(\frac{\pi}{T}+\omega)} e^{-j(\pi+\omega T)} - \frac{1}{j(\frac{\pi}{T}+\omega)} e^{j(\pi+\omega T)}\right\}$

$= \frac{k}{2j}\left\{e^{-j\omega T}\left(\frac{j\frac{2\pi}{T}}{j(\frac{\pi^2}{T^2}-\omega^2)}\right) - e^{j\omega T}\left(\frac{j\frac{2\pi}{T}}{j(\frac{\pi^2}{T^2}-\omega^2)}\right)\right\}$

$= \frac{k}{2j}\left[\frac{\frac{2\pi}{T}}{(\frac{\pi^2}{T^2}-\omega^2)} j2\sin\omega T\right]$

$= \frac{2k\pi/T}{(\frac{\pi^2}{T^2}-\omega^2)} \sin\omega T$

16.25 $F(j\omega) = \int_{-\infty}^{\infty} f(t) e^{-j\omega t}\, dt$

$= \int_{-T}^{0} k\left(\frac{t}{T}+1\right) e^{-j\omega t} + \int_{0}^{T} k\left(\frac{-t}{T}+1\right) e^{-j\omega t}\, dt$

$= Tk\left[\frac{2 - (e^{j\omega T} + e^{-j\omega T})}{(\omega T)^2}\right]$

$= 2Tk \frac{2\sin^2\frac{\omega T}{2}}{\omega^2 T^2}$

$= Tk\left(\frac{\sin\frac{\omega T}{2}}{\frac{\omega T}{2}}\right)^2$

16.26 $F(j\omega) = \int_{-2}^{2} 2\cosh t\, e^{-j\omega t}\, dt$

$= \int_{-2}^{2} (e^t + e^{-t}) e^{-j\omega t}\, dt$

$= \frac{e^{(1-j\omega)2} - e^{-(1-j\omega)2}}{1-j\omega} + \frac{e^{(1+j\omega)2} - e^{-(1+j\omega)2}}{1+j\omega}$

$= \frac{2}{1-j\omega}\sinh 2(1-j\omega) + \frac{2}{1+j\omega}\sinh 2(1+j\omega)$

$= \frac{4\sinh 2\cos 2\omega + 4\omega\cosh 2\sin 2\omega}{1-\omega^2}$

16.27 $F(j\omega) = \int_{-5}^{5} \cos\pi t \; e^{-j\omega t} dt$

$= \frac{2\omega}{\pi^2 \omega^2}\left(\frac{e^{j\omega 5} - e^{-j\omega 5}}{2j}\right)$

$= \underline{\frac{2\omega}{\pi^2 - \omega^2} \sin 5\omega}$

16.28 Since f(t) is even

$F(j\omega) = 2\int_0^3 t \cos\omega t \, dt$

$= 2\left[\frac{\cos 3\omega}{\omega^2} + \frac{3\sin 3\omega}{\omega} - \frac{1}{\omega^2}\right]$

$= \underline{2\left[\frac{\cos 3\omega - 1}{\omega^2} + \frac{3\sin 3\omega}{\omega}\right]}$

16.29 (a) $\mathcal{F}\{\delta(t)\} = \int_{-\infty}^{\infty} \delta(t) e^{-j\omega t} dt = \underline{1}$

(b) $\mathcal{F}\{\delta'(t)\} = \int_{-\infty}^{\infty} \delta'(t) e^{-j\omega t} dt$

$= (-1)'\left(\frac{1}{j\omega} e^{-j\omega t}\right)\Big|_{t=0} = \underline{\frac{1}{j\omega}}$

16.30 $f(t - \tfrac{1}{2}) = \mathcal{F}^{-1}\left\{\frac{\sin\omega}{\omega}\right\} = \tfrac{1}{2}, \; -1 < t < 1$

$= 0$ elsewhere

$f(t) = \tfrac{1}{2}, \; -1 < t - \tfrac{1}{2} < 1$

$= 0$, elsewhere

$\Rightarrow f(t) = \tfrac{1}{2}, \; -\tfrac{1}{2} < t < \tfrac{3}{2}$

$= 0$ elsewhere.

16.31 $f(t) = \mathcal{F}^{-1}\left\{\frac{(j\omega)^2}{1 + j\omega}\right\}$

$= \frac{d^2}{dt}\left\{e^{-t} u(t)\right\}.$

16.32 $4y'' + 6y' + y = 2x$

$[4(j\omega)^2 + 6(j\omega) + 1]\,\underline{Y}(j\omega) = 2\underline{X}(j\omega)$

$\underline{H}(j\omega) = \frac{Y(j\omega)}{X(j\omega)} = \underline{\frac{2}{1 - 4\omega^2 + j6\omega}}.$

16.33 $\underline{X}(j\omega) = \mathcal{F}\{e^{-(t-2)} u(t-2)\}$

$= e^{-j\omega 2} / 1 + j\omega$

$\underline{Y}(j\omega) = H(j\omega) X(j\omega)$

$= \underline{\frac{2e^{-j\omega 2}}{(1 - 4\omega^2 + j\omega)(1 + j\omega)}}$

$y'' + 3y' = 4x$, $[(j\omega)^3 + 3(j\omega)]\,\underline{Y}(j\omega) = 4\underline{X}(j\omega)$

$H(j\omega) = \underline{Y}(j\omega)/\underline{X}(j\omega) = \frac{4}{j(3\omega - \omega^3)} = \underline{\frac{j4}{\omega^3 - 3\omega}}$

16.35 $\underline{X}(j\omega) = \mathcal{F}\{x(t)\} = \frac{\sin \omega/2}{\omega/2}$

$\underline{Y}(j\omega) = \underline{H}(j\omega)\underline{X}(j\omega)$

$= \frac{j4 \sin \omega/2}{(\omega^3 \;\; 3\omega)\,\omega/2} = \underline{j\frac{8\sin \omega/2}{\omega^2(\omega^2 - 3)}}$

16.36

```
*FOURIER COEFFICIENTS
*OF A PERIODIC FUNCTION
R_R1   1 0 1
V_V2   1 0 PWL(0 1 3.14
+0 15.7 0 18.84 1 21.98 0)
.TRAN 0.1 21.98
.FOUR 0.053078 I(R_R1)
.END
*TRANSIENT RESPONSE I(R_R1)
DC COMPONENT =    1.666461E-01
HARMONIC FREQUENCY     FOURIER
NO          (HZ)       COMPONENT

1        5.308E-02     3.039E-01
2        1.062E-01     2.280E-01
3        1.592E-01     1.351E-01
4        2.123E-01     5.699E-02
5        2.654E-01     1.213E-02
6        3.185E-01     3.782E-05
7        3.715E-01     6.175E-03
8        4.246E-01     1.424E-02
9        4.777E-01     1.501E-02
****************************
*THE FOLLOWING PLOTS ARE THE
*ORIGINAL WAVEFORM AND ITS
*FOURIER TRANSFORM
****************************
```

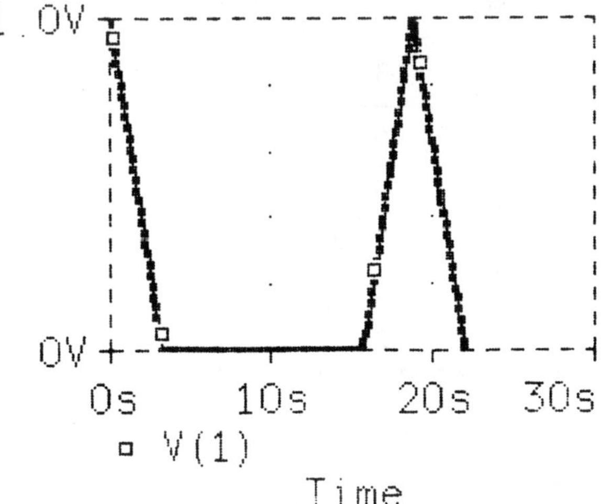

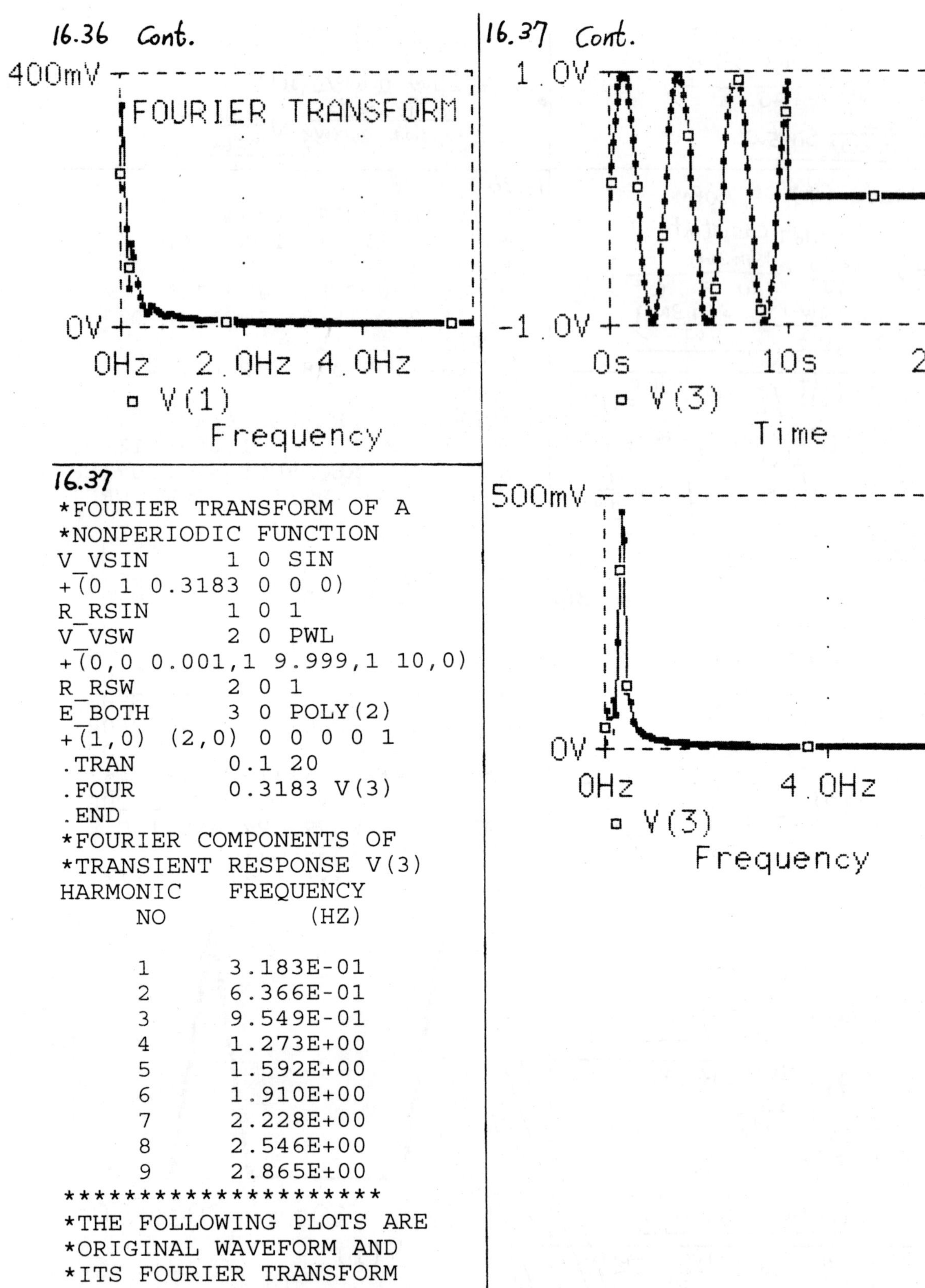

16.37

```
*FOURIER TRANSFORM OF A
*NONPERIODIC FUNCTION
V_VSIN      1 0 SIN
+(0 1 0.3183 0 0 0)
R_RSIN      1 0 1
V_VSW       2 0 PWL
+(0,0 0.001,1 9.999,1 10,0)
R_RSW       2 0 1
E_BOTH      3 0 POLY(2)
+(1,0) (2,0) 0 0 0 0 1
.TRAN       0.1 20
.FOUR       0.3183 V(3)
.END
*FOURIER COMPONENTS OF
*TRANSIENT RESPONSE V(3)
HARMONIC    FREQUENCY
   NO          (HZ)

    1       3.183E-01
    2       6.366E-01
    3       9.549E-01
    4       1.273E+00
    5       1.592E+00
    6       1.910E+00
    7       2.228E+00
    8       2.546E+00
    9       2.865E+00
*********************
*THE FOLLOWING PLOTS ARE
*ORIGINAL WAVEFORM AND
*ITS FOURIER TRANSFORM
***********************
```